Designing the Best Call Center for Your Business

Designing the Best
Call Center
for Your Business

A COMPLETE GUIDE FOR LOCATION, SERVICES, STAFFING, AND OUTSOURCING

Second Edition

by Brendan B. Read

CRC Press
Taylor & Francis Group
Boca Raton London New York

CRC Press is an imprint of the
Taylor & Francis Group, an **Informa** business

CRC Press
Taylor & Francis Group
6000 Broken Sound Parkway NW, Suite 300
Boca Raton, FL 33487-2742

First issued in hardback 2017

CRC Press is an imprint of Taylor & Francis Group, an Informa business
No claim to original U.S. Government works

ISBN 13: 978-1-138-41216-3 (hbk)
ISBN 13: 978-1-57820-313-0 (pbk)

Cover design by Damien Castaneda
Text design by Robbie Alterio
Text composition by Greene Design

**Visit the Taylor & Francis Web site at
http://www.taylorandfrancis.com**

**and the CRC Press Web site at
http://www.crcpress.com**

DEDICATION

To all the hard-working people who answer and make the calls,

do the chat sessions, and send the emails and faxes that we all depend on:

the staff in call centers everywhere.

Contents

Foreword   ix

Preface   xi

Introduction   xvii

Acknowledgements   xix

Chapter 1: What Are
Call Centers?   1

Call Centers Are Not Just for Calls!   2

Call Centers Are Anywhere
and Everywhere   2

Defining Call Centers   3

Just How Widespread Are Call Centers?   4

General or Specialized   4

People or Technology?   6

Work Environment   8

Who Performs the Work?   9

Location   10

Why You Need a Call Center   11

Call Center Challenges   13

Justifying a Call Center   14

Chapter 2: Call Center
Functions   17

Sales   17

Lead Generation and Qualification   19

Customer and Market Surveys   19

Collections   20

Directory Assistance   20

Customer Service   20

Customer Support   21

Employee Service and Support   21

Telemessaging   22

Emergency Services   22

Issues   23

Customer Relationship
Management (CRM)   27

Chapter 3: Planning
in the Call Center   31

Characteristics of Call Centers   31

Key Legislation and Regulations   37

Fraud and Harassment   38

Sizing   42

Business Continuity   45

Chapter 4: Self-Service   47

Advantages   49

Downsides   53

Technology Differences   56

Is Self-Service Right for You?   56

Enabling Self-Service   58

Chapter 5: Home Working   59

Home Working Compared to
Outsourcing   59

Characteristics   60

Advantages   62

Downsides   67

Minimal Benefits   72

Is Home Working for You?   72

Enabling Home Working   73

Chapter 6: Outsourcing   81

Characteristics   81

Functions   82

About Outsourcers   82

Program Accommodation   83

Outsourcing Program Types   83

Outsourcing Service Options   87

Advantages   92

Downsides   96

Customer Dissatisfaction
and Annoyance   97

Enabling Outsourcing   99

Chapter 7: Site Selection   105

Labor, the Key Site Criteria   105

The Life Span of Your Center   108

Labor Factors   108

Facilities Availability   116

Site Options   120

Distributed Call Centers   128

Shared-Service Centers   130

Enabling Shared-Service Centers   132

Enabling Site Selection   133

Chapter 8: Onshoring,
Nearshoring, and Offshoring   137

Site Selection in Relation to
Outsourcing   137

The Case for Onshore   138

Onshore Issues   140

Enabling Onshoring   142

The Offshore Revolution   143

The Case for Nearshoring   151

Nearshore Challenges   154

Enabling Offshoring
and Nearshoring   156

Outsourcing-Specific Factors   159

Enabling Onshore Outsourcing   160

Offshore Outsourcing   160

Enabling Offshore Outsourcing   164

Blendshoring   169

Chapter 9: Serving Hispanics   171

Characteristics and Issues   171

Outsourcing Compared with
In-House Centers   174

Onshore and Offshore Locations   175

Selecting Spanish-Speaking Agents   177

Chapter 10: Foreign Markets   181

Foreign Characteristics   181

Site Selection Options   186

Outsourcing   186

Offshoring   189

Nearshoring Alternatives   194

Foreign Market Strategies   194

Serving Foreign Markets   198

Serving Foreign Customers
While Abroad 202

Serving Foreign Customers at Home 202

Foreign Facilities and Management 203

Connecting Foreign and Offshore
Sites 208

Chapter 11: Property
Considerations 209

Property Requirements 209

Design Features 220

Property Choices 224

Existing Building Options 226

Conventional Offices versus
Conversions 229

The Recycled Call Center 234

Real Estate Strategies 234

Checking Out the Property 237

Compromises and Options 240

Chapter 12: Design,
Ergonomics, and Security 247

Effective Design is Key 247

Justifying Effective Design 249

Design Considerations 250

Effective Design Elements 250

Communications Impacts 256

Ergonomics 257

Ergonomics Appliances 263

Enabling Effective Design 268

Don't Forget Home Agents 272

Maintaining Your Call Center 273

Safety and Security 274

Chapter 13: Staffing Your
Call Center 279

Staffing 279

Designing Your Ideal Agent 281

Agent Certification 287

Qualification 288

Recruiting Channels 292

Quality is Key! 297

Targeted Recruitment 298

Chapter 14: Training
and Retention 303

Key Skillsets 303

New Hire Training Techniques 306

Cross-training 307

Soft Skills Training for Tech Support 308

Refresher Training 309

Agent Retention 317

Chapter 15: Management
Issues 325

Management Structure 325

Supervisors: Your Call Center's
NCOs 325

Measuring Performance 335

Demonstrating Value
to the Enterprise 337

Benchmarking 338

Running the "24" in 24x7 341

Unions 341

The End (Literally) 343

Getting the Affairs in Order 346

Chapter 16: Resources Guide   349

Other Books   349

Trade Media   350

Associations   351

Special Assistance to this Book   353

Appendix I—Call Center Gallery   355

Core Communications   355

Infonxx   356

Nextel   357

AmeriCredit   358

Ebay   359

**Appendix II--Locate In The Big Apple?
Don't say 'fugedaboudit!' just yet!**   361

Foreword

When Brendan asked me to write the forward for this book, I readily agreed—not just on the merits of the book alone (of which there are many), but also because I know a bit about what goes into his work. Brendan has been a journalist, writer and observer in the call center industry for many years. I have often been on the other end of his interviews—each time, thinking that his were the kinds of questions that come from someone who not only understands the industry but also cares deeply about it. Through his work, he has both covered key developments and helped to shape them.

That is the kind of thought and care he has put into *Designing the Best Call Center for Your Business*. Yes, it fulfills its promise and you'll learn a heck of a lot about the totality of designing a call center—from what it's supposed to do, to site selection, technology, management, ergonomics, outsourcing considerations and a host of related topics. You'll also gain valuable insights into leadership, management and the innerworkings of this ever-changing industry that will help you in virtually any aspect of directing, managing, supporting, or working in a customer contact environment.

There are a number of additional pluses that make this book a wise investment for your organization's library or learning center. It's well-organized and you can get to specific topics easily. As a reference, the checklists, questions and issues discussed will help you work through specific aspects of design and management, and ensure that your project plans are realistic. And as with any work that contains original thinking, you'll find yourself agreeing with some things and debating others; in the end, it will give you a great head start for making decisions that fit your organization's vision.

Call centers are a critical component of today's communication-intensive economy. A thoughtfully designed and better-managed call center not only has potential to improve your organization's bottom line now, it can also be a differentiator in an age when products and services are quickly copied—and that helps to ensure a more profitable and certain future. In short, well-designed and managed call centers make a difference—to customers, employees and shareholders. That is the most important reason to learn everything you can about them, and this book is an important part of that journey.

Brad Cleveland
President and CEO, Incoming Calls Management Institute (ICMI)
and publisher, *Call Center Management Review*
www.icmi.com
bradc@icmi.com

Preface

Plus ca change, c'est le meme chose

—Alphonse Karr

That famous quote, the more things change, the more they stay the same, applies to call centers and to this book.

On the surface the call center universe has been upheaved, in some ways tragically since the first edition came out in late 2000.

✪ THE DOT-COM ASHES

The dot-coms bombed, taking away vast sums of wealth and crippling the high-tech sector. That also went for the trade media that depended on that boom. *Call Center Magazine* is only half the size and staff compared to then. I became a victim of the shrinkage when I was downsized in June 2004 shortly after beginning work on this edition. Other publications and media also downsized and disappeared.

Yet in the ashes of that failure the Internet has grown roots as a prime enterprise and communications avenue. Vast expansion of broadband networks, both wired and wireless, have made e-commerce and Web self-service doable and popular.

The self-service trend, aided by these investments, is challenging live-agent call centers in a big way. Airline and hotel reservations, shopping and tech support have gone automated with speech-rec-enabled IVR and web self-service, permitting companies to close centers and lay off staff.

At the same time, voice over IP has migrated from a wonk technology to everyday reality. My home office phone is VoIP, supplied by Vonage. VoIP, combined with broadband and new tools like instant messaging, has made home working by call center agents increasingly popular.

Home working enables call centers to tap and retain quality workers such as stay-at-home-moms, baby boomers and those living in wired communities outside of commuting distance. I took my home working experience and research and wrote a book on it, *Home Workplace*, also published by CMP.

Technology changes continue to roil the business and trade media as search engines rather than print ads have become the advertising channel. With my downsizing I switched careers to do corporate communications for The AnswerNet Network, a fast-growing and very marketing-and-business-savvy outsourcer.

Online readership is small compared to print, but is growing rapidly. Citing a Merrill Lynch report, *The New York Times* reported January 24, 2005 that online advertising is expected to account for about 3.7 percent of US ad spending. But the amount, $9.7 billion, is expected to grow 19 percent in 2005 as the largest advertisers "shift budgets from print and network television to cable and the Internet, the report said."

Yet online circulation gives media outlets a far wider circulation than in the print-only past. For example I get the NYT delivered to me online daily every morning to the small city in British Columbia where I had been living. Ten years ago you would have had to drive to Vancouver or Victoria to see one, and then only the national edition and only in specialty newspaper/magazine stores. I found a job at Primedia in Manhattan in 1990 while living in upstate New York only after I had subscribed and specified the New York and Long Island edition.

Yes, it is more work to scan words on a screen, but the information is current, not two days or months old. Does this make paper, like this book irrelevant? Far from it. I've striven for a mix of universal points with a long shelf life and advice with reasonably current information.

9/11/01

Another more sinister bombing took place on 9/11/01 that upheaved the American psyche, perhaps permanently. People are more afraid and distrusting. Fear has displaced the sunny optimism that gives them hope, makes them willing to invest in the future and spend money that in turn helps call centers to grow.

The response: the "war on terror" is in danger of turning into the "perfect quagmire" where young Americans are being killed and maimed by enemies who have struck on American soil. I'm old enough to remember the Vietnam War and to be worried about the draft. At least the Viet Cong weren't skyjacking aircraft or planting bombs.

Since the attacks call centers are more interested in security and disaster response. The bombings/skyjackings, combined with the SARS outbreak, fears of a deadly bird flu pandemic and the northeast power blackout has encouraged businesses to look at options like home working and distributed call centers to minimize their vulnerability. I know these fears firsthand. I witnessed the terror attacks on the World Trade Center. I was on my way into the office of *Call Center Magazine*, then on 12 West 21st St. in Manhattan when the driver of my bus yelled "The World Trade's on fire!" We had just passed the twin towers only a few minutes before. My wife, myself and the other passengers rushed to the left-hand side and looked up and out at the smoke rising from the north tower.

When I got out at my stop I watched the smoke while the radios spurted reports of what was happening. Then I saw a big puff of smoke and debris from the south tower as the other plane hit.

Call Center Magazine closed its doors and we evacuated. Chief Technical Editor Joseph Fleischer accompanied me, my wife, and my sister-in-law, who worked nearby, walking to the West 38th St. ferry docks. Orderly lines stretched out for blocks, everyone scanning the skies nervously.

Unfortunately my sister-in-law had a mild heart attack. People kindly let us through to the New York Waterway ferry. On the other side an ambulance took her and my wife to a hospital. I made my way to the home of a friend of my wife in central New Jersey via bus and train. I was going to walk across one of the bridges back to Staten Island to get my car, but the authorities closed them.

In the meantime our son, a paramedic, was called in. Because the cell towers disappeared under the rubble, we did not hear from him for two days.

I also lived through the aftermath: seeing the National Guard trucks covered in dust and debris from Ground Zero rumbling up my Staten Island street, the bomb scares and security alerts. The fear.

And what I hear from my son regarding preparedness in response to the next catastrophe, does not have me feeling overly optimistic.

Offshoring

At the same time call centers, or more accurately companies using call centers, accelerated outsourcing the work offshore to cut costs in the wake of the economic downturn. Not long afterward they began to face obstacles of public resistance, union opposition, and perhaps more seriously, customer dissatisfaction arising from poor service.

Companies are belatedly realizing that having calls handled in other countries is not the same as having clothes or steel made abroad. When you buy a shirt or dress or a coil of hot-rolled steel, it doesn't matter who made it as long as they did their work well. But it does matter when you are calling, because those individuals' accents, dialects, empathy, and understanding makes or breaks the transactions. I know: I've worked in the garment industry, and I used to cover metalmaking.

Do Not Call

Since the first edition of this book the outbound telemarketing industry completed its own less deadly suicide mission with the implementation of federal Do Not Call (DNC) regulations. The telemarketing industry successfully strangled the outbound sales goose that was laying the golden egg with incessant calls, call abandons that scare recipients who pick up their phones and hear nothing, and violations of laws. The "leaders" failed to stop the problem; they were in denial, derailing potentially viable good-practices certification programs.

Claims about potential job losses from DNC made by the likes of the American Teleservices Association fell on deaf elected officials' ears that were ringing from consumer complaints. Their case was not helped by allegedly misrepresenting job loss numbers.

The fatal flaw of marketers and salespeople is not listening. By only talking and not hearing they don't get the message from customers. So when the phones slam and the regulators and attorneys-general offices come knocking there is only one set of people to blame, and that's in the mirror.

Call Center Magazine Editorial Director Keith Dawson put it accurately:

"They deserve [the heat]. It's extremely disingenuous of them to go around to mainstream media outlets like *The New York Times* and trumpet the absurd "fact" that DNC is going to destroy two million outbound telemarketing jobs. First, there are thought to be somewhere between three and five million call center jobs in the US—in total. And anyone who knows what they are talking about will tell you that only a tiny fraction (perhaps 10%) is outbound. That puts the real outbound job total at around 400,000.

"And restricting telemarketing isn't going to make those 400,000 jobs disappear. Scare tactics backfire, and now the ATA has no credibility. We should have real concerns about jobs moving offshore, but the jobs we should care about losing are the well-paying and well-skilled inbound jobs."

"DNC is good for people, good for call centers, and ultimately good for the ATA's members. The ATA does them a striking disservice through their mangling of the facts."

Spam and scams

Crooks know where the money is, and that's in the Internet: in identity theft, fraud and spam. While some still play with telemarketing scams, targeting the elderly, the real El Dorado lies online where finding and tracking you is next to impossible.

Since the first edition, the federal government did pass the CAN-SPAM Act to try and control it. Good luck. Spam and scams, aided and abetted by hole-riddled software that let this crap in have grown to the point where people are reluctant to open or preview commercial email even from trusted sources because spammers have spoofed or faked web sites. The ROI on spam is so high it only takes one or two suckers to send bundles of cash like to Nigeria to keep it going.

The CRM Fizzle

The customer relationship management (CRM) trend cooled down but is still relatively warm. Software implementations stretched months if not years at a time when Wall Street wanted payback in months. Outsourcers, stung by investing in these tools, bagged out, offering to implement customers' platforms instead.

Overriding these issues is the experience of customers. Instead of treating all people well but the better customers better—what CRM should be about—banks and other CRM buyers treated some clients well and others less well, tossing them self-service dungeons.

The customers got their revenge. They stuffed loyalty if you stuff service. They began to buy strictly on price and convenience.

Survey after survey is showing that customers are not happy. People who went through the IVR and Web self-service want to talk to someone in authority, not to some well-meaning overworked underpaid agent limited only by what's on the script and in the knowledge base.

At the same time the fundamental value of CRM still remains. It is about identifying the most profitable customers and finding ways to retain them. Which is what any smart business does.

That's why telemessaging is beginning to emerge as an alternative to this paradigm. These agents do not attempt to solve problems. Instead they answer politely and quickly, triage the calls and contacts and connect them to the professionals who can meet their needs, or if not available record, store, or transmit messages and pages.

The telemessaging providers are becoming highly capable. The AnswerNet Network, which offers both outsourcing and telemessaging has over 50 locations and 1,600-plus workstations.

The Book Has Changed Too...

The book has changed quite dramatically in size and scope with these changes, and with the additional information and knowledge I've gained in the four-plus years since I wrote the first edition. There is greatly expanded coverage of self-service, home working, outsourcing, offshoring, real estate and design. There is more detail on staffing, training and management. Business continuity i.e. disaster planning and response is worked into planning, site selection and operations.

There is less emphasis on technology than in the previous edition. There are many fine volumes out there that cover that topic, which gets dated quickly.

Yet the Fundamentals Are The Same

Yet the fundamental decisions and issues of setting up and running call centers have not changed. Organizations still need to decide whether or not they are going to outsource and manage those relationships, or have their own call centers that they need to locate and design. They also face the challenges of staffing, training, and managing employees in what is a stressful, tightly controlled and relatively low-paid yet essential line of work.

The approach of this book hasn't fundamentally changed either. I cite and quote outside sources like the journalist I am because it is their experience-and-knowledge-generated insight that I am relaying to the readers. Yes, there are touches of my humor and a few of my stories to illustrate key points and to keep you on your toes.

That's why this book is a second edition rather than a new book. The kernels of advice contained in what I wrote in 2000 still have the same value today.

Brendan B. Read
January 2005

Introduction

Call centers are a key part of our lives even if we have never set foot into one of them. We depend on their friendly live or automated voice information and assistance to repair products and acquire goods and services. Call centers are not so much an industry as a function.

This book is about how to design, plan and create call centers to supply those functions cost-effectively to satisfy both the users and your organization.

Yet there are many permutations to this task because there are so many ways to deliver call center services. With live employees, interactive voice response, or Web self-service. By voice, e-mail, chat, fax, or video. From your own organization or an outsourced one. By employees or by the self-employed. In traditional offices, former supermarkets, factories or homes. In your own country or abroad.

Whatever way you choose to have your call center you also need to ensure that it has the right people hired, trained and managed to get the most out of your investment, which can run into the millions of dollars. That's why this book delves into staffing, training, and management issues. And because not even call centers live forever, this book touches on how to close them.

There is not room in this book to satisfactorily explore all the facets of call centers. That's why I've focused on call center creation. And because volumes like these are as good as the day the author turns the copy into the publisher, there is the risk that information often becomes out of date by the time the reader picks up the copy. At the same time, there are many associations, consultants, and firms who can assist you with developing your call center functions.

For these reasons this book contains a Resources Guide that points to other books with substantial management, operations, technology, and training content. The guide also contains trade media sources for current information. At the same time it outlines trade associations that have a lot of information.

Finally, the Resource Guide contains the names of consultants and firms that have helped me in my articles and research. Which is why I've cited many of them in the content. Without these people and companies there would be no book.

I am a writer, journalist, and observer. The function of people like me in books like this is to present information obtained from doing what I do best in a clear, thorough, and concise manner. I say "concise" with a smile. This book is short compared to the vast tomes written in magazines and online on call centers.

I hope you find this book helpful, interesting, and informative in assisting you with your call center. I and others who use call centers will be counting on you.

Acknowledgements

The information for this book comes from an interesting mix of firsthand and secondhand experience. Like most people, I've used call centers and know what makes good service from them. I've also worked for call centers in small business (John Garde and Company in Toronto, Ontario, Canada), as a telemarketing fundraiser (for the King County Democrats in Washington state) and in handling corporate communications for an outsourcer (The AnswerNet Network of Princeton, N.J.).

More importantly, I've learned a lot more about call centers by meeting and interviewing many fine people. A few of these great individuals include:

Berta Banks and Kathy Dean, Banks and Dean
John Boyd, The Boyd Company
Maggie Klenke and Penny Reynolds, the Call Center School
Philip Cohen
Jim Trobaugh, CB Richard Ellis
Bob Engel and Gere Picasso
Ron Cariola, Equis
Rosanne D'Ausilio, Human Technologies Global
Jeff Furst, FurstPerson
Suresh Gupta, The Paaras Group
Jack Heacock, Jack Heacock and Associates
Brad Cleveland, Incoming Calls Management Institute
Roger Kingsland, KSBA
Paul Kowal, Kowal Associates
Linda Lauritzen
Bryan Mekechuk, Pacific Crest Consulting
Rudy Oetting and Geri Gantman, R.H. Oetting
Kay Jackson, Response Design Corporation
Jon Kaplan and Mark Schmidt, TeleDevelopment
Anne Nickerson, The Call Center Coach
King White, Trammell Crow Company

Then there are my former colleagues at *Call Center Magazine,* CMP, and Telecom Library. They are: Keith Dawson, Joe Fleischer, Richard "Zippy" Grigonis, Jenn O'Herron, Mary Lenz, Alison Ousey, Robbie Alterio and yes, Harry Newton. I learned plenty from them, and I wish them the best of success. A special appreciative smile goes to Janice Reynolds who has worked with me on past books.

I also want to thank the very patient people at CMP Books: Matt, Gail, Dorothy, Meg and Sachie for working with me on the second edition. Putting together a book: from conception to marketing and selling is no easy task and takes a dedicated team of professionals at each step of the process to ensure it is done right.

The last person, but certainly the first I should acknowledge is my wife, Christine, who spent many an evening and weekend upstairs while I worked on the book downstairs. Her support and common-sense advice kept me on track.

Chapter 1: What Are Call Centers?

When people strike up a conversation with me, they invariably ask what I do for a living. When I reply, they almost always follow up with: "What is a call center?"

I then say, "You know when you dial a toll-free number to talk to someone, like in customer service or sales?" Their heads nod. "That's a call center." Or I will add: "If you've dialed a doctor's office and got an answering service, dialed a hotline or 911, you've reached call centers, too."

The questioner smiles knowingly. Most everyone has dealt with a call center, even though they do not call it as such or know what one is.

Say you take one of those get-away-from-it-all trips to a place with no landline phones and terrible wireless connections. At some point, you need to check your airline or train reservation with a call center.

If you got a diamond-plated smart card in the mail, you probably talked to a telemarketer who called because you had a silver-plated card. And it doesn't stop there: The telemarketer was probably working off a list of names that had been matched up with data indicating an income and lifestyle to match this high-prestige (and high-interest) product.

Or if you got a call from your favorite or not-so-favorite political candidate, they're calling from a call center. If you voted and they got your number, you're going to hear from a call center.

If what you heard from that call literally made you sick, you make an outbound call to your doctor. A friendly courteous voice replies saying they will page the physician. You will have reached a—guess what—call center.

Companies now realize that friendly customer service is a vital marketing tool that differentiates them from the competition. Today, many billboards and TV ads feature smiling call center agents.

Even at work you deal with call centers. The person who called you yesterday with a deal on copier toner dialed you up from a call center. How are you going to get back to that inside software sales rep about fitting out all your employees with video-equipped laptops to cut a deal on the price? By picking up the phone or sending a fax or email to a rep. Where does that rep work? A call center. Many businesses or institutions with call centers refer to them by customer service, sales, reservations, help desk, and collections rather than as a call center.

✪ CALL CENTERS ARE NOT JUST FOR CALLS!

Call centers are not just for phone calls. They also handle other communications media such as email, chat, fax, and video. There are call centers, such as eBay's, which are nearly all email.

If you had a problem with that computer you just bought, chances are the first place you looked for help is that manufacturer's web site. If you could not find the answer you want on the web site's FAQ, then you might email your issue to a support rep who then handles it in a call center.

If you were applying for a credit card and the issuer wanted you to fax additional information like paystubs to prove your income, that fax is received by a call center. That employee who got it may call you back, or you may call them to verify that they got the fax, which means you're dealing with the call center.

Call centers are part and parcel of self-service solutions such as interactive voice response (IVR) and the web. These tools take on simple requests like account balances, bill payments and orders and reservations, freeing up employees to handle more complex tasks.

These tools supplement call centers and make them more efficient by intaking customer or employee information, which enables callers to be transferred to the right people and allows call center employees to identify them without wasting time asking who they are. Computer-telephone integration (CTI) reads the identifiers such as phone numbers or Social Security numbers, which lets software pull up and delivers the files to the employee's desktop.

At the same time, call centers supplement computer and online services. They provide support for Internet services; web site FAQs are not helpful if your connection is down. You have to buy hardware, software, and connections to get on the web before you can order on the web.

There are many occasions when people call to carry out online transactions, such as web order entry or tech support. Examples include customers or employees who are mobile or are still on dialup. The tiny screens on web phones are hard to see, especially for older people who are making up a larger percentage of the population. Also, many web sites are difficult to navigate and require expert assistance. And there are people who still prefer to give their credit card numbers directly to a person rather than keying it online, even though the agents are doing the same tasks.

✪ CALL CENTERS ARE ANYWHERE AND EVERYWHERE

Call centers are in formal workplaces, which can be in traditional office buildings, converted department stores, or warehouses. They also can be the reception desks in clinics, on the retail floors, in vehicles, and in homes. The locations can be your home country or abroad, handling domestic and foreign customers. Those customers can be buyers, clients, employees, and the public.

The call centers can be part of your organization, or they may belong to a contractor you outsourced that function to. That outsourcer could deliver the call center functions on your premises, on its facilities, or use home-based employees. The workers can be your hires, or hired by an outsourcer or staffing agency to work in facilities on property.

In short, it is not the location or the employee-worker relationship that makes call centers. Rather it is the function.

✪ DEFINING CALL CENTERS

So here's my definition: Call centers are where formal interactions and transactions such as information, service, support, sales, employee assistance, and emergency occur virtually, that is, not face-to-face. This takes into account people-to-person as well as self-service.

Why did I define call centers and their dimensions? Because there are no agreed-on definitions for call centers or their employees.

Yes, "call center" may seem to be an antiquated phrase. But the hard fact is that most contacts between people occur by phone and will continue to be that way. The research firm Datamonitor reports that over 95 percent of communication in call centers occurs over the phone.

The reason: Only the phone provides effective, low-cost, real-time, bidirectional communications between two or more parties. Gartner Group reports that live-person email interactions take longer than voice. Why? Because it takes longer to write a message than to speak it.

I learned that lesson in my high school composition class. The teacher asked us to write down exactly what he said. He finished talking before we were finished. That's why there is shorthand, formal and informal. It's why good journalists back up critical interviews with recordings.

All other popular methods, such as chat, email, and instant messaging (IM), require you to wait until the other person has transmitted their remarks, but with the phone, as in regular conversation, you don't have to, and you can butt in.

This is a lesson that too many IM users don't get and need to be reminded. I've learned to let them finish spitting out their text before responding. When they ask "are you still there" I politely remind them that I am waiting to avoid having them "overwrite" what I wanted to communicate, to avoid misunderstandings.

"Call center" is also what I call a foundation phrase, one that describes an object when it is created, becomes established, and is in widespread use.

One of my favorite examples of a foundation phrase is "carfare." Carfare refers to the money people pay for transportation—usually mass transit. But its origin dates from the streetcar era, which still lives on in many cities. Do you pay carfare when you board a bus or subway, or on the revived and upgraded streetcars known as light rail transit (LRT)? I paid carfare when I rode the King streetcar line to Spadina Avenue when I worked in Toronto years ago. Was it still considered carfare when I changed to the Spadina bus to go to my stop? Is it carfare now that the Toronto Transit Commission replaced the bus with a restored streetcar (a.k.a. LRT)?

Think of the term "dial." Do you really put your finger inside a hole over a number, yank down, and let it click back when you dial someone on your wireless or landline phone? You can use email and surf the web on cellphones; are they still phones?

I prefer call center to other descriptions for other reasons. "Contact center" sounds a little too personal, like someone poking me in the arm. "Customer interaction center" is too nebulous. I talk, email and text-chat with someone, and that's how I refer to these functions—not "contact" or "interact."

✪ JUST HOW WIDESPREAD ARE CALL CENTERS?

I've received many inquiries from public relations firms, researchers, and suppliers who ask, "How many call centers are there?" I reply that describing the call center business is like the blind men describing an elephant, to use the definitely *not* politically correct and blatantly sexist, old-fashioned, but accurate metaphor.

The research firm Datamonitor reported in a study *Opportunities in North American Call Centers to 2007* that there were 55,800 call centers in North America: 50,200 in the US and 5,600 in Canada. But it requires a call center to have an automatic call distributor (ACD) or equivalent or 10 or more workstations to be included in its research.

That's the view of only one research firm. Others may not agree.

Why 10? Why an ACD and not direct-inward-dial or key system phone systems? What about centers that do only or mostly email, like eBay's centers? Surely it is the function that counts, isn't it?

For example, when I was 18, I was a one-person call center for John Garde and Company, a small family-owned firm that assembled and serviced factory sewing machines in Toronto.

I answered the phone, took orders, arranged for repair calls, and called delinquent customers—the same functions that many call centers provide. Only I was known as an accounts receivable clerk, not an agent.

Or what if employees work at their desks for only part of the day? Sales reps make and take customer contacts, yet they are also in the field, visiting suppliers at their offices, or from home offices and at trade shows. When at John Garde and Company, I did not answer calls all day. I served customers at the counter, lugged parcels to the post office, delivered machinery, and collected on accounts in person.

Or in another example, staff at Kaiser Permanente's 13 Portland, Oregon-area clinics take calls fed via a Siemens switch from the HMO's 40-agent center when they are not handling in-person inquiries. Are these 13 clinics call centers too?

And as noted above, call centers can be in the home. Do you count their houses or apartments as call centers?

I find that knowing size of the call center industry is not very important to the industry and especially to the people that really matter to the organization: the customers, clients, and employees. Instead what is truly important is the service you provide them.

✪ GENERAL OR SPECIALIZED

As you can see, call centers handle many different types of tasks in a variety of ways.

These can be grouped into several dimensions that you examine and pick from to best meet your needs.

The first grouping is generalist, specialist, and telemessaging.

Generalist

Generalist call centers is my term for what people think when they come across them in their day-to-day experience, inbound or outbound, in customer service, information, sales, support, and collections. These are the agents that you reach sometimes immediately but more likely after losing yourself in an interactive voice response (IVR) menu maze or after spending 20 minutes on hold. Or who is calling you about an order or late bill. Or who is dialing you at dinnertime to sell you steak knives that you should have bought the last time they called, as the meat lies there impervious to your stabbing.

The characteristics of generalist agents are that they are trained to handle specific tasks, strictly scripted, and look up information in specialized databases, known as knowledge bases. Their education level is usually some college.

If there are matters that these agents cannot resolve, they call in their supervisors or escalate the calls to experts or specialists. Generalist agents rarely have the authority to act on requests outside of instructions, like offering a user a deal on a new computer because their current one has been breaking down a lot. They do not have the training or expertise to think out of the box to fix problems that cannot be remedied by drilling through the knowledge base.

Also, generalist agents rarely communicate often with the same customer or employee unless it is on a specific issue. Call centers are too busy to permit assigned callers.

Specialist

Specialist call centers are those where the employees have authority to make decisions, experience, education, and specialized training or licensing. Examples include inside sales, advanced technical support, engineering, law, medical information, and agents in insurance, real estate, and securities. Specialist employees often act as account representatives serving specific customers.

As technology improves and becomes less expensive, especially including audio, data, video, and web conferencing, expect more specialists to carry out their functions by phone and online rather than in person. Already much of medicine has moved from the clinic to the home.

Telemessaging

Telemessaging centers are an outgrowth of traditional telephone answering call centers that have been around since the 1920s. They are similar to other call centers with modern phone switching technology and can handle other channels i.e. email, live chat and the Internet. Yet telemessaging call centers are different from generalist or specialist call centers in that the employees do not answer inquiries, resolve problems, or make sales. Instead they take messages, triage, patch calls or emails through, or page specialists, professionals, or staff who can complete those tasks. Telemessaging staff also sift through email inboxes, send acknowledgements, toss out the junk, and forward important mail to recipients; they page them for urgent issues.

Perhaps the biggest difference between telemessaging call centers and others is how they are applied. With telemessaging you usually interact with a live person first, rather than undergoing an IVR menu that the center's management hopes you will complete your transaction with instead. As this book explains later, it is a lot less expensive to talk to a machine than to speak to a human.

Telemessaging is the oldest call center function, and it dates back to the invention of the telephone. Callers asked operators to put them through, say, to a doctor. With the growth of automatic phone switching, telephone answering services began to supplant operators.

Contractors, lawyers, and physicians have historically relied on telemessaging to handle calls when they are busy or after hours. But this method is gaining new uses as call center tasks and the expertise and authority required begin to exceed the capabilities of generalist agents. Telemessaging then works in tandem with specialists.

For example, telemessaging staff now function as Level zero in customer support. They answer calls, ask basic questions, triage callers, and open trouble tickets online for tech support staff. Agents can also sell paid support plans. For a detailed explanation of these levels please read *The Complete Guide to Customer Support* by Joseph Fleischer and Brendan Read, also published by CMP Books.

Telemessaging is low-skilled compared to many generalist call center positions. The upside is that such call centers can draw on a wider range of applicants, which keeps a handle on call center costs. Telemessaging interactions are typically short: 45 to 60 seconds compared with three minutes or more in a call center transaction, which makes it competitive with low-cost self-service options when you factor in the human touch.

✪ PEOPLE OR TECHNOLOGY?

The next split is whether people or technology conducts the interactions. The choices are:

Live Agent

Live agents are the people who make and take calls and other contacts. Agent is the best, most popularly used term for call center employee. Other terms that are used include CSR for customer service representative and rep for representative or operator, as in "operators standing by." Live agents employ several channels to communicate with others, both inbound and outbound. These include phone, email, chat, escorted browsing, fax, video and mail.

Live Agent Advantage

Live agents give that human touch. They provide empathy, understanding, and get to the point. They think, rather than what to be reprogrammed. You are talking to people, who can compare apples and oranges by seeing that they are fruit. People can identify needs, cross-sell, upsell and supply advice.

Live Agent Challenges

This out-of-box labor is expensive to provide per transaction compared with self-service (see below). As you will see later on this in this book, there are a lot of resources that go into supplying that interaction, be it outsourced or in your own call centers, in setup, training, and management.

For example, The Boyd Company, a leading site selection firm based in Princeton, N.J., did a study showing that a sample 200-employee call center can cost between $7.3 million and $10.4 million to operate, depending on location.

In another example, the Trammell Crow Company, said to accommodate 270 workers, requires a 35,000-square-foot building that would cost, say, $18 per square foot, or $630,000 in rent alone. Outsourcing prices, which account for all operating and capital costs plus profit margins (and they are not great in the outsourcing business), run about $27-$30 per hour in the U.S. Organizations, for-profit and non-profit alike, try to minimize live agent interactions. Having calls handled offshore, which many organizations are doing to cut costs as low as $12–$14 per hour, is still several times more costly than calls taken care of by technology.

Self-Service

Self-service encompasses automated technologies that also process contacts, both inbound and outbound. Self-service tools, depending on what they are, can screen and direct calls or handle transactions. Chapter 4 explores self-service.

Self-service includes:

Interactive Voice Response (IVR)

IVR is a both a hardware and software tool where you communicate with computers either by touchtone (using keypads) or by speech recognition (talking).

Outbound Voice Messaging (OVM)

OVM, also known as outbound IVR, entails sending or dialing out pre-recorded voice messages for people to hear directly or left on their answering devices.

The Web

The web has become the leading self-service channel. People use it for information, order placement, and problem solving, such as through FAQs and downloading software bug fixes.

Outbound Email

Automated outbound email is used to broadcast email messages, send acknowledgements to inbound emails, fax on demand, and broadcast faxes.

Fax Services

Yes, people still use faxes. Self-service fax services include fax on demand, initiated by IVR and broadcast fax messages.

Self-Service Advantages and Challenges

Self-service is less expensive than live agents. The Gartner Group reports that live agent interactions cost $5.66 apiece compared with 45 cents for IVR and just 24 cents for the web.

As technology improves and public acceptance grows, self-service tools are taking on more tasks that had been handled by live agents. One notable example is ticketing and reservations. Automated online booking has largely sidelined live agent call centers and in-person travel agents alike. .

Because self-service tools are quicker and inexpensive, they can conduct transactions immediately, avoiding waiting in queue for a person. But they are still machines, limited by what has been programmed into them.

Blended Live Agent and Self-Service

Most commercial call center applications blend live agents and self-service to get the best of both worlds. When you call a business such as a bank or credit card company you will get an IVR menu where you can do simple transactions like obtaining the balance but where you can also zero-out or reach a live agent if you have a question about holds or outstanding checks.

IVR is a great tool to identify customers. You can use it to ask them for numbers, like a phone number or password, and it will link the callers to live agents who have the customers' files on their screens, known as screen pops, through CTI.

Speech rec makes more people comfortable using self-service. Smart outfits like Amtrak are using real recorded voices as prompts that give the customer an acceptable illusion of talking to someone, when they are conversing literally with a machine.

To make this blended solution work well requires very careful planning and implementation. One of the many mistakes organizations have made is creating long scripts and complicated IVR menus that trap and frustrate callers. Some outfits stupidly omitted a zero-out option to a live agent.

I say "stupidly" because few things will want to make a customer who has a problem rip up, curse out, and badmouth a product to others more than not being able to reach a live person. I have several companies on my list.

Remember, bad news travels farther than good. Ten times as many people will hear and act on a complaint than a compliment. If you are in a competitive environment, not providing an out to a live person will cost you more where it counts than if you had implemented the service correctly to begin with.

✪ WORK ENVIRONMENT

The next dimension is the environment the work takes place. Call centers apply to the function but not necessarily the location. The choices are:

Premises Call Centers

These are the call centers that everyone thinks of, the rows of people sitting inside cubes or at tables wearing headsets in buildings paid for by the employer. The structures range from new and renovated conventional offices to converted stores, factories, warehouses, and bowling alleys. Supervision is both electronic (through monitoring calls and emails and chats) and face-to-face. Training is by live or conferenced instructors and by technology on agents' desktops.

Home Agent Call Centers

Home agent call centers or home agents are where the employees work from their own homes but are still connected to callers, called parties, and to supervisors. Home workers are monitored electronically; they stay in touch through instant messaging, email, and phone.

Chapter 5 examines home working; Chapter 6 looks at home agent outsourcing. There is also a separate book, *Home Workplace*, written by myself and published by CMP Books that explains how and why home working operates.

Having agents work from home is less expensive and more productive compared with having them on premises. You can get them up and working faster because they don't have to commute. But you have to select the right people for their self-discipline and problem-solving abilities, and you have to be flexible in your organization culture to allow it.

◎ WHO PERFORMS THE WORK?

There is another dimension: who performs the call center work. The options are:

In-house

In-house is the term for carrying out call center functions, live agent or self-service, by your organization. You own the equipment and hire and manage the people.

When you have your call center in-house, you have maximum control over what it does, how it is set up, maintained and managed, including staffing, training, site selection, technology acquisition and voice and data carriers. But those investments come directly from your pockets.

Outsourced

You can also contract out live agent call centers and self-service to third parties, known as outsourcers or service bureaus. Outsourcers can do their work in traditional premises call centers or at agents' homes. Application service providers (ASPs) are technology-only bureaus. Outsourcing is covered in depth in Chapters 6 and 8 and is touched on in Chapters 9 and 10.

Outsourcers have made the investment in the people, locations, and tools. You and the outsourcers agree and contract for specific services over a given period of time.

Outsourcing shrinks or eliminates capital costs and can get programs up and running quicker because the investment is there or can easily be made. You buy only what you need, and you can add or subtract capacity. But you have less control compared to in-house centers, less say in the people and technology. And in the long run you'll pay more, because outsourcers need to make profits.

Insourced

Insourcing is a special type of outsourcing in which the outsourcer either staffs your call center with their employees under your management or manages your entire center. With insourcing you control the location, technology, and process. But the outsourcer is responsible for the people.

Blended

Often outsourced projects are standalone. Common examples are outbound campaigns where a service bureau is contracted to call a certain number of people and after-hours, where all calls are shifted to the outsourcer.

But larger firms like credit card issuers have in-house and outsourced agents doing the same work. That limits the capital investments in call centers, gives flexibility, and keeps both the in-house and outsourced call centers honest through comparisons of performance. The message is clear: Either the outsourcer or the in-house staff can be replaced if they don't perform well.

Another example of blended is in-house traditional premises call centers and home agent outsourcers. During periods of heavy call volume you can hire home agents through bureaus to take the volume over and above the average call load during the day. That way you avoid paying for call center seats and employees who are idle for most of the period.

✪ LOCATION

But location is still another factor in the call center. Chapters 7-11 look at this issue in depth. In summary the choices are:

Onshore

Onshore is having the call center interactions and transactions take place in your home country, which, for most of you who are reading this, is the United States. But it can apply to any country. Onshore applies to in-house and outsourced live agent call centers.

Onshore is on familiar ground, with services provided by live agents who are most familiar with the people who are interacting with culturally. On the other hand it is often the most expensive location. Labor accounts for 60–70 percent of call centers' operating costs.

Offshore

Offshore is having the call center interactions and transactions with your home country's consumers, employees, businesses, governments, and nonprofits take place in another country, both in-house and outsourced.

Offshoring has developed because the labor costs are less in countries and regions in countries such as Canada, New Zealand, Northern Ireland, and in developing countries and regions like India, The Philippines, Mexico, Central and South America, Africa, and Eastern Europe. "Nearshoring" is a term applied to countries like Canada, Ireland, parts of the UK and northern Mexico, which are close to the nation's customers and head offices culturally or geographically. The downside of both offshoring and nearshoring is that no one can relate better to Americans than other Americans.

Onshore/Offshore ("Blendshoring")

To literally get the best of all worlds many larger firms are in-house or outsourced to both onshore and offshore locations. Low-touch, less culturally sensitive, high-volume interactions, like first-level customer service, order taking, and lead generation and qualification go offshore, while higher-level contacts or customers stay onshore. Services and sales issues unresolved offshore are escalated onshore.

International

This is where the call center is intended to serve foreign markets (see Chapter 10). These centers can be live agent or self-service, premises call center or home agents, in-house or outsourced, onshore or offshore. But the details such as content, how it is presented, the language used, agreements, and even the locations and buildings must conform to the cultural, legal, regulatory, and geographic requirements of those nations.

Because international entails handling foreign customers it is not be confused with offshore. But you can have in-house or outsourced agents serving foreign domestic and as well offshore (including foreign offshore, for example, British and French) customers.

✪ WHY YOU NEED A CALL CENTER

So given the complexity and costs of having call centers, what is the justification for them? The answer: Call centers are the most versatile business and organizational apparatus around.

You can use them for nearly every type of person-to-person transaction that does not require physical interaction, including sales, service, collections, business-to-business transactions or business-to-customer transactions. Call center interactions may range from something as simple as taking messages to something as complex as servicing industrial machinery.

Call centers collect a vast amount of data on individual customers and users. That enables organizations to find ways to better serve them and offer products and services at the right price. When you dial into a call center, it "knows" you if it's done business with you before. Only in the smallest retailers do you find such information and attention.

Call centers are less expensive than in-person service. The Center for Customer Driven Quality at Purdue University estimates that an in-person retail transaction costs $50 and that a field sales rep visit costs $500 compared with $5 in a call center.

Call centers are far more flexible than retail outlets or showrooms; customers do not have to wait until they open to shop. A call center's low operating costs let you turn the lights on earlier and switch them off much later. This provides customers greater convenience; many call centers never close.

But that doesn't mean call centers are open all of the time, staffed at full volume. Many companies run call centers located in different time zones, including other countries, and shift calls to these places in a practice known as "following the sun." Customers do not even realize they are being routed elsewhere.

Call centers' presence is everywhere because they provide so many valuable functions for a vast range of products and services, such as customer service for toilet paper. You can bet if this item "malfunctioned" users will not hesitate to buy another brand.

This is how it works. Say you have a travel service, such as an airline or bus charter. If inbound call volume trails off after 9 p.m. in your Scranton, Pa. call center, you can shut the center down and shift calls to your Enid, Okla. center at 8 p.m. CT. When Enid shuts down at 9 p.m., you can route calls to your Fresno, Calif. center, which is two hours ahead and stays open until midnight. If you reopen your Scranton center at 6 a.m. ET, you need to cover only the three-hour period when all three centers are closed; just use an auto-attendant with an opt-out to an on-call emergency supervisor. This three-hour period varies per time zone (for example, 3 a.m. to 6 a.m. ET, 2 a.m. to 5 a.m. CT, 1 a.m. to 4 a.m. MT, and 12 a.m. to 3 a.m. PT). Your customers are always taken care of, no matter where they are routed to or what time it is.

Call centers help your business in other ways. For example, if a retail shop has long lines or is not fully stocked, customers are stuck. The next branch may be miles away, and there's no guarantee that items will be in stock when customers come back (if they have not gone to your competition already). A call center has access to your central inventory, so it can pick and ship goods; agents can take pre-pay orders and ship goods as soon as they are in stock. If you have more than one call center, you can set up routing to send incoming calls to the first available qualified agent—wherever he or she is located.

If your customers are experiencing long hold times and it is too costly or not feasible to hire more agents, you can always outsource the overflow to a service bureau that will handle calls and contacts on contract. Or you can install an opt-out to an IVR self-service menu, invite callers to leave a callback message, send email to an automated response database, or invite callers to visit your web site. All this beats customers having to stand in line, twiddle their thumbs, and walk out the door, never to return.

Call centers provide unique benefits for business-to-business transactions. If you sell B2B through a call center, you know where your agents are. This cannot be said for field salespeople, unless you have tracking devices implanted into their bodies.

If you make purchasing decisions for your business, you may prefer to be in contact with people through a call center rather than in person; you can put a sales call on hold and tell your assistant to call people back, even temporarily ignore your email.

By going through a call center, you and your staff can avoid dealing with that well-dressed, briefcase-burdened character who is checking her voice mail and email, flashing a webcam image of the receptionists' tightly garmented physique to their home computer simultaneously, or playing Internet poker on a laptop. Your company can also save face with a call center—you will never have to tell a visitor waiting in the reception area that "something just came up" and they cannot be seen.

Thanks to their versatility, call centers have come a long way from outbound cold-call telemarketing, collections, and airline and hotel reservations—which are what they were originally used for. Now they are virtual storefronts, with sales and service shops located in buildings that may be thousands of miles away from customers. As mentioned, e-commerce firms now realize they need these human virtual sales associates to assist customers with online shopping.

Call centers may also make or break a sale. By tapping into the same software, dealers and outside salespeople can pick up leads and new customer info from agents, be it in real time off their cell modem-fitted palmtop or from a pay phone just off the Interstate highway. The call center's unsung workers—agents—can also save their tails.

Darren Nelson, founder of GWI Software, knows the consequences of not collecting up-to-date customer information. He once worked as a salesperson for several large, high-tech firms. When he made a sale, he would turn over the contracts and details to his company's implementation department. Several times he returned to clients only to discover problems that could torch future sales and commissions. "I looked stupid going in there," recounts Nelson. "I'm selling the clients better connectivity, and I didn't even understand the problems they were having with the products they were already using."

Call centers and their agents are the first lines of defense when a firm fouls up. Certain situations prompt people to call or send email more quickly than others,such as a software bug, a toxic waste spill or a defect that a safety expert alleges could injure a child. How you instruct your agents to respond determines whether or not the public and its government agents will tear off your hide.

Call center agents are corporate windows to the customers. The customer service they supply is now a vital marketplace differentiator. It is no coincidence that many TV ads and billboards show a smiling, headset-wearing agent engaged in conversation. Call centers have become customer and senior management showrooms for service bureaus and in-house firms alike, with corporate-style furnishings. My cable company, Shaw Cable of Calgary, Alberta, shows off their call centers.

✪ CALL CENTER CHALLENGES

As noted, earlier call centers are not an inexpensive investment. The cost and time involved does not include the substantial IT, HR, legal, and other corporate resources needed to commit to making a center work. The following chapter discusses technology, outsourcing, and home agent alternatives to owning a call center.

Also, call centers have their limitations. The main one is that they can't fix inherent problems with your product or service. They can't turn a sow's ear into a silk purse. Call centers can't make software easier to use, remedy production flaws (like introducing better-grade materials), make the planes run on time, or lower prices. Many telcos have excellent call centers, but the fine agents can't replace the bad wiring from the central office to your house; they can only relay the problems to the repair department.

After a while customer complaints will get to the agents; that is human nature, and customer service will deteriorate. The solution: Make sure the product or service that you make, sell, and are responsible for is the best quality it can be at the pricepoint your customers will accept and that your agents will be proud of it.

To find out what type of call center is right for you, examine your existing and future needs closely. You do not necessarily need a center to sell, survey, or service customers. Customers can come to your store, office, movie theatre, or other physical place of business, and buy from (or yell at) you and your staff in person. You can obtain customer response information through focus groups, market through brand advertising, and sell through dealers and retailers.

For example, few people today place orders for their cereal, coffee, or Thanksgiving turkey through a call center (except for gourmet foods like Balducci's Kobe beef and the pre-packaged clambakes found in airlines' SkyMall catalogs.) Most shop for food at supermarkets, but increasingly, shoppers in urban areas are beginning to going through delivery services like Peapod, which are now coming back after the dot-com bust and which need call centers.

Yet at some point you need to communicate directly with your customers. Through small in-house or outsourced call centers, food manufacturers provide information, such as recipes, about their products. Butterball's popular bilingual Turkey Talk Line, staffed by a team of home economists, handles calls from Thanksgiving to Christmas on topics ranging from the best way to thaw a turkey to knowing when one is done.

Some of these calls can get weird. A trucker, for example, wanted to know if his turkey would cook more quickly if he drove faster—he planned to roast the bird on his 18-wheeler's engine block. A California restaurant owner wanted to know how to roast a turkey—for a vegetarian meal. When one woman called to find out how long it would take to cook her gobbler, the Talk Line economist asked her how much it weighed. The woman replied: "I don't know. It's still running around outside."

If you sell through a dealer channel, chances are you will need a small call center for service, as a supplement to an outside sales team. But as Jon Kaplan, president of consultancy TeleDevelopment Services discovered when he was hired by B.F. Goodrich over 20 years ago, your customers may prefer doing business with call center agents rather than sitting down with a field sales representative.

B.F. Goodrich had Kaplan develop an inside sales call center to obtain dealerships where the tire giant did not have them. The firm franchised 16 dealers through the inside sales call center within the first 10 months. The inside salespeople (for example, agents) would say, "We don't have distribution in your area, but you're a big player in the market. We know you carry X, Y, Z brands, and we'd like you to carry the B.F.Goodrich brand for these reasons."

"The company had a philosophy where, if a franchisee reached a certain size, it would turn them back to the field sales rep," recalls Kaplan. "At this point they deserved a field sales rep's time and attention. Yet guess what? We gave dealer customers the option of having a field sales rep or continuing the inside sales relationship they had. Nine times out of 10 they wanted to continue the relationship they had. Why? If the company turned them back to the rep, the rep would visit them maybe once a quarter. In the grand scale, they were still small fish in a big pond. But for me they were a large account. And they could call me on a toll-free number at the B.F. Goodrich building world headquarters, where I had access to product managers, R&D, and price people, and get answers right away."

✪ JUSTIFYING A CALL CENTER

So given the information, benefits, costs, and challenges noted earlier, how do you make a business case for a call center, in-house or outsourced?

Most organizations now recognize that effective customer service from a live agent phone and online center is essential, if only to have people there to apologize for the company when it screws up. There are many studies showing a strong correlation between customer service delivered by call center, customer retention, and revenues.

But obtaining the funds to build, expand, and improve call centers or pay for quality outsourcing is another matter. It is difficult to quantify the results from investing in a call center or buying call center services from an outsourcer. On the other hand, some activities are easy to justify, such as outbound sales, inbound order taking, collections, and cross-sells and upsells on customer care or service calls.

Attributing a sale is more difficult where there are many channels, such as in B2B, where the call center generates and qualifies the leads, then forwards them to a field sales rep or a dealer, and in business-to-consumer, where the call center literally drives callers to the sales floor. Which person in the transaction process convinced the customer to buy the product—the agent or the dealer? Or did the customer already decide before contacting the company?

Justifying a center becomes more challenging, especially if it is for customer service where no money is involved in the transaction. Such call centers become an expense, or a cost center, even when they are outsourced. In today's hyper, bottom-line-sensitive environment, anything listed as an expense is usually marked for a slow or a quick death.

That fate may also befall companies that treat call centers like cost centers. Rudy Oetting, president of marketing consultancy R.H. Oetting and Company, points out that the cost center mentality—encouraging shorter calls, lower pay, high supervisor ratios, and high agent turnover—leads to higher customer acquisition costs, fewer dollars per sale, and lower retention rates caused by bad customer service. In short, you could be strangling your own business neck if you don't give a call center enough air.

The telcos are a good example of this. These are wireline and wireless carriers that have been taken over, gone belly-up, or could go that way in the wake of number portability, increased competition, and voice-over IP, which have allowed the cable firms to enter the phone business. Why? Because too often the service these carriers provided stunk and their customers had a choice. Good riddance. You know who you are. Enjoy your forced retirement. That's how the marketplace should work.

If this rings a bell, or alarm bells, you need to show how customer service contributes to overall profitability. You will need to draw a relationship between customer satisfaction and retention and sales. If senior management doesn't listen well, you've done your best. Now get that resume updated and out there and call the headhunters.

"If a call center is not a direct revenue generating department, then management should move toward calculating their value based on customer retention," says Kathryn Jackson, associate with Response Design Corporation.

But there are pitfalls to moving too far in the profit-center and sales-center direction. Steve Murtagh, principal with consulting firm North Highland, says that treating a center strictly as a profit center ignores vital activities such as customer service or support that benefits other direct revenue-generating functions like manufacturing, product development, and sales.

"A call center that is only a selling channel is not living up to its potential," says Murtagh. "I often find it best to treat the call center as a cost center, but with specific revenues objectives and responsibilities."

There are of course call center functions that can't be openly quantified on a profit basis, like 911 and corporate hotlines. But they are justified by lives and property saved or damage minimized.

Mark Schmidt, vice president of TeleDevelopment Services, adds that you should not neglect cost center functions like an external help desk. "Even though these functions are cost centers, the service they provide is important to customer retention over the long haul," Schmidt says.

If you want to help your center—even one engaged in sales—get credit where it is due by establishing sound cost and revenue allocation methods. Jon Kaplan, president of TeleDevelopment Services, advises creating predefined standard operating procedures and key performance measurements to allocate revenue targets and operating expenses.

To work in that framework, you must set up back-end measurements (for example, closed-loop lead-tracking processes and cross-selling activities) to track results to recognize what is driving the business, and break down all process components to clearly allocate credit. Once done, you can determine how each sales and service step contributes to revenue, then allocate a percent of revenue to the appropriate department. "It is much easier to get time, attention, and resources from top management when you can demonstrate a corporate contribution," says Kaplan.

CHARGING FOR SERVICE

In the services-that-are-no-longer-free department are travel agencies, airline reservations, and customer service. Travel agents add fees to tickets. Airlines charge service fees for tickets bought by phone.

This trend follows on to what has been happening for years in customer support for higher-end hardware and software (see *The Complete Guide to Customer Support*). Individuals and companies typically pay per minute with credit cards or prepay support plans.

There is good reason for this. Experts calculate support costs to be anywhere from 5–25 percent of corporate revenues. Al Hahn, president of Hahn Consulting, estimates that the support costs for packaged consumer hardware or software often begin at $149 per unit, equal to or more than the cost per unit to develop and sell the product.

Also, many companies have turned to charging for support initially to recoup costs but eventually to create a profit stream. Just like when you buy a razor. The money is in the blades, not in the handles. Many high-tech companies such as Oracle have come to depend on support as a revenue stream.

Charging for service and support keeps list prices low and shows customers exactly what they're paying for. But it risks creating the perception of nickel-and-diming. Banks have long maintained this practice, which has led to some legislators calling for abolishment of ATM fees and other such charges.

"What typically happens is that a company launches a new product or service that is innovative, grabs great market share and prices it so that it has a high enough profit margin to absorb support," explains Hahn. "Then as more companies enter the market, prices come down, which squeezes support costs. Eventually they are forced to break out services like delivery and support with separate charges."

Will you be able to charge for your customer service like you can for support? It depends on what your customers are willing to tolerate, but only if you give them a user-friendly self-service option.

Chapter 2: Call Center Functions

You know what call centers are, the size and scale of the industry, what a center could do for your organization, and why you may need them, either before you began reading this book or after going through the previous chapter. That's the overview. Now let's take a hard look to see what functions they provide for you to determine which of them that you need for your organization and how to justify their costs.

✪ SALES

Any time you buy a product or service over the phone or online, you are engaged in a sales transaction. You can conduct it inbound or outbound, via live agent voice, email, chat, fax, or video, with IVR or Web self-service.

Inbound Sales

Inbound sales occur when anyone contacts you to buy your product or service, consumers or business alike.

That includes order desk, where buyers place orders by any number of means: phone, email, a Web site for an Internet service provider. Customers may be responding to a print, radio, TV, Internet banner ad, a direct mail piece, faxblast, outbound email, message burst, or were flipping through a print or online catalog, or found out about your goods or service from whatever marketers and technology wizards dream up next.

It also includes reservations. When you call to make airline, Amtrak, hotel, or rental car reservations, or if any of these companies contact you with an offer and you accept, you're in sales too. You made a commitment to purchase services, and the tightness of that agreement depends on the contract terms.

Outbound Sales

Sales also comprise communications such as outbound calls, outbound voice messaging, emails, IM and wireless "bursts" and faxes from marketers and contact centers to targeted prospects and existing customers. They include communications to consumers, such as for Internet services, steak knives, and Hello Kitty hood ornaments and for credit cards and low-interest loans to pay for them.

They also include communications to businesses and organizations such as pitches for leases in office towers, the steel and concrete used to build them, janitorial and security services, and furniture, computers, printers, copiers, toners, phones, and fax machines for the occupants.

Lists Are Key to Sales

The keys to driving sales are lists of consumers and businesses. These lists include both existing customers and those likely to buy products and services similar to yours. They have also been put through sophisticated analysis. By examining past buying patterns, you can determine what customers might acquire and at what price point and when. Outbound and inbound marketers then contact customers and prospects from lists. Marketers may also pop up offers if you visit their Web sites, asking you to visit a certain page, email, chat, or call.

There are two types of lists: compiled and response. The American List Council, a big list marketer, explains the differences between the lists and their pros and cons. Compiled lists are general lists. They are pulled from sources such as directories, phone books, public records, retail sales slips, and trade show registrations. Firms can use them to reach entire markets.

With compiled lists, marketers usually know your name, family, lifestyle, and neighborhood characteristics, and they match them against demographic and geographic profiles. They give comprehensive, specific market coverage, like when a firm is opening a new store and wants to drive business into its catchment area—the zone where it expects the bulk of its customers to come from.

Response lists are those that list consumers who buy regularly through a certain channel, like by phone. The marketer may have rented this list from a magazine or even from that online garden center where you bought the pink flamingoes for your front lawn, after being drawn in by the free John Waters movie offer.

Response lists generate higher response per name. They are more psychographic, as they include attitudes and buying patterns. They are more expensive than response lists.

Marketers append information to their lists. They will add phone numbers if they have only your address, and vice versa. They also append buying data and demographic and geographic changes to your file. If your wife just had a baby, you'll get showered with offers for goodies like infant-safe toys and teddy bears. Marketers know when you move; they obtain public-knowledge U.S. Postal Service National Change of Address lists.

The marketers' reason for gaining such information is to find out what, when, and how much you want to buy, because you will buy something, sometime. They want to target only the offers that you are mostly likely to accept. Smart marketers don't want to waste time and resources buying lists and paying agents, along with ancillary facilities and technology costs, trying to sell you something you don't want or need, for example, sending you promotions for baby clothes if you don't have kids (unless wearing them is your thing and your name is on a targeted list of consenting adult buyers and prospects that are into such "variations"). It is a truism of human nature that for every itch there is a scratch and today that means relying on call centers to connect the two.

Cross-selling and Upselling

Cross-selling and upselling is an indirect form of sales. This is when you dial into a center to get information or obtain a service, such as activating a credit card, and an agent suggests or offers a product or service like an awards plan.

With outbound sales fading away, more companies are doing upselling and cross-selling every time a customer contacts the organization, for whatever reason. For example, you could be calling in about your checking account, and as you are talking to the agent their screen pops up a suggestion to offer a new credit card. The system, set to business rules your firm created, popped that offer based on a marketing push they were making at the time. It also analyzed your transaction history and any income or other personal data on file to select you as a likely prospect.

This function also occurs online. You can be browsing or shopping on a web site, then suddenly see an "intelligent agent" icon (similar to "Clippy," the venerable infamous animated paper clip in Microsoft Word) that asks if you need assistance, tells you of sales, or makes you an offer.

Say you're vroompersonalizations.com. An intelligent agent shaped like a car pops up and blurbs: "I noticed you're looking for hood ornaments. We have a limited-edition Darth Vader that fits all SUV models for $59.99."

Smart companies offer their customers who have difficulties with their products or services discounted or free items, like a month of free Internet access, 40 percent off of their next software purchase, or an upgrade to business class. This way they can retain customers by giving them a little in return for the time-sapping inconvenience.

✪ LEAD GENERATION AND QUALIFICATION

Lead generation and qualification is inbound and outbound. Here is how it works: You generate interest in your product or service by advertising, sending a direct mail piece, or sending an outbound email, and enclosing your phone number, email, Web site, or mailing address. Or your agents call or email from targeted lists.

When customers and prospects reach out to your firm or when agents make outbound calls and contacts, agents ask questions about job title, if they buy or specify products, and what they spend in order to qualify them.

If the customers or prospects ask your agents to send them literature, the agents make an offer and close the sale or pass the lead to your inside or outside account representatives, depending on your program. If the prospects have any technical questions, you can conference in your experts. The call center should then follow up to see if the inside or outside salespeople and dealers contacted the lead and find out if the contact had been satisfactory.

Lead generation is also accomplished through OVMs and outbound IVR. You leave pre-recorded messages on called parties' answering devices with your pitch, either to call or email the prospect or go to their premises. With this technique you can reach people who are not at home or work or who are just screening calls.

✪ CUSTOMER AND MARKET SURVEYS

These outbound and inbound calls and contacts are important business development tools. You can use live agents or OVMs for them. Businesses use them to find out what customers and prospects are looking for, what their attitudes would be to the firm's products and services, and what their experiences have been with the company. Call centers try to implement surveys when a customer calls in for service or sales; they usually take place at the end of the call.

The same caution for outbound cold call sales also applies to customer and market surveys. People don't want to get bugged. When I was researching a transportation venture with a friend, outsourcers had told me that consumers wouldn't take kindly to getting market survey calls. Instead I should place survey ads in publications and hope for the best.

❂ COLLECTIONS

This is the essential call center function that recipients always dread, both outbound when they get the call and inbound when they have to respond to an OVM or make an arrangement to pay a debt. These types of calls traditionally have required agents to be sharp, tough, and to the point, to tell debtors that they owe money and to obtain promises to pay to avoid further action.

With credit ratings key to buying homes and cars, and in some cases obtaining jobs, consumers especially do not need much prodding to remember the consequences. Collection agents are becoming credit advisors, helping people stay on the right side of the bureaus and ensuring that the money keeps coming in.

The key in collections is obtaining the promise to pay and to make sure that promise is kept. It requires therefore agents with cool firmness and persistence. Keep in mind there are laws to follow here, such as the Fair Credit Reporting Act.

❂ DIRECTORY ASSISTANCE

Directory assistance is one of the oldest call center functions, dating from the development of the telephone. It is used when people cannot get their hands on a printed or online directory or if the number has been changed.

❂ CUSTOMER SERVICE

Customer service is arguably by far the most common call center function. This includes inbound from businesses and consumers who have concerns, questions, or a problem with products or services. Inbound also includes IVR or Web self-service.

Customer service is also outbound, with agents calling customers to see if the product or service was satisfactory. OVMs and emails also deliver important messages to customers, like location changes and product recalls.

Delivering excellent customer service is essential before all other call center functions. Your agents need to be polite, friendly, eager to listen and assist whenever your customers call for whatever reason (sales, lead generation, market surveys, collections), and to be efficient at it so that they don't waste your time (and money) and the customers'.

My wife and I used to live in Staten Island, the most suburban of New York City's five boroughs. When we had fuzzy signals from Staten Island Cable, I called in to point that out. About an hour after they fixed the problem, the center called back to see if everything was fine.

Very often but not always, customer service can help you retain the customers you've attracted. A customer who is satisfied will likely stay with your organization, may recommend your organization to others and may be receptive to cross-selling, upselling, and outbound calls and emails. The rule of thumb is that it costs 10 percent of the expense to retain customers as it does to acquire them through advertising, direct mail, email, and telemarketing.

✪ CUSTOMER SUPPORT

Customer support is similar to customer service. The key difference is that in support, highly trained agents diagnose and often fix problems rather than entering the matter and forwarding it on to others to resolve.

Customer support handles complex products and services like appliances, computers, cable, and Internet services. Reps offer suggestions, get into the equipment to find out what is causing the problems from their own computer—or perform a combination of the two—and if need be, fix the problems remotely such as zapping bug fixes into the gear. A customer support agent resolved the previously mentioned Staten Island Cable problem. The calls and contacts can get hellishly complex and sometimes take hours if not days to resolve.

Customer support has unique issues. It is more expensive to provide than customer service or sales. Wages are higher: by at least $2-$3/hour more than service or inbound sales. Support agents often require much more training; they may need certification such as for Microsoft certification to fix applications that use that software. To compensate for these expenses, support desks grapple with issues such as paying for support, either through pre-paid or as-you-go plans.

Support managers also deal with setting and escalating problems between support levels and setting up, updating, and working with knowledge bases. Depending on the product and service, these executives may need to integrate call center support with field support, sometimes provided by third-party firms.

To learn more about customer support I highly recommend that you read *The Complete Guide to Customer Support*, written by Joseph Fleischer and myself. It too is available from CMP Books.

✪ EMPLOYEE SERVICE AND SUPPORT

Call centers provide internal service and support for your employees with dedicated individuals and teams for functions including accounting, HR, and IT. They handle calls and emails on such matters as employees wondering when they are going to get their expenses reimbursed, vacation time, and getting computers fixed. In large organizations, these teams are housed in "shared service centers," often with customer service, support, and sales.

One key employee service and support function is internal support, also known as the help desk, where IT specialists diagnose and repair computer, phone, and network equipment. While this is similar to customer support, internal support reps and their departments are also responsible for other tasks, such as monitoring equipment like computers to make sure they don't have viruses or unauthorized software.

Internal support teams also manage employees' computers and phones, known as asset management. They track assets' lifespan, remotely install bug fixes and upgrades, and switch out equipment. For example, when I worked for CMP, its IT department told me that I needed a new laptop; my old Dell machine was obsolete. I picked up my new computer, an IBM ThinkPad, and the IT staff transferred files and settings from one to the other.

Employee and Corporate Hotlines

Another specialized employee call center service is corporate hotlines. Workers use them for many functions, including sign-in/out, accidents, report alcohol/substance abuse, theft, fraud, harassment and violence.

Corporate hotlines also let employees know what to do or where to go in emergencies. They are also good news tools, providing information such as new openings, town meetings, and office parties.

Employers can also use call centers outbound, with live agent or OVM to notify staff at home, on the road, or in satellite offices. Examples include a sudden surge in customers that require staff to be on hand and disasters.

✪ TELEMESSAGING

Answering calls and emails are the most basic functions call centers provide. Answering services have been around since the 1920s.

At its most fundamental, telemessaging, which answering services have evolved to, entails just that: an agent picking up the phone and taking a message and then repeating that message to the intended recipient. But telemessaging can be more than that. Agents can screen and then forward calls directly, convert them into emails that are playable as .wav files, or place them into voicemail.

If the calls are urgent, agents can make a warm transfer or send a page, if the intended party is not available. Agents can also dispatch calls, such as to field sales or service reps.

Sending acknowledgements to emails and sorting them helps organizations deal with this electronic deluge. Spam filters are not foolproof. You can waste just as much time sorting through the recycle bin as going through the inbound box.

Telemessaging functions can easily be integrated with other call center functions. Here are some examples:

* A client calls in with an urgent request but the center is closed. An agent providing call answering can process, forward or store the message for an on-call staff member.
* A caller wants to speak to a specific person who dealt with them before. If that person is not at their desk the agent can page them.
* The call is about a hardware problem, and the agent can't fix it. They can then contact dispatch to send out a technician.

✪ EMERGENCY SERVICES

Call centers are vital tools for handling crises promptly. The most common example is 911. But they also include hotlines for abuse, violence, and suicide. Call centers can also process nonemergency calls, with agents having a good ear and training for urgent issues, like words indicating domestic violence, that they can escalate to authorities.

FORMALIZING YOUR CALL CENTER

Nearly every business has a call center. But there is a difference between call centers where there are a dozen people who make and take contacts and supply sales and customer service, and call centers where there are over 50 agents in a room or two, technically chained to their workstations.

Some consultants and publications use figures like 25 or 50 agents or seats as the benchmark between formal (50-plus seats) and informal (under 50 seats). Site selection experts say that it is easy to find people and places for informal call centers, but the task becomes more challenging for formal ones.

Andrew Hewitt, principal with consulting firm Pittiglio Rabin Todd and McGrath, recommends that you ask yourself several fundamental questions before choosing to open a formal call center:

- Does your product or service require any level of pre- or post-sales support?
- Have you sold, or are you going to sell, a sufficient unit volume to require dedicated resources to address customer issues?
- How important is it that your customers have an avenue to get their questions answered or get their issues resolved in a consistent and efficient manner?

"In today's hyper-competitive marketplace, attracting and retaining customers is a constant challenge," says Hewitt. "For many businesses, the call center is your face to the customer. It's responsible for maintaining an ongoing relationship. Call centers should be in the business of enhancing the overall customer experience and value of the product. A call center can also provide a wealth of data about your products and services."

✪ ISSUES

Sales

Outbound and inbound sales face several issues worth looking at and preparing for when putting together a call center program, either in-house or outsourced.

Outbound cold calling has been on the way out, and call centers have had to change with it. As an example, outsourcer LiveBridge closed its outbound Wilsonville, Ore. center, seen here from the first edition. At the same time the firm opened new centers in Washington State, Canada, and Argentina. Credit: Brendan B. Read

The Decline of Cold-Call Business-to-Consumer Outbound

Be especially careful when planning business-to-consumer outbound, especially cold-calling campaigns. The advent of federal and state Do Not Call (DNC) legislation (see Chapter 3) has limited outbound to existing customers and those people who have not removed themselves from DNC lists.

Many consumers have been porting their home numbers to cellphones. One reason is to avoid telemarketers. States have begun to ban cold calls to cellphones. Louisiana enacted one such bill in 2003, with an exemption to wireless providers.

Business-to-consumer advice

Scrub your lists against government DNC databases and the Direct Marketing Association's Telephone Preference Service, which some states now reference. There are services from vendors such as Call Compliance and Gryphon Networks that automatically block federal, state, wireless, third-party, and in-house DNC lists.

Customers and prospects open to outbound calls are gold. Make sure you have targeted offers and respectful scripts that seek to find out customers' needs rather than simply push products.

Take every step you can to avoid reaching cell lines inadvertently. If the called party is annoyed, be sure your scripts have the agents apologize, hang up, and have the number immediately scrubbed from the database. The DMA has a couple of services to assist you. The Wireless Block Identifier Service, which identifies numbers that have been assigned as cell phone numbers—numbers they may not call without express permission from the holder of a number. The Wireless Ported Number File identifies land-based phone numbers that have been ported to cell phone numbers

Otherwise if this becomes a problem like cold calling to landlines, expect pressure for the federal government to enact "DNC Part II." If that happens you can bury the business-to-consumer teleselling industry altogether: the second self-inflicted blast of the barrel, the first being the greed and stupidity that caused the industry to fire the first shell into its netherparts that led to the state and federal DNC legislation.

Looming B2B Problems

Be careful how you plan your business-to-business outbound campaigns. Businesses are also becoming annoyed with cold calls, especially for nondurable items such as ink, toner, paper and cleaning supplies. Many decisionmakers also work from home and on the road from time to time. You may find yourself afoul over the business-to-consumer and cellphone-calling issues. Lawmakers are beginning to crack down on B2B outbound. DNC Part III? Will such telemarketers learn from their b-to-c brethren? Or are they going to once again sit back and do nothing, hell-bent on making the money and running, and destroy that sector too, before legislation and litigation padlock doors?

B2B advice

Learn all you can about your customers and prospects before selling to them, such as what they buy, when, and who makes the decisions. Make your offers as targeted as possible. Their Web sites often contain that information. Do quarterly filings with the SEC.

When planning the program, be respectful and courteous. Time is money to the people you're selling to. In B2B the first person you're likely to encounter are the gate-keepers—in-person or virtual receptionists and executive assistants. Marketing professionals, including outsourcers, offer excellent tips how to get past them to reach the people who sign the checks.

* Email, OVMs and Junk Faxes

Email started out as a great marketing tool, but the human vermin got a jump on the marketing industry who were in "we can self-regulate" denial—and inboxes became stuffed with spam. Consequently many consumers and businesses won't look at, let alone open or respond to, emails from firms they don't do business with for fear of the very serious consequences: porno, fraud, identity theft, viruses and "zombiefication" of their computers. Web site spoofing is so good that you can't tell if the merchant is real. Spam is so cheap that offenders need next-to-nil response rates. Anti-virus software and spam filter vendors work ferociously hard to keep on top of the scams. Furthermore, filters often trap legitimate commercial email, and many people don't draw the distinction.

OVMs, outbound IVRs, and unsolicited faxes fall into the same category. These methods are easier to trace and address with legislation and prosecution. Even so there are lowlifes who abuse these media, which have brought on tough laws. As Chapter 3 outlines, the laws are very strict on these avenues. The message is the same: Don't contact either business-to-business or business-to-consumer customers unless you have a *proven* business relationship.

Email, OVM, and Fax Advice

If you're going to use outbound email, especially unsolicited email, make sure your organization's name, address, and phone number are front-and-center. Also go beyond what regulations like CAN-SPAM say. Be solicitous of prospects. Request receivers' permission to continue emailing to them.

With OVMs and faxes, make them targeted and clear. Leave your address and telephone number for them to call you, and have live agents ready to pick up the calls.

With OVMs, preferably identify your business. People screen calls; if they see "number unavailable," they *know* it is an unwanted caller.

* Cross-Selling and Upselling

Be careful how you do your cross-selling and upselling. If the matter the customer calls about is unresolved to their satisfaction, then they'll be in no mood to buy your product or service. Also, in tech support, especially where customers pay for their support either per-incident or on a contract, they might be annoyed to get a sales pitch.

Many agents do not like doing cross-selling and upselling because they feel it is violating customers' trust. Also companies have been often clumsy in how they write the scripts.

Typically, the transition from service to sales is unnervingly abrupt. Agents are popped sales scripts even before the customer's original issue is resolved to his or her satisfaction.

"Too often I hang up when I hear the agent say, 'Is there anything I can help with today?' because I know that when I answer 'No' the agent will still reply with two or three completely unrelated, heavily scripted pitch lines for cross-selling and upselling," says Todd Beck, service portfolio senior product manager with training firm Achieve-Global.

Too-aggressive cross-selling threatens to trigger consumer hostility to inbound. Could there be a "DNC Part IV"? Are the culprits in this industry that deaf and short-sighted?

The key is training the agents correctly and have the proper scripting. Teach staff how to initiate and build relationships through engaging customers in a dialogue.

Kathy Dean, principal of consultancy Banks and Dean, uses the example of a customer calling a golf gear company to inquire about a brand of club. The old cross-sell strategy has the agent launch into a pre-scripted pitch at the end of the conversation on something like golf bags, which were completely irrelevant to the conversation. However, in relationship selling the agent knows enough to ask the customer about his game and what equipment he uses. If the customer responds, the agent continues to build the dialogue. The agent then thinks like the golfer, listens, and identifies needs.

"If a customer reveals a need such as an edge in accuracy in their putt, then the agent supplies information like 'We just got in this new version of putters with tungsten inserts that provides better roll. I could send or email you a fact sheet on them,' " says Dean.

Customer Service

Call center customer service has limitations. No matter how good it is, it can't guarantee business. There are many reasons why satisfied customers won't buy again, such as better pricing and features on the same products or service the next time they shop.

I am not an airline loyalist. I pick the carrier that gets me to my destination when I want at the best price with the least amount of hassle. If I have a roughly equal choice I know which ones I'll stay away from—they know who they are.

The same goes for shopping. My wallet is packed with loyalty cards from several different supermarkets. But we don't stay at one and do all our shopping there; instead we shop around for the best deals. Fortunately we live in a small city where all the stores are no more than 15 minutes from each other. The local gas stations offer loyalty cards, but with prices the same wherever we go, we don't bother. What's the point?

Personal needs change. When I began home working instead of driving and commuting, my car usage dropped. Net result? I have less need for tires, tune-ups, parts, and repairs.

There is also a limit to how many instances of one item or service you need. After all, how many plastic pink flamingoes or other tacky lawn artwork will someone want to buy? I have two, and my wife barely tolerates them. One is discreetly in a planter in the back of the deck and the other is in my office.

Lastly, if your product or service stinks and there are alternatives, your customers are going elsewhere. If it becomes outmoded in favor of newer, better, more-effective products and services, like when men and women literally dropped their garters for stay-up socks and pantyhose in the 1960s, all the fine customer service in the world won't keep those customers.

The women's underwear manufacturers managed a comeback in the late 1970s when the younger generation that did not have to wear garter belts as necessities discovered that they were appealing to the partners not wearing them. Their male counterparts are not as successful.

A woman in bra, panties, and garter belt is a sex symbol. Yet a man wearing an undershirt, boxers, or briefs and sock garters is a comic target. The reasons are for that are beyond the purposes of this book to explain.

✪ CUSTOMER RELATIONSHIP MANAGEMENT

Customer relationship management (CRM) is a blend of sophisticated management and/or software, and/or scripting that delivers the appropriate level of service to every individual customer based on their present or future value to the organization. CRM also enables you to determine the profit per sale based on volume and price so you know whether to make the deal or see if the buyers wish to pay more to get the goods or service. CRM therefore allows you to use your resources, like your live agents (which represent 60–70 percent of a call center's operating costs), to their best advantage.

The marketing adage (mistakenly called the Pareto Principle) is that 80 percent of one's business comes from 20 percent of the customers. You give that top tier the best service, like no on-hold times, no IVR, and assigned account representatives, when they call in. Others may first have to go through an IVR menu for automated responses, which are less costly than live agents, offering them to zero out, that is, dial zero for a live person if their enquiry can't be met with a computer.

CRM systems' records that contain customers' data are used by organizations to generate targeted cross-sell and upsell offers by phone, direct mail, or email. The software can better coordinate, aim, and get the most value out of your sales and customer acquisition and retention programs. You can also use what you've learned to devise new products and services or remodel them to make them more profitable.

The CRM approach benefits mass-market B2B and business-to-consumer firms that sell and contact different channels and need to keep, update, and dynamically use data gathered from customers and develop relationships with them. Examples include catalogs, financial services, healthcare, high-tech firms, Internet service providers, telcos, and other utilities. One company, Community Playthings, saved $600,000 in catalog mailings yet increased sales by 10 percent by analyzing who were their best customers and sending targeted mailings to them.

CRM is especially valuable to firms that sell to a limited customer base, where the buyer and the product or service has high value and whose sales loss would cause considerable pain to the balance sheet. This is especially important where there is limited or slow growth in the number of customers and prospects.

The CRM concept is not new. In 1994 I wrote an article for *Metal Center News* about a Canadian steel service center that used "gross margin index" (GMI) to help determine high-performing and low-performing customers. The GMI was calculated by gross margin dollars per invoice x gross margin percentage. The firm sets a GMI threshold based on minimum profitability per order. If an order came below the GMI the salesperson could either reject it or look at ways to boost it, such as seeing if the customer would accept higher service charges.

Here's the example published in the July 1994 issue:

$120 [gross margin on a $500 order]

x 24% [gross margin percentage]

GMI: 28.8

Now let's boost the GMI:

$120

+ $35 [delivery fee]

+ $22 [skidding]

$177

= = = =

$177

x 24%

GMI: 42.2

Personal Experience—CRM at Work

Essentially, CRM duplicates for medium-to-large businesses what small businesses take for granted: customer knowledge. I first saw this at work in the mid-1970s when I worked for John Garde and Company, a family-owned factory-sewing-machine dealer in Toronto, Canada.

When a customer entered our showroom, I got up from behind my desk, walked to the counter, and asked how I could help them. Then if a second customer sauntered in that I knew or who asked for Mr. Garde, I would yell back or page him if he was upstairs in the repair shop. He would talk to that second customer, taking his time while I took care of the first.

The same went for the phones. When a customer who owed us money dialed in for a repair, I would remind them after I wrote out the repair ticket. If it was an important caller, someone I knew who had been trying to reach Mr. Garde or who insisted on talking to him, I'd try to find my boss.

When Mr. Garde prepared to leave on a repair call, he would ask me for a copy of all outstanding invoices (if the customer owed him money), grab a couple of blank invoices, plus his tools, parts, and his cheap, foul-smelling cigars. With a smile through his stogie, he flung open the wooden door and waved. A few hours later, he'd come back with a bigger grin. He'd have an order for me to process and a check, which I would write up in the musty black bindered ledgers and deposit at the Royal Bank.

What Mr. Garde and I had done single-handedly was customer service, help desk, sales, and collections, with a literally three-dimensional 360-degree view of the customer. As the owner, he had the authority to try to meet the customer needs. I escalated calls and customers to him for special treatment.

CRM Weaknesses

Unfortunately the term CRM has become discredited after too many companies ham-handedly implemented the concept and the enabling software to create two classes of customers, the elite and the hoi polloi. The elite got live agents, and the hoi polloi spent eternity on hold or in hellishly complicated IVR menus. Not surprisingly, many of the latter weren't thrilled by that treatment. And you can bet that when they merited elite status they took their business elsewhere.

"How can you have a relationship with your customers, if your agents are trying to get them to hang up?" points out marketing expert Rudy Oetting, senior partner with R.H. Oetting and Associates.

The CRM business rules that rigidly created the classes followed the fallacy of history: What happened in the past or present does not determine what will happen in the future. The past and the present is just that, reporting. It is not necessarily a precedent for actions in the future. There is no way business rules can be written to predict the myriad of factors affecting individual wealth such as the economy, an employer's business and individual job performance, who they meet and partner with, births, deaths, and divorces.

In a letter published in the February 8, 2000, *Financial Times*, Don Cook of State College, Pa., says that while customer segmentation tools are excellent for providing personalized service to top-level customers, they are not yet sharp enough to chop bottom-ranking customers. He cited a bank (that will go unnamed) that has made that mistake. "Companies that use it as such do so at their peril," says Cook.

The software itself, which costs up to millions of dollars to license, became a demon to install, integrate, and train staff within 18 months. Wall Street demands a return on investment in 12 months, or ideally 12 weeks. You also had to spend the time and resources of your IT department.

There are also privacy concerns with CRM. Despite the popularity of Caller ID, some people are still unnerved when you know their identity when they call your organization, so have your agents very respectful and "innocent" when the customer dials in.

You may also not *need* CRM. If you still have a natural monopoly, depend on walk-in traffic, do not have multiple sales and service channels, or have low customer acquisition costs, you are wasting your time and money on CRM methods and software. This is especially true for smaller consumer products firms. The software is also not worth your while if you have little contact with end customers or sell low-value products to a small customer base and have a tiny number of contacts.

Chances are you also won't receive much payoff from your investment if you do not have repeat business from customers or if your product or service market consists of customers who primarily buy based on the lowest price or the sharpest cutting-edge technology. If your business functions in islands, designated by product line or function and not connected by fixed links, you will also not benefit from this approach.

Your customers may not want a relationship with your organization. As illustrated in the discussion on customer retention, buyers are loyal until the next deal or innovation, depending how much money they have or want to spend.

Here's a great example. My wife and I live in a small Canadian city. Canadian businesses have bought into loyalty cards and clubs big-time. I have in my wallet cards from QFC, Save-on-Foods, Safeway. All the major gas chains such as Esso and Petro-Canada have "points" plans.

On the groceries side, we know which stores carry what items at the best prices and when the specials come out. All of the outlets are within 10 minutes of each other. There's no difference on the gas prices. So guess what we do? Shop on price and convenience. So much for building a "relationship" with us. If the business does not have what we want when we want it at the price we want, then it doesn't get our business. It's that simple.

The same with airfares and hotels. They have become commodities like food and gas.

Price and convenience ... price and convenience ... price and convenience. Maybe if you repeat it enough it will displace "CRM" as the new marketing mantra.

As *Call Center Magazine* Editorial Director Keith Dawson put it: "Customers don't want to have a relationship with you, and never did. All the wishing and marketing in the world won't make it so."

CRM Advice

Be very careful when considering, devising, and implementing CRM. See if you need it, or more accurately, do your customers care about it when making buying decisions to begin with. If you do go this route, make sure that you treat *all* your customers well. No long hold times. No IVR jails. No one minds seeing people go to the penthouse floor or towers if they are all staying in a four-star hotel, but they will complain if they're assigned rooms with lumpy mattresses, scratchy blankets, and the do-it-yourself maid service.

As *Call Center Magazine* Chief Technical Editor Joseph Fleischer pointed out in a December 2003 article on CRM ("Relationships Redefined"): "CRM isn't only about giving special privileges to the most valuable customers; it's about striving to make all customers more valuable to a company. A company earns repeat business in return for learning, and acting upon, what its customers want. In time, the company knows its customers well enough to cross-sell and upsell to them."

Chapter 3: Planning in the Call Center

You've decided that you need a new or expanded call center. You know what functions you want it to provide. But do you know what is entailed in incorporating a call center into your organization?

Call centers are expensive, resource-demanding entities. They can cost up to millions of dollars to set up and maintain. For these reasons you need to know the characteristics of call centers and determine the size and number of centers you actually need so that you have fair idea how much of an impact they will have on your organization.

From there you can begin to explore your strategies. You will then be left to look at alternatives for live agent handling, including self-service (IVR, outbound voice messaging or OVM, fax, automated email, or Web self-service, Chapter 4), home agents (Chapter 5), and outsourcing (Chapters 6 and 8). Once you've sifted the volume with those meshes you will be left with the contacts to be accommodated in your own call center. You may find you need a far smaller call center or none at all.

Deploying, expanding and consolidating a call center may be a great opportunity to take a new look at how you do things. There are new strategies, methods and technologies that enable organization to carry out their goals better, for less money. Why not take a look at them now?

✪ CHARACTERISTICS OF CALL CENTERS

Call centers are different from other functions in an organization, in ways that impact facilities, design, telecommunications, staffing, training, management, and business continuity (disaster) planning. Here's how:

Cost

Call centers are not inexpensive to provide. To accommodate employees costs $20,000–$40,000 per workstation per year, depending on the location and size of center, furnishings, equipment, and maintenance expenses. The figure includes property, amenities, cabling, heating, air conditioning, power, voice, data and furniture and equipment amortization.

Wages and benefits and allocated HR (staffing, training, and management) costs per employee run from $7 per hour for basic customer service and order desk to $20 per hour for third-level tech support, says John Boyd of site selection firm The Boyd Company. Team and group manager salaries range from $57,400/year to $74,700/year, reports Mercer Human Resource Consulting.

Density

Call centers squeeze more people in a given floor area than other uses, typically one-third more. While a typical office may have 4 workers per 1,000 square feet, a call center may have 6 or 7.

There are many implications to this reality. You get more people for your real estate dollar, but you need more air changes (hence higher-capacity heating, ventilating, and air conditioning [HVAC] units), additional break room space, washrooms, and parking. More people on the premises mean more workstations and voice, data, and power cabling. You will need higher quality, more-durable flooring. Greater density means more workers per floorspace who have to be evacuated in an emergency.

Call centers located in multiple-tenant office buildings, especially multi-story office buildings, have caused problems because there are more people using common facilities like elevators. There have been parking issues caused by more people per square foot and by shift changes as workers arrive before others have left.

On the other hand, more organizations are becoming more like call centers. They are finding ways to squeeze more people in the same space. Check out the densities in your non-call center offices and you might find them similar.

Smoking

Many call centers apparently have a higher ratio of smokers to nonsmokers compared with other functions. That has serious cost, health, and environmental consequences. Smoking poisons both the smokers and others exposed to the fumes. Smoking-related illnesses increase sick days, decrease productivity, force up medical costs and jack up the odds of fires.

Organizations must either accommodate smokers with outside or pressure-ventilated smoking areas or ban smoking anywhere on the premises, especially around the building. Call centers have earned a bad reputation in offices because of smokers congregating around entrances.

Noise

Call centers are much noisier than other offices, about 60 to 65 decibels, not simply because there are more people, but because in most cases they are continuously on the phone. Higher densities mean more workers yapping in the same space. Dr. John Triano of the Texas Back Institute likens the din to power tools. Research by Professor Gary Evans and his student, Dana Ellis at Cornell University, showed that a typical office noise level of 55 decibels is high enough to cause problems.

Chapter 13 looks at methods to curb noise. One of the most common tools is the headset. The universal logo for call centers is an agent wearing a headset. You will need to buy them. High-quality, very durable sets run into hundreds of dollars and are worth it.

Stress

Call centers are high-stress environments. To make sure calls and contacts are handled without long wait times and obtain maximum productivity, agents are confined to cubicles except for pre-determined breaks. Centers are staffed to meet expected inbound or outbound volumes, leaving little idle times between calls, emails, and chats.

To relieve stress, midsized-to-larger call centers provide break rooms where agents can chill out. Some of these rooms have videogames or old-fashioned pinball to permit staff to unwind.

24x7 Operation

Many times call centers are open evenings, nights, weekends, and holidays. Buildings and operations must be set up to accommodate those employees.

HVAC and lighting must be kept on. Provision must be made for food service because even if it is just vending machines, most cafeterias and restaurants will be closed.

Security is a prime consideration. You will need electronic keypasses, desk guards, and patrolling guards.

24x7 operations must be taken into account in site and building selection. An area that is reasonably acceptable during the day can be a nightmare at night. You must pay much closer attention to local crime rates, incidents of vandalism, massive amounts of graffiti, broken glass, youths hanging around, nearby rowdy bars and liquor stores, and lighting and visibility.

Short Breaks and Need for Amenities

Call center workers take shorter breaks and lunch hours compared to other office and industrial employees. That means the facilities must located near restaurants or minimarts, or food must be provided on site.

Labor Cost and Availability

Call centers have particular location considerations and requirements. A city or town that is ideal for a regional or headquarters office, or for its factory or warehouse, may not be so for the call center.

Labor accounts for 60–70 percent of the operating costs. Wage rates vary by city, state, and country, affected by factors such as other call centers in the same labor market. Specific needs such as Spanish speakers also restrict labor supply and raise costs.

Because the workforce is generally lower-paid than other staff, they are not as willing to commute as far. As noted earlier, those employees may need mass transit. Also, break times tend to be short, usually 30 minutes, unlike the hour or so for salaried and other office employees. That means you have to site your call center close to restaurants or have onsite catering.

Communications Needs

Call centers, practically by definition, are communications-intensive. Their needs are easily double or triple that of conventional offices. An accountant, programmer, or writer may be on the phone 15 or 20 minutes per hour—if that—but a call center agent is on for 45 minutes per hour.

A conventional office, factory, or warehouse can easily use direct-inward-dial (DID) phone system to individual departments or individuals. They also could use key systems, where the system hunts for an open line and then rings it. The state-of-the-art offices use voice over IP (VoIP) routers over their data networks. Either way, the function is basically the same: taking incoming calls and connecting them to others.

But a call center needs to more efficiently route larger numbers of calls and or emails, chat, or video contacts to agents. Routing can be based either on availability or specific attributes, for example, for Gold Card accounts. It can also be based on skills such as fluency in French or Spanish or knowledge of Apple Macintoshes, known as skills-based routing. That is done with automatic call distributors (ACDs), which queue and forward contacts to agents. The ACD functionality is hooked into network-hosted or network-routing, telco central offices or Centrex, proprietary computer or PC-based private branch exchanges (PBXes), or voice and data IP routers that you own and set up.

If there are home agents, you may need special off-premises extension (OPX) cards in switches, standalone OPX boxes, or IP gateways to convert calls to VoIP. Where reliability is critical, such as for emergency or utility customer service call centers, you may need to be fed by two local carrier central offices.

Call centers suck up and spit out vast amounts of data, and that information must be received and transmitted reliably. That requires fat pipes like T1s into buildings, backed up by extra connections in case the line is cut, like by a backhoe.

Once inside the call centers, you need a lot of cable to connect the switches and routers to agents' desktops, either computer softphones or separate handsets, both with headset jacks and headsets. If you go with IP, you cut down but not eliminate cabling.

Don't even *think* of wireless in call centers. The data volume is intense. You also don't want to have to have people calling you back because the information got lost, especially in critical apps like 911. There is also the danger of outside hackers ripping off data. If it is found that you didn't do enough to protect the information, you could be held liable.

Contact Processing

Because call centers receive and make high volumes of expensive-to-handle calls and contacts, they rely on hardware and software that automate the contact processing. To get those labor-saving, productivity-enhancing benefits, you need to invest in these tools or contract with outsourcers.

The key tools are:

Interactive Voice Response

Interactive voice response (IVR) is available in touchtone or optionally with speech recognition. IVR automates inbound calls through:

* An auto-attendant ("if you know your party's extension enter it now...")
* Menu selection by either touchtone or voice response to different sources
* Taking identification to send calls to the agents designated to handle them, such as by product or service type and by caller-designated language

❋ Receiving data such as name, address, account, or credit card numbers and transmitting them to live agents, saving time and money by avoiding having the caller repeat that information

Email Templates

Email templates are tools that speed up live agent processing of inbound emails by identifying keywords and coming up with set responses that staff then quickly check before sending. That saves time and enables uniform replies to the same issues compared with writing customized non-templates replies.

Predictive Dialers

Predictive dialers are sophisticated computers that dial numbers pulled from databases and if answered, connect the called parties to live agents. They are far faster than dialing the numbers directly or by popping the numbers on agents' screens for dialing. Predictive dialers also detect whether there are answering devices on the lines and do not connect, also saving time and money.

Computer Telephone Integration

Computer-telephone integration (CTI) refers to software that takes caller information and matches it with the record you have on that customer or prospect and transmits them to your agents. CTI saves them from putting your customers on hold while they look it up.

Customer Relationship Management

Customer relationship management (CRM) software at its most basic records customer interactions, no matter the channel (see Chapter 2). The information is then used to profile customers, determine their individual value, and target them with products and offers. The data collected by CRM systems lets you deliver the level of customer service —i.e., direct to agent, IVR first—appropriate to the value of the sale and customers' existing and future worth to maximize profitability, both short-term and long-term.

Management Tools

Call centers also use management tools. They include call monitoring and recording, scripting, call logging and reporting, and information displays, like calls in queue. A key technology is workforce management software that enables you to forecast and schedule agents to meet service requirements and make changes when absences happen.

Training Technologies

Call centers also make intensive use of what I call technology-based training (TBT) or eLearning tools to initially or refresher-train agents. This topic will be covered in depth in Chapter 14.

Computing and Technology Requirements

Call centers have become hardware and software-intensive . In addition to phone switches and routers, they need heavy-duty fault-resilient servers to support costly applications such as CTI, CRM, TBT, and workforce management.

Call centers are also information-intensive. You may need databases and supporting servers to house customer and employee information, and knowledge bases from which agents can pull information to answer questions and solve problems. You may also need middleware to link databases, like checking, savings, and credit card accounts.

Chances are agents will need computers. The size and cost depends on the function. If your employees will be storing data on their PCs or printing it, off-standard sets are fine.

If they are just talking calls and working with your databases and knowledge bases then all you need are storageless PCs, which are essentially glorified dumb terminals. Storageless PCs are cheaper to buy and operate and provide much better security— because they lack user-accessible data storage memory means. Because they can be used only on your network, they are less likely to walk out the door. But if your agents have to save or download data, then you need plain PCs.

Call center technologies are not exactly plug-and-play despite so-called "open standards," unless you're prepared to have it hosted. Be prepared for integration and support costs.

For example, CTI is easier said than done, especially if your firm is big and established. Let's look at a banking application. A customer who lives in San Diego just got a big raise and decides to move their money and calls your toll-free number. You may have their checking account on a HP Integrity server running Linux in a Tampa data center, their savings account on an IBM A/S 400 in the bowels of a regional office in downtown Los Angeles, their IRA on a Windows NT machine in Las Vegas, and their credit card accounts on a Sun Solaris in Ottawa, Kansas. And your call center is in Port Angeles, Washington. Somehow all of these computers have to be made to talk to each other.

For an in-depth look at phone systems and call center technologies, I highly recommend several CMP books including *The Call Center Handbook* by Keith Dawson, the *Telecom Handbook* by Jane Laino, and *A Practical Guide to Call Center Technology* by Andrew Waite.

IT Support

Call centers need IT staff to maintain, upgrade, and troubleshoot the expensive technology. Installing new hardware and software tools may require the assistance of costly consultants. That personnel can be your employees, contract or outsourced in separate call centers.

Staff Is Intensively Managed

Call centers, like retail and hospitality businesses, depend on having adequate numbers of trained employees on hand to deliver the expected level and quality of service to deliver optimal results. Call centers also often get sudden and unplanned demand spikes, like after the release of a hotter-than-expected product or a storm that floods the switchboards with calls about power outages.

That means, unlike much other office work, you must pay very strict attention to absenteeism, late arrivals, early departures, personal time offs, sick days, and vacations.

Call center managers may need to ask people to come in early, stay later, or work on their days off.

Because call center employees directly interface with customers, clients, or other employees, their activities need to be checked up on. This is especially critical when agents are making sales offers or other tasks where there could be lawsuit-generating disputes between your organization and other parties. To enable this checking your center will need tools like call monitoring and recording and email and chat session reading and buffering, that is, reading email and chat interactions before they go out.

Because what these agents do impacts your organization's performance, their performance needs to be measured. You will require means to define and track performance such as customer and employee satisfaction surveys taken immediately after interactions so that experiences are fresh in the surveyees' minds and the results can be quickly acted on to have the most impact. You will know what you're doing right and wrong and how to improve on both inputs.

High Turnover

Call centers are marked by high voluntary turnover, 30 percent to over 100 percent annually. The staff changes are similar to those in retail stores and restaurants, which are also revolving employment doors, rather than offices, which call centers outwardly resemble.

That means your recruitment or HR office needs to be equipped to efficiently recruit, screen, and select large numbers of people and have them trained. Alternatively they can seek out staffing agencies to do that for them or hire these agencies or service bureaus to bring on their own employees to work in your centers, a practice known as insourcing.

◎ KEY LEGISLATION AND REGULATIONS

Call centers interface with the public, selling and providing services. They also house data, and the agents handle data. Both of these factors leave centers open as channels for crimes like fraud and identity theft, harassment, and annoyance. In response, the public has demanded that elected officials take action through legislation, regulation, and enforcement.

There are many federal and state laws that impact call centers. If you have one or outsourced one, you, your legal department, and your management teams must be aware of them.

Important note: The information provided below is reasonably current as of press time. Laws and regulations change. Make sure you stay on top of them.

Federal Rules and Regulations

Most of the U.S. laws and rules that apply to call centers are administered by either the Federal Communications Commission (FCC) or the Federal Trade Commission (FTC). The FCC and the FTC have different areas of jurisdiction. For example, the FCC covers banks, insurance, telcos, and transportation firms that the FTC does not. Therefore there are examples, as you will read later on, where there are two acts that have the same intent but which reflect these jurisdictional differences.

Here are some of the key U.S. federal laws and rules to remember:

* Telemarketing and Consumer Fraud and Abuse Prevention Act

The Telemarketing and Consumer Fraud and Abuse Prevention Act authorizes the FTC to make rules to prevent telemarketing fraud.

* Telephone Consumer Protection Act (TCPA)
* Telemarketing Sales Rule (TSR)

These laws and regulations are similar, but the seeming duplication is necessary because of administration and jurisdiction. The TCPA is administered by the FCC; the TSR by the FTC.

Important note: Many of these rules also pertain to inbound sales. For example, customers must be told information, such as price, quantity, terms, conditions and limitation, no refund policies, prize promotions and negative options that affect their decision whether or how much to buy or donate. Also inbound calls exempt by the laws automatically come under them if those interactions become cross-sells/upsells.

Key Provisions
Do Not Call Lists

The rules set out a national blanket Do Not Call list (DNC) that consumers can place their names on so that they won't get cold-called by or on behalf of businesses. Companies must also maintain individual DNC lists. Call centers must scrub their lists against the DNC lists and pay fees for doing so.

At this writing, B2B calls, market surveys, charitable calls, and political calls are exempt. But for-profit firms, like outsourcers that call on behalf of charities, must honor individual-specific DNC requests.

The FTC and FCC allow marketers to call consumers who have given consent to be called and who have established business relationships with those firms, defined as transactions within the last 18 months and inquiries within the last three months. However, if consumers tell companies that they don't want to be called, even if there is a business relationship, the companies have to honor it.

Also, individuals can explicitly express written agreement to receive calls, even if their names are on the DNC registry. The signature may be a valid electronic signature if the agreement is reached online.

No, this doesn't mean "no telemarketers." This is the stamp of PhoneBusters, a Canadian police-based effort to stop destructive, fraudulent Canadian and US telemarketing that gives the function of outbound calling a bad rap.

Remember, IF YOU HAVE WON A PRIZE IT DOES NOT COST A DIME.

PHONEBUSTERS

1 (888) 495-8501

✪ FRAUD AND HARASSMENT

The TSR has several provisions to limit fraud and harassment. The rule bans misrepresentation for the purpose of getting someone to pay for goods or services or make a charitable contribution. It requires express, verifiable authorization and bans unauthorized billing.

It is also illegal to give or receive unencrypted account numbers to others for consideration. That is to stop people from illicitly trading unencrypted account numbers that enable fraudsters to place unauthorized charges against consumers' accounts.

The TSR also prohibits credit card laundering. It restricts payment for sales of credit repair services, prize recovery services, and advance-fee loans.

The TSR require telemarketers using pre-acquired account information and free-to-pay conversion offers to obtain consumer assent to charge a set account number, ask consumers to recite the last four digits of their account numbers, and record the interactions.

Telemarketers are banned from calling consumers who placed their numbers on individual DNC lists, especially with the intent to harass them, like making rude remarks or hanging up on them. Telemarketers cannot make threats, intimidate, or use obscene language.

The rule also bans "substantially" assisting sellers or telemarketers while knowing or trying not to know that they are violating the law. You can't take steps to ignore violations or look the other way.

Predictive Dialer Abandonment Rates

The rules regulate predictive dialers' abandonment or dead-air calls. For dialers to operate efficiently they cannot connect every agent each time it connects with the called party. So they must abandon some of the calls.

The TCPA and the TSR say call centers must connect live agents to called parties in four rings or within 15 seconds, otherwise they consider the calls abandoned. They prohibit companies from abandoning more than 3 percent of their outbound calls answered by live people. The dialers must record and report data such as the actual abandonment rates and compliance.

Caller ID

Telemarketers must transmit certain Caller ID information.

Calling Hours

The federal rules allow outbound calling between 8 a.m. and 9 p.m.

Faxes

The TCPA regulates outbound faxes, including those sent to computers or fax servers. You cannot send faxes without express permission. You also cannot send customers faxes asking for permission. You will need to maintain a do-not-fax list. There are no exemptions for nonprofits. When faxes are sent, they must include the legal registered company name of the sender.

Outbound Voice Messaging

OVMs are strictly regulated by the FCC, which calls them autodialers, and by the FTC. You need prior consent or a business relationship to use them to pitch to consumers, unless you're using them on behalf of a tax-free charity.

The FCC bans you from calling emergency lines, the lines of any guest or patient room at a hospital, health care facility, home for the elderly, or similar establishment, or to numbers where the called parties are charged, such as pagers, cellphones, and radio-phones. While you can call businesses, you can't tie up two or more lines of a multiline business at the same time.

The messages must state at the beginning the identity of the business, individual, or other entity initiating the call. During or after the message, the caller must give the telephone number (other than that of the device of the business, other entity, or individual who made the call. The number can't be a 900 number or any other number for which charges exceed local or long-distance transmission charges.

The FCC requires that the OVM devices release the called party's telephone line within five seconds of the time that the calling system receives notice that the called party's line has hung up. In some areas there might be a delay before you can get a dial tone again. Your local telephone company can tell you if there is a delay in your area.

The TCPA and TSR count voice messages as abandoned calls unless they meet certain criteria. These vary by rule including prior consumer consent, existing business relationships, and inclusion of a non-extra-fee phone number, such as a 900 number, with the name of the seller.

Paperwork

The regulations have stringent record-keeping requirements. For example, the FTC requires that you keep records for two years on advertising and promotional materials, information about prize recipients, sales records, employee records, and all verifiable authorizations or records of express informed consent or express agreement.

Tracking Employee Names, Numbers, and Addresses

The FTC requires that you keep track of the names, phone numbers, and home addresses of any employee directly involved in telephone sales or solicitations, outbound and inbound. This is more of an issue offshore than onshore.

"Many offshore telemarketers use fictitious names for their agents to make them sound like Americans," says Larry Mark, chief technology officer of dialer maker SER Solutions. "The FTC allows this; however, each fictitious name must be traceable to only one specific employee."

CAN-SPAM Act

CAN-SPAM regulates outbound email. It prohibits deceptive or misleading headers, transmission information, and content, and nonfunctioning or false return addresses.

Senders must also have valid remove features on the emails and keep them functioning for 30 days after transmission. They must not send emails to such addresses once the remove requests have been received. They cannot forward or market "do not spam" requests, long a favorite trick of spammers because the receipt of remove requests are gotchas that the email addresses are live.

Email includes text messages. You are not allowed to send unsolicited mobile service commercial messages that reference an Internet domain associated with wireless subscriber messaging services. But you can send customer service or billing information, subject to FTC regulations.

Debt Cancellation Contracts and Debt Suspension Agreements Rule

The Debt Cancellation Contracts and Debt Suspension Agreements Rule requires that if financial products covered by federal law, like credit cards, are sold by phone, customers can authorize the transactions. Banks must maintain sufficient documentation to show that customers have been told of the terms.

Financial Services Modernization Act

FSMA requires financial institutions to provide initial and annual privacy policy notices. It also gives consumers the right to be notified when a firm wishes to share data on them with other companies, and it gives consumers the right to opt out of such arrangements. Firms cannot share account numbers and similar data.

Health Insurance Portability and Accountability Act

HIPAA requires you to trace who handles data at every step. It makes health care providers, plans, and clearinghouses (that is, "covered entities") responsible for handling consumer health care data. That includes outsourcers.

Call centers must obtain signed authorization from patients before they can use or disclose protected information (name, admission or discharge dates, or Social Security numbers) for marketing and purposes not related to health care or payment.

Outsourcers, known as "business associates," must comply with HIPAA at the behest of covered entities, such as by developing privacy notices, training employees on privacy requirements, and protecting data. They may have to sign compliance agreements with covered entity clients.

Application

Most of these regulations and provisions apply to business-to-consumer calling. But be careful. Charitable and some B2B calling are also regulated.

Charities must make certain prompt disclosures in every outbound call, get express verifiable authorization if accepting payment by methods other than credit or debit card, and maintain records for 24 months. They cannot call consumers who asked not to be called. But they are exempt from the National Do Not Call Registry provision.

Also, B2B outbound calls to sell nondurable office or cleaning supplies such as ink, toner, paper, and pencils, are covered by the TSR. But these marketers are exempt from the DNC and record-keeping requirements.

State Legislation

States have their own legislation in many areas. Some federal regulations pre-empt state requirements, others do not. Examples are requirements for telemarketers to register with them and use state DNC lists. Some states have calling hours and define existing business relationships more restrictively than federal rules do.

Foreign Laws and Regulations

When you locate in or sell or serve customers in other countries, you fall under the jurisdiction of their laws. Ask your counsel about this. There are in particular two sets of foreign laws that have direct impact on call centers. They are:

Data Protection Directive [Europe]

The Data Protection Directive, devised and approved by the European Union, has been adopted into law by the EU's Member States.

The Directive gives European consumers the right to opt out of corporate access to their personal data. This includes names, addresses, and phone numbers. They can opt-in to collection of sensitive data such as ethnicity, political preferences, health information, philosophies, religious beliefs, lifestyle, and union activities. There are limited exceptions to the opt-in requirement. Check out the U.S. Commerce Department's web site for more information (www.export.gov/safeharbor). Also, consumers must be told why the data is being collected and have the right to object, and to access and rectify incorrect information.

The Directive imposes restrictions on moving personal data to nations outside of the EU Member States. This has led to potential disputes with the U.S., which has a much different, more private and self-regulatory privacy protection system. To get around this issue, the EU and the U.S. Commerce Department have a Safe Harbor agreement that allows U.S. firms to voluntarily adhere to the directive. This provision also comes into play on offshoring from Member States to non-member countries.

Because Member States can alter Directives when making them into laws be sure to have your legal staff check into those in each country where you have operations.

The Personal Information Protection and Electronic Documents Act [Canada]

PIPEDA, administered by Canada's Privacy Commissioner, requires businesses to obtain consent from consumers before collecting personal information that can identify them. Examples include credit card numbers, financial, health, and income data, and identification numbers.

The law requires different types of consent, depending on the sensitivity of the information. For example, you can have implied consent to obtain the individual's address for the purposes of delivering a product. But you need their express consent to obtain sensitive information such as financial and health care data.

PIPEDA allows businesses to collect some publicly available information such as published phone books. Businesses will also need consumer consent before transferring the data to other users, with the exception of debt collection, government or police investigations, or emergencies. If businesses have no use for the data, they must dispose of it in such a way as to prevent improper access.

PIPEDA affects American data handled in Canada, such as in call centers. Check with your legal staff.

Note: Provinces have passed privacy legislation. Ask your staff to look into that as well.

✪ SIZING

The impact a call center will have on your organization depends on the contact volume you expect to receive or distribute. I recommend that you assume that all of your contacts will be handled by live agents. Pretend that self-service doesn't exist. That way you get the complete view of your contact volume.

After you establish the total volume and costs, then you look at self-service and other strategies to pare down the demand that has to be handled by live agents. You then look at options like home working (Chapter 4) and outsourcing (Chapters 5 and 6) to cut that inflow and outflow further. At the end of this diverting process you get the actual flow that you need to carry with your own call centers, and scale and design them accordingly.

Call center programs are typically measured by number of inbound and outbound calls and contacts per minute, per hour or by the number of calls or contacts received with allowance built in for offline work like looking up information and additional training. You can then equate the minutes to agents' compensation, which is almost always hourly, or hourly with bonuses and commissions for sales work. Very few agents are paid salaries.

The "unit of production" in a call center is the workstation or seat. That is where the call center agent works. The call center's base output is based on how many workstations there are. Additional output comes from the same workstations with second, third, weekend, or holiday shifts, just like the number of sewing machines in a garment firm or buses with a bus company or transit authority.

Factored into the program measurement and per-minute costs are supervisory and managerial expenses. Call centers require one supervisory staff member for every 8 to as much as 20 agents.

Calculating Demand

The key to figuring out how big a call center or project should be is to determine how many agents you will need to handle expected demand. From there you calculate how many workstations you need or minutes or hours to rent from outsourcers.

Assume that the expected demand over a given period is the highest you expect. That way you will avoid making your center or program too small, leaving you with annoyed customers, employees, and members of the public.

Also budget for idle (7-9 minutes) and wrap-up time, such as for online entry and paperwork (6-10 minutes), and short rests (a 2-3 minute rest per hour. Figure from 10 minutes to 20 minutes in total.

The actual amount of wrap up time can vary by function and by industry. Brad Cleveland, president, Incoming Calls Management Institute (ICMI), a leading industry professional organization, reports that insurance calls require 3 to 6 minutes per calls, or 15 to 20 minutes per hour. At the other extreme are reservations, with 1 to 2 minutes per call.

The rest period is important. It is a quick recharge of agents' batteries to permit them to continue to work productively. Outbound is easy. You calculate how many minutes you need to complete each call with buffer time in case it goes longer, call wrap-up time, and how calls you will be making. Each agent usually makes 15–18 outbound calls an hour if using predictive dialers.

Inbound is something else. It is a very complicated exercise beyond the scope of this book and is best worked out between you and your experts, based on your experience. If you already outsource, your partner may help you there.

To get a handle on staffing issues, I recommend reading *Call Center Management on Fast Forward* by Brad Cleveland and Julia Mayben of the ICMI. The ICMI offers a wealth of information through conferences, consulting, seminars, and media.

The Erlang Formula

Here are the basics of inbound call calculation. In an ideal world there will be an agent to answer every call that comes in. But that gets very expensive because odds are that most of your staff will be idle; why pay for agents doing nothing? Also, you may be paying for telephone trunks into your call center that are just storing matter.

There is a well-known mathematical formula, Erlang C, that enables you to roughly predict how many agents you will need based on queue or hold times and the capacity of the phone trunk lines terminating at your center. But Erlang C assumes no busy signals or lost calls, which do happen.

Cleveland and Mayben report that to make Erlang C work you need to input four variables. The formula works for 30-minute intervals:

1. Average talk time, in seconds
2. Average after-call (wrap-up/rest), in seconds
3. Projected number of calls, which is based on factors like historical experience
4. Service level objective, which are the seconds in which you plan to answer a certain percentage of calls, for example, in 10 seconds an agent needs to answer 85 percent of calls.

Simple? Not quite. Different programs and customers have different tolerances for being put on hold. For general customer service is it 3.45 minutes, for help desks 9.75 minutes. If you're taking reservations for a five-star hotel, you know your clientele's patience for waiting 5–7 minutes on hold is in the range of nil to zip. But waiting 5–7 minutes is an instant for packaged or downloaded software consumer customers seeking technical support.

Average talk time, also known as average handle time (AHT), also varies. Some calls like telemessaging, dealer locators, and simple orders are short, while other contacts like replacement of lost credit cards, flight cancellations, reticketing and tech support are long. Too often call centers follow AHT too closely, having agents shove callers off before they are finished, annoying them. Yet having AHT that is too long degrades productivity.

When planning, err on the side of allowing for a little more but not much more volume than your calculations state that you need. One of the worst decisions you can make is grossly underestimating the workstations your call center will need. However, you don't want to waste money buying space that you don't need, oversized hardware and software, or paying for agents who will spend most of the day playing solitaire on the Web, buying swampland in Florida, or looking for more challenging jobs elsewhere.

While most people understand that they have to wait, there are limits to their patience, especially if they contacted your live agents after trying to order or solve their problem through your IVR or Web site. After all, it's not the customers' problem that you don't have enough agents and workstations. They paid you for it, in a portion of your product or service's price, a service contract, or paid technical support.

Usually the less sophisticated the call type, such as outbound sales or inbound order-taking, especially direct response, the more control you have over it, therefore, the easier it is to calculate call load. You know from experience how long it will take for each agent to pitch and the customer to ignore or swing back in an outbound call; you can set your predictive dialer to throw a set number of calls—say, 10 to 15 each hour—to your agents. In inbound order sales and response you will have some idea from your records of the average call length; if you use infomercials, you will know precisely when the volume spike will hit your centers.

Blended and Multimedia Contacts

Complicating these matters further is the decision whether or not to have blended agents, inbound agents who make outbound calls when inbound demand is slack and vice-versa. You will need to train the agents in both skill sets. Another decision is whether or not agents handle multiple media: email, fax, chat, white or "snail mail," and video, as well as voice.

If your call center will also be handling email and chat, as many increasingly are, you must account for the fact that it takes longer to receive and reply in writing than by voice, which means that each contact will have longer handle times than with voice. You can easily demonstrate that by comparing how long it takes to write out a question or answer compared with saying similar words, allowing for mental composition time.

This is a lesson this old journalist learned in English class eons ago. That's why there is shorthand, formal and informal, and recorders.

Though there are email response templates to speed answers to where it could surpass voice, chat and escorted transactions can be two or three times as long or more than voice, according to ICMI president Brad Cleveland. This depends on the nature of the dialog, the customer's equipment and typing proficiency, the quality and speed of the voice and Web connection, the tools and resources available to the agent, and the proficiency of both the customer and agent in handling these dialogs.

On the other hand, in many instances you can set aside email responses or pre-arrange for chat at less busy periods. But don't be too long. Nothing annoys a customer who is emailing or chatting more than a company that will not respond promptly. E-commerce researchers report that the lack of timely response to emails ranks in the top tier of customer complaints.

Your best bet, say experts, is to measure your existing multimedia transactions, look for averages, come up with the equivalent of AHTs in voice, and plot from there. Then you apportion your voice and voice AHTs into your planning based on the experienced percentage of both.

○ BUSINESS CONTINUITY

Call centers have particular business continuity needs that affect location, design, operations, and costs. The more your customers, clients, employees, and the public depend on the services you provide, and the more your bottom line requires that your organization be available, the more essential it is that you plan to ensure that your centers keep functioning no matter what.

Business continuity means ensuring that your operations will continue and survive in the event of a disaster. Business continuity planning is assessing the risks and looking at ways to avoid or minimize them and how best to respond to those events.

Your call center, or if you outsource the service bureau you deal with, should have business continuity plans to protect your investments. That includes methods, techniques, and technologies from prudent site selection, such as away from floodplains, to backing up data and to having calls and contacts handled by outsourcers, home workers, agents in other centers or by agents or other staff in offsite recovery rooms. It also includes having an evacuation plan and drilling your employees on it.

The center should be designed to minimize disaster risk and recover from it. That includes additional voice and data lines, multiple power feeds, onsite power generation, and locating computer and switching equipment to minimize damage. The investments, planned right, are well worth the cost.

Your organization should have a business continuity plan. If it doesn't, they should develop one. Make sure you fit your call center, existing and new ones, in the plan. Several later chapters touch on business continuity in real estate selection (Chapter 11) and day-to-day management (Chapter 12).

Also, call centers can and should be a key element in any business continuity program. As Chapter 2 noted, you can have in-house or outsourced employee or corporate hotlines to inform them what is going on, with a mix of live agents and self-service.

COMPARING SWITCHING STRATEGIES

Call centers, by definition, handle large volumes of calls. To get them to agents, they are routed through switches, either network-hosted or network routing, telco central offices or Centrex, private branch exchanges (PBXes), or voice and data IP routers that you own and set up.

There are advantages and disadvantages to all four routing schemes. In summary they are:

- Network and Centrex routing drop capital costs, but you are dependent on the carriers for the technology and upgrades, and the systems are not customized to your needs. You are also locked into those carriers to handle your calls and contacts.

- Premise-based PBXes give you total control, enable you to be carrier-independent, but they are costly (hundreds of thousands of dollars), take years to depreciate, and you have to find a safe room for them in the building. If you move you either pay to take them with you, sell them for cheap, or throw them into the property deal.

- IP routers treat voice as data. Circuit-switched calls are converted to data packets at gateways, put into data streams, and routed like emails. The equipment is less costly than PBXes; you can have ASPs or carriers host the gateways and routers. You need only to wire your centers with one wire rather than two. If you have satellite centers or home agents living outside of the local calling area, you won't get dinged on the long-distance charges. IT staff understanding routing, but only the telecom person understands phones, and little else.

Chapter 4:
Self-Service

In determining the size and cost of your call center you will need to calculate how much of that volume: calls, emails and faxes you can automate with self-service. That will determine how many workstations and people you will need either in-house or outsourced.

Here are the leading self-service methods:

Interactive Voice Response
IVR is a very sophisticated technology that captures and recognizes input from callers and responds to them with answers from databases. In essence callers are communicating with computers.

IVR interacts in two key ways: touchtone and speech recognition. With touchtone customers reply to IVR prompts by pressing numbers and symbols on telephone keypads. They make menu selections, answer yes-or-no questions and spell out names or words. With speech recognition, callers literally talk to computers running highly sophisticated programs that recognize words and convert responses into data.

Both methods complement and compete with each other. Touchtone is quicker and more accurate for vital functions like entering ID numbers, selecting menu items, and yes/no questions, and it is less expensive. Speech recognition avoids long menus and is more natural. It is best for names, which are a nuisance to key in. To make IVR more acceptable to users, call centers are now employing recorded voice talent from real people rather than synthesized voices.

IVR works directly with voice, anytime, anywhere. Where your customers have a phone they have access to IVR. You can't (or shouldn't) surf the web while driving, but you can talk to a machine, through hands-free on your wireless device or on the vehicle's dashboard. As companies improve design IVR with easy-to-use menus and top-notch speech recognition, more people will use it.

IVR works best for basic tasks. These include address changes, account information, arrival and departures, order taking and status, and reservations.

IVR is also good at and commonly deployed for letting callers identify themselves to the call center by entering information keys such as telephone, Social Security, and PIN numbers, account types and preferences such as language. IVR, coupled with the ACD and CTI, routes the calls to the agents best able or designated to assist them. CTI pops the customers' files to the agents' desktops, which avoids the annoying and time-wasting hassle of repeating the same information such as name and account number.

Outbound Voice Messaging

Outbound voice messaging (OVM) systems, also known as outbound IVR, dial numbers from databases and deliver pre-recorded messages. Some OVM systems connect called parties to an IVR menu or to a live agent. OVM gets the word out and saves time spent with live agents or avoids them altogether.

OVMs are great tools to deliver sales pitches. They are especially helpful for customer notifications, like software bugs and reminding them to activate their credit cards.

Fax Services

There are two principal types of fax services. The first is fax-on-demand, where prewritten faxes are stored on servers and sent when callers request faxes. That can be done from an IVR or web self-service menu. The technology enables customers to leave messages to have the faxes sent at another time or to a specific location, such as a hotel room or business, like a Mail Boxes Etc. The other is broadcast fax, where faxes are sent to numbers loaded into computers.

Web Self-Service

Web self-service is where customers access your company via the Internet or internally through your intranet to do such things as obtaining assistance from the help desk or buying products and services. Users point and click and key in information. They can also have information products delivered, from documents to media to software to tickets. The web is also replacing faxes for document transmission.

Web self-service can save time and hassle in transactions that must ultimately be handled by live agents by recording identifying information about customers or employees and tracking what they did online. When the system can no longer help them and the customer needs to email or talk to a live agent, the trail of what has been done has been captured for easier handling and resolution.

Web self-service is so prevalent now, and a majority of homes having computers and broadband Internet access. For many people, especially the younger generations, the web has become the first interaction channel. The phone has become secondary except for voice over IP. With the phenomenal growth of high-speed wireless connections, web self-service is there anywhere your customers are.

Email

Email self-service consists mainly of sending acknowledgements and messages. Customers can respond by sending an email back or by calling, faxing, or visiting in person. All outbound email can have links to your web site, thereby driving traffic to it.

Email at this writing costs virtually nothing to send, compared with white or "3D" mail (but there is a movement at this writing to impose costs on email). Acceptance and rejection is immediate, instead of waiting several days for a fax, phone, or white mail response.

You can easily customize outbound email messages. To do the same with "white" or "snail" direct mail is expensive and sometimes impossible since it is dependent on large mass volumes of the same pieces to keep printing costs down. Changing and dropping in text and zapping them out electronically is far faster than redesigning and laying out new copy, plates, and cranking out separate press runs. Outbound email is especially important if you're targeting niche markets with small numbers of potential customers. Using direct mail may eat up your entire profit margin.

OWN OR OUTSOURCE SELF-SERVICE?

In many cases it is probably advantageous to own your self-service hardware and software. You pick the technologies that meet your needs and requirements. Most proprietary or PC-based PBXes and network routing services come with IVR for example.

But in many other cases you may want to outsource self-service partially or completely. Those occasions are usually short-duration, high-volume programs like customer notifications with automated email, faxblasts, OVM, customer surveys, limited-time offers, and disaster response with IVR.

These outsourcers can get the additional capacity available for you probably faster than you can do it yourself. You also save money and time by not having to invest in hardware and software.

Outsourcers like SafeHarbor, based in Satsop, Wash., say you can save upwards of $2 million compared with buying, installing, and integrating the complex software internally. You also slice your implementation time to as little as eight weeks compared with seven months to one year for an owned in-house solution.

Says Geri Gantman, senior partner, R.H. Oetting: "Outsourcers provide huge economies of scale in technology. They've made the investment and integration in IVR ports, email management systems, and interactive web site back ends. Because they've done that, it costs very little for them to add or remove a client from their system."

But outsourcing self-service is not for everyone. You will pay more over the long-term than if you invested in the technology yourself; after all, the bureaus have to make profits. You're also tied to the bureaus' technologies and their upgrade and implementation schedules, which may not exactly fit your needs.

"Whether you should outsource self-service depends, like live agents, on your capacity and capabilities," Paul Kowal, president of Kowal Associates, points out. "If you've got IVR ports to spare and a well-functioning interactive web site, and IT staff on hand to regularly update them, then you probably don't need to outsource. But if you don't, or you have a large short-term campaign coming up, then you should consider outsourcing."

✪ ADVANTAGES

Here are the advantages of self-service:

Lower Costs

Self-service costs just pennies per transaction compared with dollars with live agents.

A self-service transaction costs just 5–10 percent that of a live agent voice or email interaction.

A May 2001 study by research firm Gartner Group, 'Contact Center Self-Service Costs," noted that the average cost for a transaction involving a rep was $5 per email message, $5.50 per call and $7 per text chat session. By contrast, automated transactions average 24 cents online and 45 cents by phone if callers only receive information through IVR systems.

Self-service enables customers to perform many of the same tasks as with live agents, such as placing, changing and canceling orders, managing accounts and transactions, requesting and obtaining information, solving problems, locating dealers, entering contests, and filling out surveys. The self-service applications are growing increasingly sophisticated.

But you don't have to talk to the experts to see how prevalent self-service has become. Look at travel. Who *hasn't* bought a ticket or reserved a hotel room online or through an IVR on a hands-free cellphone while driving? Who goes to a ticket or travel agent and not to a kiosk unless they have to?

The net result of deploying self-service is smaller and fewer call centers, which saves money. The more work handled by machines, the less work handled by people.

"Julie'" Amtrak's IVR agent, has saved the cash-strapped passenger railroad more than $13 million, reports *The New York Times* in an article that appeared appropriately enough at Thanksgiving, the busiest travel period of the year. She handles about five million people, or 25 percent of Amtrak's call volume.

Moreover, Julie, who is the voice of Julie Stinneford, reached an approval rating of more than 90 percent from customers. The *Times* reported that many riders said she sounded and acted so real they didn't know they were talking to a machine.

Improved Customer Service

As the Amtrak example demonstrates, self-service, if implemented correctly, can improve customer satisfaction and retention by doing away with long hold times. Customers get what they need quickly, anytime, anywhere.

The web, coupled with high-speed landline and wireless access, has enabled people to obtain vast volumes of information instantly, such as to help them buy products and services. They have through their screens more data than they can find out in a phone call or flipping through a catalog.

If a caller is on hold, you can invite them to go through the IVR, sometimes without losing their place in the queue, should they wish to return and wait to speak to a live agent. OVMs, automated email, and faxing also deliver acknowledgement, marketing, reminder messages, and customer care confirmations. You can also use self-service media to drive callers and contacts to your live agent call center.

Assisting Live Agents

By definition, self-service assists live agents, by providing simple information and answers to questions. That leaves the more complex apples-and-oranges-are-fruit to staff. If customers are trying to get problems solved, you can capture the information and trace what they have done on the self-service. Then when the agent gets the call they don't waste time, and money, and cause more aggravation by asking callers what they did.

IVR systems especially cut down on total transaction time and reduce costs by enabling call centers to identify customers and send them to the right agents. It is superior to automatic number identification (ANI), which relies on capturing the callers' phone number. ANI does miss calls, requiring agents to ask basic information from callers, eating up time and costing money. But IVR captures callers' account identifiers, such as PINs, which are much more reliable, that they enter themselves by touchtone or by voice.

There are also self-service applications that enable shorter and more-effective customer service. Web co-browsing cuts costs by having an agent on the hook only for the period when the customer needs help.

Another example is speech-to-text software that records and digitizes what customers say and transmits it to live agents. Those technologies cut costs by shortening call time, which enables your agents and reps to handle more calls, which in turn can also result in smaller, money-saving call centers.

For example, outsourcer Teleperformance USA offers clients integrated speech-to-text and speech rec-enabled IVR technology. The product, Enhanced Screen-Pop (ESP), gathers data at the beginning of each IVR-handled call and delivers that information, plus the customer's file, to the agent's screen in real time. The outsourcer claims the solution can cut call-handling costs by 25 percent.

"ESP saves valuable seconds off the agent having to repeat the same questions and key in data, and it increases accuracy," explained Dominic Dato, Teleperformance USA's president. "Those savings add up for our clients. [The solution] also saves end-customers time and aggravation from having to repeat themselves."

Better Live Agent Utilization and Retention

There is a second key advantage of self-service: better live agent utilization. Given that people are expensive, you want to use them as efficiently as possible. The people who work for you feel the same way. Basic tasks like order taking and dealer locating does not tap into the creativity, intelligence, and customer-service skills of your agents. Many of them get bored quickly, and that usually means lowered performance followed by door-slamming "byes!"

Also, when faced with growing call and contact volumes and static agent supply because you can't find enough people when you need them, self-service can fill the gap.

Customer Preference

Customers may prefer self-service because it is quicker and easier. Why be on hold for 10 minutes when they can find out what they need with an IVR or on the web in two minutes? With the web they can also see the product and find out more information about it. With the aid of video, the web can give the consumer an experience that is the next best thing to being there.

Self-service is especially important if you are marketing personally sensitive items, such as clothing and personal hygiene supplies like incontinence pads. Online shopping at sites such as Fredericks of Hollywood and Victoria's Secret is the best thing for guys who'd choke if they tried to order the same allegedly sexy flimsy or excruciatingly tight garments over the phone, let alone walking into a store.

And many women like to buy clothes online or use an IVR, for the same reasons. According to my wife, no woman (notwithstanding the fact that many are objectively in good shape) likes to reveal her size to another person, even if it is over the phone.

As self-service technology improves and as people get smarter at deploying it, more people will use it. Speech recognition has increased acceptance of IVR by make it easier to communicate with. The growth of residential high-speed Internet connections has made web self-service even more powerful. Both the younger and older generations have become very web-savvy.

Outsourcers have seen remarkable success in diverting live agent calls to self-service. West, a leading live agent and self-service outsourcer, has seen diversion rates around 30–40 percent and as high as 80 percent. Speech recognition-enabled IVR could boost diversion to 50 or 60 percent.

Safe Harbor says companies find they can improve customer service by switching from live agent to web-based self-service because their customers can download the information they need, rather than scribbling notes as they speak to live agents. The method also frees agents to handle more-challenging support problems. One of Safe Harbor's clients, a large multinational high-tech firm, now handles 70–90 percent of all contacts by web self-service, up from 30 percent prior to signing on with Safe Harbor.

"Intelligent" machines versus "intelligent" workers

While self-service technology has been improving, the same cannot be said, sadly, for many potential live agents. A study by American Diploma Project (ADP) said more than 60 percent of employers rate graduates' skills in grammar, spelling, writing, and basic math as only fair to poor. The report also reveals that 53 percent of college students take at least one remedial English or math class.

Jeff Furst, president of training firm FurstPerson, reports that in many communities, applicants have fewer skills than in the past. Reading, writing, and comprehension abilities are diminishing along with the work ethic.

Transaction Traceability

Lastly, with email and web-based self-service, you have transaction traceability. You can keep a permanent, instantly accessible record of what was said, promised, ordered, and delivered, and the price. It's faster compared with the slower, clumsier method of listening and reviewing recordings with call monitoring, which also may be legally restricted to protect privacy. You will find fewer "he said/she said" disputes. The records enable easier escalation of leads and problems to higher-level agents and management.

Business Continuity and Disaster Recovery

Self-service has another benefit: disaster recovery, also known as business continuity. By locating IVR ports, OVM devices, fax servers, and web servers away from your call centers, they can deliver messages about what is happening with your organization and provide a limited amount of service. But as these technologies improve the amount of interactions they can handle are growing.

✪ DOWNSIDES

Up-Front Costs

Self-service technologies like IVR and a fully interactive web site, with ordering and problem resolution, are not cheap and take months to implement. Then you must have the funds and the financing in place when you acquire the system along with bearing the costs of the inevitable installation headaches.

To buy IVR with speech recognition costs 10 cents per minute, including capital and installation costs (around $500,000) spread out over four years. The e-commerce software and servers cost tens to hundreds of thousands of dollars and take months to install and debug. According to Kowal Associates, at the lowest end a company could spend $100,000 for an email-enabled web site over the first year and more than twice that for an e-commerce site.

The more features you add, the higher the price. Experienced web programmers don't come cheap, especially those who are experienced at back-end integration of databases and fulfillment.

And it isn't a one-shot deal; web technology becomes obsolete very quickly. To keep your web site fresh and attractive, you have to redo your site and your offers often.

Poor Implementation Risks

IVR (and to a lesser extent the web) has suffered from poor implementation. Few aspects annoy customers more and prompt them to zero-out more often than long and complicated IVR menus—people cannot remember more than a few options at a time. Web sites that have broken links and Javascript errors will also make surfers reach for the phone.

Despite success stories like Amtrak's Julie, self-service is still getting beat up, rightly or wrongly in many satisfaction surveys. One study by Connell+Associates said only 24 percent of respondents were satisfied with IVR and 38 percent for web self-service; 59 percent found IVR frustrating to use. Jupiter Research showed web dissatisfaction jumped to 49 percent in 2003 from 31 percent in 2002.

Yet live agent voice does not look good either. Connell+Associates said just 63 percent were satisfied with voice and 65 percent with chat.

IVR Challenges

IVR is also the most easily abused of inbound self-service means. In the wrong hands, that is, incompetent or lazy managers and executives, IVR can destroy customer satisfaction and retention. Unfortunately those results bite long after these parasites have left their host and moved on their next victim.

It is very tempting to shove as many customers as possible to IVR, such as by making customer access to live agents difficult if not impossible, to reduce costs. If all you care about is cost and your customers are suckers, then why should you take the take the extra time design good, easy-to-use menus?

Empirix, which provides testing and management solutions for voice and the web, found an inverse correlation between poor IVR performance and customer satisfaction. The higher the IVR failure, the more annoyed customers will be. The firm also warned, in an article in February 2003 *Call Center Magazine*, "Poor application design can also increase agent turnover, as customers take out their frustrations on harried agents."

OVM Challenges

Outbound voice messages are strictly regulated because companies have abused them, like cold-calling at all hours of the day and not disconnecting when customers hung up (see Chapter 3 on legislation). For example, you can make OVMs only to existing customers; you can't use them for cold calls.

Unfortunately OVMs are becoming overused where they are allowed, by collections. Debtors know that they're getting a collections call when the phone rings and they hear "stand by for an important message" Click! Therefore, you might be selling a product or service but the caller hears the computer-automated voice, and click! ...because they think they're being hounded by a debt collector.

Web Challenges

Like IVR, it is easy to screw up the web site if all the organization cares about is cost reduction. Even today, when you think people would have completely learned, there are many examples of sloppy, fancy, and obscenely counterproductive web site constructions.

Too many firms are still wasting money on paying design geeks for whom the expression "keep it simple, stupid" is about as familiar to them as Mongolian-accented Finno-Urgic. Some sites reportedly take as long as 10–15 minutes from entry to completing an order, by which time customers who have a choice have gone elsewhere.

Web self-service is a lot more complicated than it looks, especially if the customers are conducting transactions online. The data can sit on many different computers, from servers to mainframes, using different operating systems and databases, strung out over thousands of miles connected by complex networks. Often the computers belong to other companies, like a bank or finance firm. Somehow they must all communicate to each other. If they don't, then there are problems and customers will get annoyed because they don't care about how complex your systems are; it isn't their problem.

My wife had been a well-paid project leader and manager on the web team for a large healthcare company. When she shows me the complex coding behind one seemingly simple web page, such as a member enrollment form, I understood why she made the money she did. This isn't kid's stuff. Programming is painstaking rigidly logical math. Make one error in the equation and the answer will be wrong. Even the simplest error will cause a program to fail or a web site to crash. As I was writing this section my computer died, a "fatal error" in Windows.

Email Challenges

Outbound email, especially commercial email, has been virtually spammed to death. Internet users have put up not-impregnable virtual fortresses with spam filters that chuck out the good with the bad, because the bad is so evil—spam is often infested with identity-stealing viruses. With practices like spoofing, where the fraudsters fake names and web sites, even commercial email from legitimate companies is rejected. But there is hope in new sender-verification technologies and standards that will sift legitimate firms from the spammers. Regulators and others have been taking action against spammers, including lawsuits and prosecution, aided by tools like the CAN-SPAM Act (see Chapter 3). (The term *spam* comes from an infamous Monty Python restaurant sketch where a hapless diner has been told that everything on the menu has Spam® in it. Then a chorus of Vikings chimes in, singing "spam, spam, spam, spam, spam, spam, spam, spam....")

As with nearly everything else that sounds too good to be true, the mutts and skels have occupied the email medium, aided and abetted by marketing industry denial. Spam would be far less of a problem today had the industry got its act together in the late 1990s when the matter first began to fry on the servers.

Fax Challenges

Alas, the Internet and email are replacing fax systems, making the technology increasingly less worthwhile to deploy. There are very strict regulations governing to whom and how you should send broadcast faxes. Junk faxes have become a major problem because they tie up fax machines, costing owners money in paper, ink, and computer time.

Take a hard look then at how many customers prefer to use fax, survey them on why, how often, and whether they will make more, less, or the same use of the feature. Look at options like outsourcing (see box) to avoid tieing up money on hardware, software, and lines you seldom need.

No Live Agent Opt-Out

It is very tempting for organizations to think: "Gee, we can do without people altogether." Wrong.

The worst mistake you can make with self-service is not allowing callers and surfers access to a live agent; it will probably ensure that they'll never call or shop on your site again. Yet I've come across IVR menus with no zero-out; there are also too many so-called interactive web sites that lack phone numbers, email addresses, or even white mail addresses. Web self-service may lull you into thinking that's all you need. Wrong—you're going to need live agents.

That's the trap the dot-coms fell into that most never got out of, dragged down also by dubious business models. Studies poured out of the research firms in 1999 and 2000 about customers abandoning online shopping carts because they could not get customer service.

Paul Kowal cites studies that show that 65 percent of e-commerce site visitors who initiate live agent contact complete the sale. You'll also need web-enabled agents who can read and push web pages and undertake text chat. An e-commerce site offering only automated self-service will annoy consumers; you can't program an automated email response or FAQ as well as you can train a truly interactive agent.

Natural Limits to Self-Service

Self-service makes for poor call screens. You can't program them to solve every known problem. The knowledge the systems have is limited to what has been written in. Garbage in, garbage....

These systems also don't show empathy. You can't ask "Julie" about her kids; neither can she say she's available Saturday night. She's coming into Penn Station on the 5 p.m. Acela Express that is due to arrive at 7:59 p.m. However, *The New York Times* reported that National Public Radio set her up with Tom, her counterpart at United Airlines.

"Unfortunately, the programmed conversation between the two computers over candlelit dinner quickly descended into bickering when the human voices behind the computer personalities argued over whether rail or air travel was better," said the *Times*.

On the other hand, many live agents are equally soulless, judging from their tone of voice. Lack of empathy has been a major issue with offshored agents, as I will explore in Chapter 8.

❍ TECHNOLOGY DIFFERENCES

There are some differences in how various self-service technologies operate. If a new customer and the company they're contacting don't have a file on them, they must fill out a form, which can be via an IVR or a web form.

In some cases with an IVR (such as dealer locator), their phone number is picked off by ANI, giving all the information a system needs. By matching the number, such as the exchange (the first three digits) with the area the exchange covers, the system can locate the nearest store and tell its location, hours, and contact information.

If the person has called before, the company will have a record on them. An IVR system will recognize the caller based on their phone number and can present customized menus or offers, or pop their information to an agent. If a person has previously visited a web site, a file known as a cookie will have been placed at that time on their computer, allowing the web site to recognize them the next time they visit.

ANI methods are not foolproof. It doesn't work, of course, with cellphone numbers and cell and pager exchanges that are also used by landlines. That's why organizations have been going with unique number identifiers such as an account number, PIN, or Social Security number (better a PIN than SSN for data privacy reasons).

This book and this chapter are not about the details of the technologies that enable the self-service and automation functions used to support customer service and sales. There are other books like the *Call Center Handbook, The Complete E-Commerce Book,* and *Computer Telephony Encyclopedia,* also offered by CMP, plus numerous articles in CMP magazines that can give you that information. I will instead briefly describe them, their applications, advantages, and weaknesses.

❍ IS SELF-SERVICE RIGHT FOR YOU?

Yes, in many cases. But take a hard look at your inbound and outbound functions. IVR, OVM, web, automated email, and fax self-service work best with set information—where the requests and responses are straightforward. But they can also intake more complex inquiries for live agents.

Of all the choices, the web is the most sophisticated. You can use it for account history, FAQs, and troubleshooting much more readily than with IVR. You can download and retain information, like tickets, confirmations, and software and follow links to other sites.

But you have to be on the web to use it. That means, for all intents and purposes, on a computer. But an IVR or an OVM can reach phones anywhere.

THE AGENTLESS CALL CENTER

Are call centers doomed? Will the vision of the dot-commers, peopleless service, actually come true?

The cost savings with self-service are enormous, whether you buy or install the technologies yourself or pay to have them hosted, which is similar to buying versus leasing a car. Either way you get a vehicle that offers significant benefits.

There is no place on Earth, even the dingiest prison call center, that you can conduct a transaction as inexpensively as self-service. A study by Datamonitor, "Voice Business in Regional Perspective: The Americas," says a speech-rec-enabled call is 15–25 percent of the cost of a call handled by an agent in India.

As self-service improves, more people will be comfortable with it. The hard fact is that the information available to a live agent is the same as to self-service. They tap the same knowledge bases for answers.

More organizations have finally gotten the message that they need top-notch IVR menus, such as with speech rec and quality web sites to keep customers on self-service, not by making it hellish to access live agents. But some that do are also charging for live agent service, especially in customer support.

Consultants like Jon Kaplan, president of TeleDevelopment Services, believe non-high tech firms will soon start charging calling fees, just as banks and travel agencies have introduced service fees. While buyers have been known to balk and complain to government officials—like when the airlines raise their fares—banks add fees and telcos do play with phone rates, so customers may acquiesce to the new reality of calling fees.

"Reaching a live agent is a value-add service," Kaplan points out. "Therefore it is not unreasonable for them to pay."

Yet would customers be getting value for money? Many call centers' rules do not permit agents to deviate from the scripts and knowledge bases. If that is the case then what is the value of having live agents?

Especially as declining birth rates, coupled with demand from other employers and a substandard education system, are shrinking the pool of people willing and competent enough to work in such call centers?

There will always be a need for live agents, only many fewer of them than at present. The agents that remain must be empowered to make decisions, like business-to-business sales reps and second- or third-level tech support. Or they perform tasks like basic customer service or order entry that requires some but not a lot of human intervention but strong empathy with others.

Alternatively, you can have a blend of low-cost and low-skilled telemessaging agents answering, triaging, directing, and taking messages for experts. Telemessaging staff can also sell support plans.

Only people can think out of the box, outside the programming to solve problems. Only people can provide support and warmth. Otherwise why have people who are no more enabled than machines?

✪ ENABLING SELF-SERVICE

You will need expert advice from consultants and from your vendors to do self-service right. Here is some advice to make self-service work.

* Make self-service user-friendly. Write IVR scripts and FAQs like you are interacting with the contacts, which you are. Keep it simple.
* Consider speech recognition, especially with voice talent. The more natural the conversation, provided the design is right, the fewer the zero-outs.
* Provide an easy route for your customers to access a live agent. Customers want that security. You also hold the self-service teams to account. It is up to them to design, deliver and maintain quality automated interactions. They can't hide behind the moldy fig leaves of making live agents hard to get
* Always ensure that everyone is on the same page. Capture every self-service interaction so that if and when customers need to speak with someone, the information is right there. That voids time, money, and potentially sales-killing annoyance from having customers repeat themselves. You also find out where the blockage with the self-service system occurred and if it is a problem, then you can correct it.
* Consider a telemessaging interface. Customers may be more willing to use IVR or the Web if they first briefly speak to a live agent who offers self-service as an option to a 10-15 minute wait in queue.

Chapter 5:
Home Working

Before you start planning or expanding your call center, with visions of happy agents saying, "Thank you for calling America's Classic Septic Service, how can I help you today?" take a hard look at whether you need to service your customers with such an in-house investment. The previous chapter examined how much of this volume you can, ahem, pump through self-service: IVR, the web, automated inbound and outbound email, fax, and voice.

You are looking at anywhere from $20,000 to $40,000 per workstation per year to set up and run a call center. The smaller the center the higher the cost per seat because of reverse economies in scale for facilities and technology expenses. You're also facing from thousands to millions of dollars in annual operating costs, mostly wages and benefits, depending on how many agents you have.

Labor alone accounts for more than 60 percent of your operating costs. Finding enough qualified people who will work in your center, when you want them, at the compensation package your firm has budgeted, over the period you expect to keep it operating, is the biggest challenge you face.

There are several alternatives to expanding and opening call centers that you should examine before going forward. They will help you spend your resources wisely while ensuring that there is the flexibility necessary to handle fluctuating customer contacts and demand.

Home working is one of those means. Chapter 6 reviews outsourcing.

✪ HOME WORKING COMPARED TO OUTSOURCING

I ordered these chapters this way because home working, for the most part, entails managing people who are still your employees. They directly access calls, contacts, and data through your servers. If done correctly, you will still maintain the same level of control and obtain equal but better performance with home agents as you do with conventional center agents. (I highly recommend *Home Workplace*, a great how-to book on home working that I wrote, also for CMP.)

But the decision to outsource involves managing an independent third party to handle your contacts. You don't have the same legal or managerial say as you do with your own employees. There are contractual issues involved.

Also, that outsourcer may have the work done in conventional call centers locally or offshore, or by agents at home. Those home agents could be their employees or contractors. I have included all discussion about offshore, contracted, or outsourced home agents in Chapter 6. I look at the cultural, political, and regulatory issues entailed in going offshore in Chapter 8.

You can use all of these methods in various combinations to deliver the best balance of service and cost. For example, you can have your senior customer service and sales staff work at home; outsource your customer service, Level Zero or Level 1 support, order entry, lead generation/qualification; and have contractor home workers cover for you on call spikes, evenings, and weekends. To minimize your costs further, outsourcers may offer to handle high-volume commodity customer service, support, or outbound calls handled offshore.

All of these methods affect your strategy and costs because they remove agents from your centers. Depending on how you deploy them, you can make a new center smaller and less costly , or avoid having one altogether or shrink or eliminate existing call centers.

✪ CHARACTERISTICS

As scoped out earlier for the purposes of this chapter home working uses your own employees. Agents take and make calls and contacts from their homes, on computers and phones that are usually monitored in a traditional or premises call center. Supervisors and managers can often also work from home. Here are the main features:

* Agents work at home exclusively. They could be full-time, part-time or on-call employees.
* Agents could also work at home part-time or in emergencies, in which case you will need to have either dedicated or shared workstations for them in your center. But those shared workstations could be formal hot desks that other visiting home agents also use, training room desks, or informal desks assigned to agents or supervisors who are out of the office.

   I look at hot desking in Chapter 11. I strongly recommend that if you go that route that you deploy software or contract for service firms to schedule and manage your workspace to avoid two people trying to work from the same desk. AgilQuest, which is based in Richmond, Va., is a highly experienced company that can do that for you.
* Home-working agents can be on your payroll or be independent contractors (see Chapter 6).
* Home working relies on eLearning, which I call technology-based training (TBT), to minimize the need for home agents to come to premises call centers. More about TBT in Chapter 14.

   Alpine Access, a service bureau with nothing but home agents and no formal call center or training facility, uses ePath's web-based solution to train 3,000 agents. It tracks class schedules and progress, where learners are in training and identifies when the agents can go live in support of a client.
* The most successful home agents and supervisors are those who are self-starting, disciplined, motivated, who take the initiative, and who solve problems. The same qualities *every* employee should have.

✱ Call centers may buy agents computers or have agents use their own. Attributable operating costs such as voice and data charges are typically expensed. Employees typically provide their own furniture, but employers should insist that the chairs be adjustable and ergonomically sound and that the computers are set up to prevent glare so that the staff can work productively without injury.

I *strongly recommend* that if your agents are working from home on a regular basis that you buy the computers. The reasons? Security—you own it, you control it, including the software and data on it. That limits hacking, viruses, and data theft, such as when the employees sell, recycle, or toss out their computers. You also ensure consistent performance across your organization. Lastly, you may run into software-licensing hassles, especially for part-time home workers who also work in your centers. You may have to pay fees per machine, as opposed to per-worker, which means you may have to shell out twice for the same agent.

I *also highly advise* that you deploy data security methods and technologies. That includes virtual private networks (VPNs) if server access is needed and strict data access procedures such as ID authentication, log in to applications that enable access to data, log out and record what they did with that information; agents must authenticate their ID. Data can be partitioned to give access only to what agents need, such as credit card numbers but no names. Wireless home networking should only be allowed if the equipment is to corporate-level standards. For example all certified products made after 2003 included Wired Equivalent Privacy (WEP), a technique that encrypts all data that passes between an access point and a wireless card. While not perfect, if WEP is enabled (and it must be) it will block casual snoops from reading intercepted data.

✱ Home agents and staff are responsible for dedicating quiet spaces for their offices and must keep away family members, pets, and neighbors. Callers and called parties *must not know* that they are reaching home offices. Agents at home must work to the same hours, including break and lunch times, that they would in premises call centers. No petting the kitty or playing with Mr. Rugrat when the calls are backing up in queue.

✱ Home offices must mimic the conditions in premises call centers, so that the performance of home agents *meets or exceeds that of premises call center agents*, to justify taking the call center work home.

That includes following privacy policies to help comply with regulations such as the Financial Services Modernization Act and the Health Insurance Portability and Accountability Act. You can insist that your home agents install locks on their doors, install lockable filing cabinets, shred unwanted documents, or buy biometric computer keys that read fingerprints or irises.

The last key characteristic of home working is that you need to get the assent of home workers to visit them and require any changes to their premises. It is their home and property. To handle these issues and others at home, you will have to spell them out in a formal home working policy signed by your employees.

Home Workplace contains detailed advice on setting up and managing home-based workers of all occupations, including home working policies. I touch on a few of those enablers later on in this section.

Who wouldn't want to work from home with a view like this from their office? Sure beats the next set of ribbon windows in some "banker's box" or of the latest fashion in dumpsters. Home working also permits call centers to attract employees living in relatively remote areas like the Comox Valley on Canada's Vancouver Island. The area is 1- 4 hours' road or road/ferry travel time from major cities and has very reliable high-speed cable and DSL Internet access. There is also a fair-sized French-speaking population thanks mainly to an air force base. Credit: Brendan B. Read

✪ ADVANTAGES

Let's look at some of the key advantages of home working.

Reduced Costs and Higher Productivity

Research by call center and home working consultant Jack Heacock reveals that a 100-workstation call center, staffed on one shift, can save over $10 million across five years from real estate savings and lower turnover. Few organizations can afford to leave that kind of money on the counter.

Even if you are in a long-term lease, or got the building for "free" from a local economic development agency, you can still incur savings. For example, home workers mean not buying and installing workstations, not laying out cabling, not paying (and waiting for) contractors and their high-paid help to move or modify work floors. You also gain from other benefits, like lower turnover and healthcare costs and greater access to quality labor, as demonstrated later.

Jack Heacock, who as project manager built and outfitted Amtrak's award-winning and ergonomically sound Riverside, Calif. call center, says it costs *over $33,000* per workstation to accommodate employees in premises offices.

In contrast, based on the model developed by Heacock, it costs just *$7,100* in one-time setup charges for home workers. The home working expense includes $5,900 for furniture, equipment, and telecommunications (those costs can be less if employees pay for some of it).

Less Tardiness and Fewer Early Departures

Call centers require agents to be there on time to maintain service quality. But that's becoming increasingly difficult thanks to traffic delays caused by too many vehicles and accidents. The average urban driver now spends 62 additional hours annually—the equivalent of 1.5 working weeks—stuck in traffic. That's up from 44 hours in 1990, reports the Texas Transportation Institute *2002 Urban Mobility Report.*

By eliminating the commute, home working limits early departures for appointments and picking up family members such as children from schools and increasingly elderly parents from clinics and seniors' centers. Employers have long struggled with both problems. With commuting times getting worse, expect these matters to deteriorate.

Here's one of the best reasons for home working: having your agents avoid the commuting mess, which isn't getting any better. With your staff safely at home you don't have to worry about meeting service levels through tardiness, early absences to pick up the kids at school, or facing increased downtime and medical costs through diseases caught and spread at workplaces.

Improved Productivity

Home working liberates you and your agents from the commuting tyranny, from wasting time in transit. That's time that could be spent getting and making calls and contacts instead of waiting until the next day, which swallows up the time one had planned to allocate to other tasks.

Procter and Gamble saw productivity climb between 5 and 10 percent when it implemented home working for its consumer products call center in 1999. The gain also allowed it to close one of two call center floors.

Greater Flexibility

Home working permits employers to meet call demand spikes. Employees can be on their computers and phones in 60 seconds, rather than the 60 minutes or more that it typically takes an employee to commute to a premises office.

People who work part-time and split shifts are especially sensitive to commute times. If they have to spend the equivalent of half their shift commuting they won't want to work for that employer. The benefits of working are rapidly eaten up by gas, vehicle wear-and-tear, or carfare.

IntelliCare, a health-care outsourcer based in Portland, Maine, adopted home working to improve productivity by offering split shifts. Home-working agents also handled regular shifts. Prior to allowing work from home, staff arriving at the call centers for a full shift would invariably be idle for a portion of it due to wide fluctuations in call volumes.

The company says home working enabled the split shifts, which allowed it to reduce its total workforce. That move and other cost-saving innovations saved it over 15 percent on labor and infrastructure costs.

* Free (to You) Employee Benefits

Home working gives agents and supervisors indirect pay raises and bonuses that cost you little on the salary, benefits, and tax side. Your home workers save money on cars, carfares, ties, pantyhose, cosmetics, toupees, dress shoes, lunch and break money and dry cleaning bills. Jack Heacock estimates home working gives each employee $4,000 to $5,000 annually.

Staff Attraction and Retention

Home working is a powerful way to attract and retain top workers. The Information Technology Association of America says in its "Anytime, Anyplace, Anywhere" survey that 54 percent of American voters felt home working would improve their quality of lives, 36 percent would choose home working over a pay raise, 43 percent felt they would be a better spouse or parent if they were able to home work, and 46 percent think that the quality of work would improve if they were able to work at home.

Attracting and retaining employees is going to become a critical issue over the next several years as the labor supply of experienced workers dries up rather than gushes forth. The U.S. and Canada are facing the likelihood of a diminishing labor pool as the baby boomers retire and die off. That generation used modern birth control, which left fewer progeny in their place. Potentially counteracting this trend to some degree has been the 'natalist' movement: right of center families having three or more children and focusing of their incomes on them. *The New York Times* op-ed columnist David Brooks reported in a December, 2004 issue that the natalists live by and large in the politically "red" or conservative areas: the exurbs and in the Plains and Sunbelt areas.

Employee turnover, especially of highly skilled employees, costs employers more than they realize. To hire, train, and bring a replacement person up to speed costs as much as 150 percent of the replaced worker's annual salary. With call centers demanding increasingly higher-skilled employees, recruiting and retaining them will be increasingly difficult and costly. P&G's at-home agent turnover is just 8 percent compared with 22 percent for at-center agents.

Many employees prefer to work from home. Only one of P&G's home workers reportedly returned to the call center because he missed the camaraderie.

Carla Meine is president and CEO of call center outsourcer O'Currance Teleservices. She reports that home working enables her firm to tap into a high-quality and more-loyal workforce. More than 80 percent of her agents work at home.

Meine's employees or agents are in their late 30s, compared with the early 20s typically found in call centers. She says they are better able to deliver excellent customer service and sales compared with younger, inexperienced agents.

Freed from the straitjackets of dressing for and commuting to work, her agents are likely to stay longer, reducing staffing and turnover costs. O'Currance's turnover is 100 percent per year, compared with 200 percent to 300 percent for typical inbound and outbound sales operations.

"Our agents are those who would never dream of driving to and working in a call center," says Meine. "They are a different caliber of people."

Larger Available Quality Labor Pools

Home working expands labor pools by freeing workers from the constraints of commuting distance and time, just as call centers themselves liberated customers from having to travel to an office or store to conduct most types of business. Working Solutions, a contracted home working outsourcer in Plano, Texas, has over 22,000 agents throughout the U.S. and Canada. It had agents in places like Florida and upstate New York working on a program for Ensenda, a San Francisco delivery firm.

Home working removes geographic limitations to serving customers with region or state-specific products and services, such as insurance where there is state licensing. There may not be enough buyers in that state to merit a formal premises-based call center but you may have sufficient demand to hire on several home-based agents. This is also a great way to build your presence in a market without heavy investment in facilities and staff.

You have fewer geographical restraints with home working even where you require those agents come in once a week or so. People don't mind traveling farther if the trips are infrequent.

For example, IntelliCare agents work at the centers once a week. The outsourcer discovered that home working more than doubled the distance agents are willing to travel to the call centers, from 15 to 20 miles to 30 or more. That increased radius takes in more people and has enabled the firm to snag more qualified agents.

Accessing Quality Staff

These restraints of commuting distance could corset your call center's performance by forcing you to hire substandard employees because not enough quality agents live within commuting distance.

Sadly, in parts of the U.S. and other places, the public education system has become part babysitting service, part juvenile prison, whose "products" would be hard-pressed to read this sentence. A study released on Feb. 9, 2004 by the American Diploma Project (ADP) and reported widely in media outlets like *The Washington Times* said that more than 60 percent of employers rate high school graduates' skills in grammar, spelling, writing, and basic math as only fair to poor. The report also reveals that 53 percent of college students take at least one remedial English or math class.

Employers have had to pick up the slack. One study estimated that remedial training costs one state's employers nearly $40 million a year.

Jeff Furst, president of call center staffing firm FurstPerson, reports that in many communities call center applicants are less skilled than in the past. Reading, writing, and comprehension abilities are diminishing; so is the work ethic.

In a typical community, only 30–45 percent of candidates tested will meet the abilities and behaviors to do the work, "which really impacts the ability to hire employees that meet the job requirements," he says.

Out of 100 interested job candidates, typically only 10 will be hired. This takes into account pre-screening, selection testing, and background checks. But, the new hires will stay longer and perform better.

Home working, however, enables you to compensate for declining education standards by widening the labor pool. The Internet is an excellent recruiting tool that lets you find people and lets them find you from anywhere. You can use the web to pre-screen, screen, train, and supervise people, all without seeing them.

Call centers need to target recruiting and retaining strategies such as those aimed at baby boomers, because they are mature, more responsible, often better educated, and more reliable than younger employees.

But many boomers won't put up with the commute and would rather work at home. Furst recommends call centers look at home working and satellite working in call centers with 10–25 workstations to draw boomers and expand the labor pool beyond commuting distance from the call centers.

Home working also enables you to recruit stay-at-home caregivers and the mobility-impaired. Both groups find commuting nightmarish.

"Traditional 200 to 500 workstation [call centers] can burn through a community's labor pool quickly," says Furst. "Today those supplies are being replenished by mediocre labor. With tightening supplies of quality labor, call centers need to take every step they can to deepen the pool."

Reduced Health Costs

Home working minimizes exposure and transmission of diseases that, at best, cause sick days and resulting productivity losses and, at worst, serious illness and death. The suffering also extends to the bottom line in the form of higher employee and employer healthcare costs.

The SARS outbreak in 2003, which forced many firms in Canada to home work and teleconference, may be the beginning of a new wave of quickly-spread diseases. Health experts are nervously watching avian flu outbreaks because they fear the virus could easily mutate into an easily transmitted version that could cause deaths on the scale of the 1918 influenza pandemic.

Premises call centers are "bug spreaders." One person coughs and sneezes, and in a short period of time so does everyone else. A 2003 survey of CEOs by PriceWaterhouse-Coopers reveals that health-care benefits cost companies nearly $5,000 per full-time employee. Any step employers can take to lower those costs, like home working, goes right to the bottom line, thus enabling organizations to do more.

Business Continuity

Home working is the best strategy when it comes to business continuity by dispersing your workforce across a broad area. So if a bomb, a power outage, or a tornado hits one locale, the others will still survive and function. If a snowstorm hits, and the electric and telco lines are up, employees can work without sliding into a snow bank.

I worked exclusively at home after the 9-11-01 attacks on the World Trade Center; we were living in Staten Island then. The destruction, security precautions, and major cuts to subway service made commuting next to impossible.

Agents will also not be trapped on the roads, in transit vehicles, in airports and train stations trying to get out in case of a disaster. Such sudden evacuations could lead to panic often only make disaster response worse by clogging roads that impede emergency response.

In fairness, people will sometimes do the right thing. My wife and I evacuated Manhattan along with my sister-in-law. We hiked 15 blocks to get to the ferry to New Jersey; we were hoping to get back to Staten Island. On the way my sister-in-law had a mild heart attack and when we told others in the long lines to catch the boats they let us go ahead of them to get on the next sailing; there were ambulances waiting on the other side. I ended up staying with friends in New Jersey for two days as the police had closed the bridges to Staten Island. But I was still able to work as I had my laptop.

If the power goes out in one neighborhood where one group of workers lives, chances are it will be on in another community or city where another group resides. Better to have some of your workforce live than none at all. I work from home, far from my office. Whatever outage hits them or me is unlikely to affect both of us at the same time.

Employee Safety

Commuting during the day is hazardous, with road ragers, subway creeps, and just plain @#%^& happens. But the odds worsen at night. If you have afternoon, evening, and weekend shifts, you are placing your agents at greater risk from robberies and violent crime in parking lots and garages, walkways, bus and train stations, and on premises.

If your agents are driving at night they face accidents, especially in poor weather. A rainy afternoon can turn into an icy, treacherous evening.

Home working limits these occurrences because your agents are safely at home. You also save on maintenance, power, security as well as potential health-care and other benefits by having your call centers closed down for the nights and on weekends and holidays. You and your employees will have peace of mind.

The Benefits Are Real

These benefits aren't theoretical. Call center outsourcer ARO saves over $1 million from lower overhead, lower turnover, and higher productivity with home agents compared with premises call center staff. Staff churn plummeted from 60 percent to just 7 percent, sickness and overtime dropped, program reject rate went from 15 to 1 percent with quality calls the first time, while output jumped by 15 percent.

ARO also reduced total staffing costs. The supervisor-to-agent ratio widened from 11:1 to 20:1, while the lower turnover enabled it to use fewer trainers and training time.

By offering home working, ARO was also able to hire better-quality agents, many of them baby boomers. They are more dependable an responsible than the 20-somethings who had worked in the premises call center. They also relate better to the clients' customers, which improved sales and satisfaction.

"I could not believe that I could get this caliber of people working for us," says Michael Amigoni, ARO's chief operating officer. "These are people who would never have stepped foot into a call center."

✪ DOWNSIDES

Most call center tasks can be done by home working. But there are costs and issues with home working that you should examine.

Need for Premises Working and In-Person Interaction

There are a few instances where the call center work cannot be done from home. One example is hardware support where the agent is physically manipulating a computer or printer to duplicate a customer's problem. It is often not cost-effective to supply all home employees with every make and model of equipment you make and support.

You may need to have new hires people on premises for training. Your call center may also rely on face-to-face team meetings. I say 'may' because you can work around both issues, by hiring employees who have worked successfully independently before and by using communications tools such as audio, data, video and web conferencing. Some call centers, such as Alpine Access, never see their employees at any time, and they function very effectively. Some centers with home working, such as O'Currance Tele-services, also have agents who prefer not to work at home even though it is offered to them, because they want in-person interaction with colleagues and supervisors

Careful Agent Selection

While most call center tasks can be done at home, many employees may not be capable of handling them from there for personal reasons.

To work successfully from home employees must be highly disciplined independent self-starters, able to solve minor problems with their computers and phones. They need to have both the maturity as well as the skills to work unsupervised. At minimum your home workers must also have space in their homes where they can work uninterrupted by outside and inside noise.

Labor Supply Limitations

If the ideal person for your call center is in their 20s—and you're not willing to change—then home working may limit the supply of people eligible to work in a call center.

Young employees often lack the maturity, the space in their residences, or the desire to be home-bound. This last point is important. Like it or not, offices are social centers, for making friends and lovers, with the obvious downsides to employee relations and productivity when these relationships end.

When I was a single-again 30-something I liked going to a workplace because of the opportunity to socialize. But when I met the wonderful woman who became my wife I no longer had any need for that. My life now revolves instead around my family, her friends and mine, as well my outside interest group activities.

Training

Training is more challenging with home agents because they are not on premises. You can have them work in your premises call centers until they are brought up to speed, which means for space and sizing reasons you have to calculate carefully how many agents will go out into the field. You can have agents come into the center periodically, which means hot-desking. Or you can create audio- or videoconferenced instructor-led and/or TBT solutions so they can learn from home offices; agents who are working in premises call centers can also use them, which avoids having them leave their cubes to go to training or seminar rooms

* New and Different Costs

Equipping agents and supervisors to work from home is not a free ride. If you are replacing premises workspaces with exclusive home workspaces, then you and your employees must agree to equip them with voice and data equipment and furniture. Your per-employee voice and data costs are higher than that in premises, although as we will see later, the technology gap between premises and home call centers is rapidly shrinking.

You may need to provide hot desks (see Chapter 11), which are shared workstations. They cost about the same as regular workstations, but you spread out the expenses by carefully scheduling your employees to use them. You can make them smaller than regular cubicles because no one is keeping files or family pictures on them. You also need to manage these spaces through workspace management solutions. AgilQuest, based in Richmond, Va. is leading provider of these tools.

You also will need to spend time putting a home-working program together and selling it to senior executives and other departments such as HR and IT. You may need a separate IT home support team and a dedicated home working program manager to oversee the program. You also will have to select, qualify, and possibly train people to work from home.

Existing Property

If your organization has just signed a long lease with a landlord for a premises call center, then there may not be any real estate savings. Or if there are too few employees qualified to exclusively work from home, you may see little savings. Also, some landlords will penalize you if you let space go vacant, even when you're still paying rent (see Chapter 11).

However, you may be able to sublease if you are in an attractive location with a good building in a prosperous market. There may be deals that could offset real estate savings with home working. Hungry landlords may offer zero-net-leases where you're paying just for the overhead, for very little or no profit to them. In some cases local economic development agencies are willing to give buildings and other incentives that reduce taxes and training costs.

You may be able to negotiate leases as short as five years and exit in three years. But you may not be able to exercise such options if you received government incentives requiring you to be in a building for a certain number of years, unless your lawyer is better than the government's and you don't care about the bad publicity.

Supervision

Since time immemorial, employers have instructed, corrected, rewarded, and punished employees to their faces. Employers have measured employee performance with their direct senses against what their supervisors expect.

Home working requires that you implement a different way to supervise workers, and that is by performance. That means supervisors have to trust employees and exercise people skills like interacting with them rather than talking down to them. Some supervisors may get with the program; others may drag their heels to sabotage it.

You will also need to invest and deploy tools like instant messaging (IM) and use email, audio-, data-, video-, and/or web-conferencing to stay in touch with your agents. They in turn need to feel comfortable using them to contact you and their colleagues.

This isn't as difficult to do as you might think. You may be using these tools already. A 2002 study of Canadian executives by International Communications Research that said 94 percent of managers often send email rather than meet one-to-one; 67 percent said it happens very often.

In a story in *The Calgary Herald* in February 2003, Bob Schultz, professor of strategic management at the University of Calgary's Haskayne School of Business, said that emails "give managers the ability to respond to more people than before. Managers don't have enough time to do face-to-face meetings with everyone."

Face-to-face communication shrinks productivity. It takes time away from your job and your co-workers, and when you do speak to them in person, it can derail their train of thought, and it can take a while to re-rail those cars after you've spoken to them.

Internal Communications

There is still nothing like talking to someone face-to-face or hearing others talk first-hand. Employees who work at home are "out of sight and out of mind" unless supervisors and employees make it a point to communicate with each other. When you need to talk to employees, either one-on-one or in a group, it is often easier, quicker and more impactful in-person than it is over the phone or via the web.

Also, when employees are on premises they usually feel more connected to what is going on. They hear the gossip and rumors, and they ask questions. There are greater opportunities to chat with individuals outside of their own workgroups. For employees wishing to advance, they often need that kind of information to plan their career moves.

Home Environment Distractions

Distractions like children and pets are another concern of employers. The last thing you want is your customers, clients, and colleagues to know that your employees are working anywhere but in a quiet, professional environment. In reality, many premises are anything but, judging from high noise levels when I've talked to people in offices and call centers. And yes sometimes people bring in their kids and pets into offices.

Fortunately, you as the employer can set out the conditions and terms of home working. Most call centers that have home working do so. Employees comply by signing by home working agreements, so if there are complaints, employers can act. Few of these incidents arise because employees very much want to continue to work from home.

Data Theft

There is a risk that home workers who have access to customer or other confidential data could steal it. Family members and visitors could see what's on the screens. Careless home wireless networking could leave data open to theft by crooks that drive by looking for computers to break into.

Data theft is becoming a very serious and widespread crime, and there are now laws to stop it, including provisions that penalize companies that don't protect it adequately. The risks increase when the data leaves the premises. Employers have less control because there is no one on site to make sure rules and regulations are complied with.

On the other hand, at-home call centers, unlike premises call centers, the home office worker typically has a great deal of control over who gets into their equipment and information. They can lock the doors and secure the computers, including disconnecting the laptop and locking it in a file cabinet.

The same protection level is difficult to achieve in premises call centers. Agents don't have control over visitors and other employees, especially if they share workstations in the case of multiple shifts.

Career Development

Out of sight is, for many people, still out of mind. People are, for the most part, reluctant to promote or to hire someone sight unseen, especially when there are tasks that involve interacting face-to-face, such as on a business trip.

But this is changing, mainly due to the growth of online recruiting and hiring, more careful screening, stricter background checks, and employers placing more focus on performance.

AT&T found that working from home actually enhanced careers because employees worked longer and were more productive. Instead of using the time eliminated by commuting for leisure, they put in more hours.

Corporate Culture

Many executives shut their minds when you mention home working because they feel that employees who work from home miss out in being part of that organization's corporate culture. Countless outfits have built an identity around their corporate culture and use that to attract and keep employees. They place a high value on company events such as office parties, picnics, field trips, getaways, and community works.

But many of these trimmings of corporate culture, along with many employees have been dispensed with to lower costs. Employer and employee loyalty are a part of history.

Agents know that they and their call centers are dispensable. Employers reorganize, downsize, outsource, and move operations to lower-cost locations. Employees go job-shopping on the web. So don't use corporate culture as an excuse not to have home working.

If you are still concerned about corporate culture with home working, I have in reply one word: JetBlue. The economies and flexibilities of home agents have enhanced the airline's culture and bottom-line results. While other carriers have been going into bankruptcy, JetBlue continues to be profitable with a loyal customer base.

You may find downsizing and restructuring staff easier with home working because there isn't the same degree of personal trauma and morale, and consequently productivity losses compared with seeing agents and supervisory staff physically leaving the buildings. There are also fewer opportunities for those who are upset to take it out on you, your other employees and your property. You also have fewer workstations to dispose of and floors to reorganize.

Staff may find it easier to be downsized if they are not on your premises. No long goodbyes, the distant "glad it's you not me" stares from soon-to-be former coworkers, or frogmarches out the doors by security staff.

I know I was glad to be working from home when I was "made redundant" by *Call Center Magazine* in June 2004 after seven years there. I then got on to the tasks at hand, writing my resume, recalculating our finances, and finding a job from my own home.

I had done the terminal ward routine when Primedia shut down a newsletter on electrical codes that I was assistant editor on in late 1991. Fortunately I was professional enough to maintain a very profitable freelancing relationship with that firm that lasted into my entry into the call center field in late 1995.

✪ MINIMAL BENEFITS

There may be cases where the benefits of home working do not exceed the costs. If you have a small (under 15 workstation) call center and you have very good deal in a building or you own it look hard at the business case, especially if your site is in a rural area where traffic jams aren't an issue.

Telemessaging call centers tend to be difficult to make the home working case for. They are often very well established with low cost structures. Employees often have strong personal ties to each other. They may live fairly close by.

Even so consider home working as part of your strategy. You may have employees who can't come in because they are ill, have to take their child to the doctor or are waiting for the electrician or plumber to show up. Home working enables them to continue to work for you, taking on the volumes that would have added to their colleagues' load and pushing your service level, potentially annoying customers.

And if you need to add staff, but don't have the room, it is easier and quicker to handle that demand at home than by trying to add office space and go through the expensive and time-consuming hassle of rewiring the new premises for your gear. Planning and having home working-supporting technology and procedures ready ahead of time equips you to handle whatever comes next.

✪ IS HOME WORKING FOR YOU?

Home working works best for most types of call centers, the exceptions being customer or employee support where the agents have to physically handle the equipment, either to replicate the problems or as field repair.

Also, tasks where there is teamwork get more challenging with home working, but they are far from impossible. Tools like instant messaging and web conferencing enable sharing of screens, data, and information. *Home Workplace* discusses methods like IM and conferencing.

The only real limitations are the employees and management. Home working is most suitable for agents and supervisors in their late 20s and older who have the independence, self-discipline, and ingenuity to work by themselves and who have their own suitable space. It is not for younger, undisciplined agents who need constant supervision.

But you can change the labor force. Outsourcer ARO began replacing its younger agents with older, more responsible workers when it began offering home working, and gained over $1 million a year.

Management is a much more difficult issue to deal with. Managers and executives still like to see people.

The hard reality that in call centers, face-to-face supervision and management is like Schrodinger's Cat, the famous thought experiment in quantum mechanics to illustrate how the act of observing something changes what is being observed.

In the experiment, a cat is inside a box with a vial of poison gas. You can't see the cat. You don't know if it is dead or alive. But the vial is rigged so that if you open the box, you either release the gas, killing the cat or the vial stays closed, allowing the cat to live.

Face-to-face management in call centers is very similar. If you interrupt the agent, you kill some of your performance because you've stopped the person from working. They spend time with you, and then it takes them awhile to get back on track.

Productivity is not the only element is killed. So is the environment that we depend on. Commuting, mainly by car is a leading source of pollutants.

✪ ENABLING HOME WORKING

There are many methods and technologies that are making implementing and managing home worker easier and less costly. Most call center technologies are or are becoming location-independent. Examples include network and Centrex routing, IVR, predictive dialers, call/contact monitoring and recording, workforce management and headsets.

Here's a brief look at them.

The Broadband Revolution

When I wrote the first edition of this book in 2000, home broadband was for early adopters. "Bleeding edgers" like myself were willing to undergo the hassles of incompetent installers, faulty connections, speed drops when other users piled on in the case of cable, and poor cooperation between telcos and other providers in the case of DSL. We also paid high prices for the privilege.

Most people relied instead on dialup, which at best pumps out data at 56 kilobytes a second, adequate for short emails and small data bursts, but not fast enough for data-heavy applications that are becoming the norm in call centers.

If you require your agents to process high volumes of data it is rapidly becoming easier for you to support those staff at home. Broadband is the norm, especially in Canada, where broadband penetration rates are much higher than in the U.S. The technology has improved and prices have plummeted.

Home users now have roughly the same performance as they do in offices. With broadband I could login to CMP using the bandwidth-intensive Lotus Notes email program when I was working there.

Intelligent Network Routing

Network routing saves you money in up-front capital costs compared with installed switches, IVRs, and CTI on your premises. It also saves you on your long-distance charges. With network routing there is no disaster-vulnerable phone room.

For example, UCN's inContact suite provides skills-based routing, IVR, CTI, and universal queue and callback to home agents, at substantial savings. It does not differentiate between a home agent and an agent sitting in a call center, except that there is another leg of long distance if agents use their home phones to take calls.

A variation of network routing is Centrex routing, which is a generic term for contracting with your local exchange carrier to handle the inbound call routing rather than using the long-distance carrier's switches on your premises. Centrex routing supports automatic number identification, CTI, and skills-based routing.

The downsides of network and Centrex routing are like leasing versus buying a car. You buy what they offer, with little customization, and you ultimately pay more in the end. You're also locked into that carrier, though resellers like UCN source from different telcos.

Switching Features

There are many new features on premises and network switches that make supporting home agents easier. Telephony@Work bases the price for its CallCenter@nywhere switch on agents' interactions with customers rather than using the more common model of per-seat licenses. The payment method avoids your being charged twice for agents who split their time between home and premise-based call centers.

The switch can track agents' absences from their desks. If the agents do not answer contacts routed to them, it automatically logs the agents out and sends reports to the supervisors; it then routes customers to other agents.

CallCenter@nywhere's seamless database switching at the server end avoids having to train agents to use different systems for different tasks. It also provides greater security; only the data relevant to customers is pushed to agents' desktops.

CallCenter@nywhere has monitoring which removes supervisors' fears that agents may be rude to customers if they believe that no one is listening to them. It also enables managers to listen for background noise, which can be an issue in home offices.

The supervisor can make the agent aware of a problem by whispering to them, and if the problem isn't resolved, a supervisor can join in or take over the call. If the agent is misbehaving, the supervisor can lock them out of the communication system altogether.

VoIP-Enabled Routing

Voice over Internet Protocol (VoIP) cuts costs because voice and data are carried over the same wire into buildings. With VoIP voice technically *is* data, and because VoIP uses the Internet, there are no long-distance charges. Many employers with at-home employees are looking at or deploying VoIP.

Most major switchmakers have installed IP cards in their phone switchgear, called "IP-PBXs." There also are PC-based phone systems that are IP-enabled. In addition, there are IP-enabled off-premises extensions (OPXs) that run off IP. And there are VoIP-equipped aftermarket OPX hardware and software.

There are VoIP-only independents like Vonage that seamlessly provide voice services off your public switched telephone network (PSTN) or carrier networks, and cable companies are about to include it as part of their service packages at this writing. My employer, The AnswerNet Network, reroutes PSTN calls to my home over Vonage, and I dial out through it. Vonage also has a cool unified messaging feature where I can see and click on my voicemails.

The local and long-distance telcos are getting into the market. Why beat 'em; join 'em. Here's a bold prediction. Expect VoIP to become the standard for residential services because it is cheaper, with circuit-switched PSTN offered as a premium dedicated line.

VoIP is becoming extremely reliable. It offers increased bandwidth and gradually improving Internet, data switching, and ISP networks. IP phones, PBXs, and gateways are also steadily improving their ability to compensate for latency, jitter, and lost packets, while maintaining good audio quality.

I dropped my dedicated home office PSTN line after Vonage proved itself over my Shaw cable broadband. Vonage works like a charm on my Panasonic cordless phone.

But is it good enough? Check out some VoIP installations and hear for yourself. O'Currance Teleservices did. It has a client in Iowa—far from the bureau's home in Salt Lake City—and handles the program with a dozen home employee agents linked by VoIP over DSL lines, with no PSTN failover.

"The supervisors actually have more control with home agents than they do in a traditional premise-based call center," says Telephony@Work ceo Eli Borodow.

There's another factor to consider. If your agents are predominantly using email or chat then why have a separate voice line? Use VoIP.

Off-Premises Rerouting

If you prefer a PSTN for quality reasons, you can buy premises-based switches with special cards that reroute calls from your premises center to your home agents, including with CTI screen pops. The routing technology to home workers also can be software that you develop and program into the switch, with thin-client applications at the home workers' computers.

For example, Alpine Access, a call center outsourcer, wrote such software to enable its Avaya and Eon switches to route calls to home workers. Employees log into the switches from their PCs, which they own, and the switch activates their lines, no different than in a traditional premises call center. The employees take or make calls with standard phone sets that they own.

Alternately, you can acquire OPXs. OPXs are separate boxes located on your premises or your home workers' to reroute calls from premises or Centrex switches. OPX vendors include GemaTech, MCK, and Teltone.

Either way, if these agents live outside of the local calling areas, you may need to pay long-distance charges. Those inbound calls become outbound calls.

Scripting

Deploying and modifying scripts for home agents has become easier. Digisoft's web-based eTelescript enables you to develop campaigns and publish them on the Internet quickly and easily. Prior to this product home agents had to get onto the call center's network.

Agent Screening

With IVR and web self-service handling more of the routine calls, the calls and contacts left to agents require them to problem-solve and to think on their feet. You will have to pay them more money accordingly.

That means those employees must be independent self-starters who can work without supervision, the same characteristics you demand of home workers. You will also recruit from ignored labor pools like baby boomers as the labor shortages start to dry up employment pools. Therefore you will eventually have a workforce that is home-workable.

There are now tools to help you screen for home workers. Staffing firm FurstPerson has a hosted assessment module in its screening and selection program that enables call centers to determine whether the prospective agents have the ability to successfully work from home.

The pre-screen portion asks agents, either online or through an IVR, whether they have the clients' requirements, such as broadband connections and second phone lines. The advanced screen asks applicants indirect questions or uses role-playing to determine whether they have such attributes as superior cognitive ability, problem solving, and initiative.

You can also test agents' technical skills, such as presenting a computer or phone line malfunction to see how they respond during role-playing.

"Assessing agents' personalities is vital," says president Jeff Furst. "When they work from home they are on their own—they do not have an in-person support mechanism, and they have to think on their feet."

Email

You may think you need to see your employees. But look at how often you do so, compared with email. Email remains probably the most-used management tool. So much so that even in premises offices, managers prefer to use it instead of face-to-face communication. Speaking in person takes time away from your job and theirs, and it can derail their train of thought.

Email, coupled with corporate Intranet sites and webinars, enables all employees to stay within the loop. CMP has a weekly email newsletter that it sends to all employees, with company news and individuals' weddings and births. When I worked there I felt like part of the company even though I was hundreds of miles from the nearest office.

Instant Messaging

Instant messaging (IM) is one of the most versatile communications tools available. IM allows an agent who has a question for their supervisor to get the supervisor on the call while speaking to the customer. It also enables agents and supervisors to network. IM is part of Teltone's OfficeLink ReVo OPX.

I rely on IM on my job, running two or more dialogs at once. I like the ability to put in links and attachments that provide additional information.

Conferencing

Audio, data, video, and the web connect home and premises agents and supervisors in meetings and in collaborations. For example, WebEx's conferencing service provides three types of services.

WebEx Meeting Center provides audio, data, or video capabilities that can be set up instantly in collaboration. WebEx Support Center hosts home agents' computer diagnosis, bug fixes, and applications monitoring and ensures that computers don't contain unauthorized software. It can also link home agents doing customer support with customers. WebEx Training Center lets instructors give lectures and provide online testing. The service can be recorded, and agents can take automatic tests and polls after they listen to the lessons.

"Let's say you have eight people that needed to be trained but only six people could be there; the other two work a different shift," says Jack Chawla, WebEx's director of product marketing and support services. "With the option of recording and then taking tests when they are available, the two other agents would be on the same page as the first six."

Training Through Conferencing

Technology-based training (TBT) makes home working more feasible. It removes or lessens the need for agents to be at a premises call center by training them at their desktops—where those desktops are is irrelevant.

The tools are becoming more potent. ePath Learning's ASAP, one of the first TBT solutions to combine the power of a database management system with the global reach of the Internet, enabled home agent outsourcer Alpine Access's training team to create and deploy courses and tests to agents quickly, in some cases in as little as a few days.

These tools save time and money and improve retention by teaching agents at their workstations. They also permit agents to undertake or refresh their training at their convenience.

Bolstering this method is conferencing. Instructor-led audio, data, video and web conferencing provide real-time interactive instruction with agents no matter where they are. After the sessions you can have all your agents, home-based and premises-based, conference with each other.

Data Protection

Recognizing the risks, call centers have written up home working policies (see *Home Workplace* for a sample policy) that address data protection. For example, P&G's home working policy prohibits non-employees from seeing customers' data.

You can also take steps such as mandating lockable filing cabinets and home office doors, or having the screens located so that they can't be seen from the street. You can also have agents connected with virtual private networks (VPNs) that encrypt data, supply them with computers so you can control what's on them, and specify only high-quality wireless networking to prevent wireless data theft.

If the data is critical, consider storageless PCs, which prevent agents from storing or downloading information. These units are also less expensive to own and maintain and are worthless to thieves. You can also buy biometric readers hooked into the PCs to prevent anyone other than your agent from accessing your computer.

Real-Time and Historical Data Transmission

Spectrum Corporation provides real-time reporting and historical data to home agents' PCs via thin-client or browser-based applications, which have little or no application logic and depend on a central server for processing activities. Home agents can also see if there are potential buyers–customers that have called before—and use instant messaging to alert supervisors that they can let these callers queue-jump.

Spectrum's system also lets supervisors know if they need to bring more home agents on the lines. By tracking call volumes, it warns supervisors of possible spikes. The managers can use that information to determine whether there are any home agents available, call them up on their days off to see if they can log in, or dial up their outsourcing partner to see if they can help.

Trained IT Staff

Because home agents are out of sight they may be out of mind for your IT staff. But they have unique needs and requirements that your personnel must be aware of, such as laptops, VPNs, and VoIP. You will need to educate your IT people, or better yet have one person or a team dedicated to home support. CMP has such a team, which I relied on when I worked there and which did (and still does) an excellent job.

In call centers, more so than other functions, the software is centrally hosted, which means if it crashes, everyone crashes, no matter where they are.

Setting Standards

You will need to draft up a formal home working policy to ensure success of this program. The policy, signed by your employees, should cover:

* Voice and data equipment
* Data backups and battery-powered uninterruptible power supplies
* Data protection such as lockable cabinets, doors, and VPNs
* Facilities and ergonomics so workers don't get injuries like carpal tunnel syndrome and tendonitis.
* Background noise, interruptions, and visitors
* Insurance
* Monitoring
* Sign in, sign out, and absences
* Site visits by supervisors
* Termination

For more details on these areas and issues I once again highly recommend that you read *Home Workplace*. The book goes in depth into home office characteristics, requirements, setup, administration, and management.

Conclusion

If after reading this chapter and looking at the pros and cons of home working you may come to the conclusion that premises call centers are unnecessary or can largely be dispensed with, you may be right. Service bureaus like ARO and airlines like JetBlue think so.

Having been covering and working in this industry since 1995 and having worked in both homes and offices in three countries for nearly two decades, I'm inclined to agree, as long as you have the right people. Except when agents need to physically manipulate equipment, I can't think of one application that can't be done from home.

I would start doing an analysis comparing working from home with traditional premises call centers and outsourcing. Every organization and call center is different. You need a strategy that's right for you.

COMMUTING IS <u>YOUR</u> PROBLEM

While commuting technically may not be your problem—it is the employees' responsibility to get to work on time—you must face reality. Commuting *is an issue for you* because it is your organization that pays the price through lost productivity or lost employees who are tired of working for you because the commuting stinks.

Commuting also may hurt your chances of recruiting the top performers you need to make your outfit the best it can be. Many potential employees won't want the hassle. They would rather work for less nearer to their homes, or from home. With call centers facing high turnover that eats into your budget and profits, should you do something about it?

Commuting also hits you and everyone else in the pocketbook. More traffic, more wear-and-tear on roads, and more packed buses and trains means more new roads and transit systems and more land ripped out of the tax base for asphalt, concrete, and steel.

The more demand for transportation, the more money spent on law enforcement and emergency services due to accidents and crimes. These include collisions, derailments, fender-benders, flipovers, hit-and-runs, impaired driving, derailments, assaults, break-ins, carjackings, drive-bys, murders, rapes, road-rages, shootings, theft, and vandalism.

Justice, health-care, and insurance costs climb up the charts, which means more of your taxes goes to pay for the commuting mess. And, in turn, that means, guess what, more taxes that you and your employees must pay.

Home working helps remedy the problems. The fewer people commuting, the better the call center performance and productivity and the less congestion and delays for those workers who have to be there in person.

Fewer vehicles on the roads mean less pollution, fewer fatalities, and fewer injuries. That could also mean reduced taxes and vehicle fees because transportation demand will be lower.

Higher performance, better competitiveness, lower taxes, and superior quality of life. Who could beat that?

Chapter 6:
Outsourcing

When planning or reviewing your call center, take a very close look at whether you want to outsource some or all of your work. *Outsourcers* typically manage contacts after business hours; supplement existing call centers during peak seasons; and take on non-core assignments like outbound customer acquisition, collections, first-level service and support, order-taking, telemessaging and dealer locators.

That strategy affects your site selection and design strategy and costs because outsourcing removes agents from your centers. Depending on how you deploy them, you can shrink or eliminate existing call centers.

These outsourcers, or service bureaus as they are also called, have their own range of environments. Many now handle their U.S. or Canadian, European, and Australian customers in lower-cost countries.

Most bureaus operate conventional call centers but an increasing number of them have home agents (addressed later in this chapter). These home workers are usually employees, but in some cases they are independent contractors. Some outsourcers specialize in home agent contracting; others offer it as an adjunct to their traditional call centers. Chapter 4 has a full discussion of home working; there is a separate book on this topic, *Home Workplace*, also published by CMP Books.

There are benefits and costs to outsourcing. It may or may not be right for you.

✪ CHARACTERISTICS

Here are the key features of outsourcing and outsourcers:

✱ Outsourcing is a contract between the client and an independent vendor. You set out what you want that outsourcer to do, at your offered price. If the outsourcer agrees to the terms, you and they are obligated to them, nothing more, nothing less.

Unless the agreement specifically calls for it, you cannot dictate to the outsourcer the how they should run their business, such as hiring, firing, management decisions, and deciding where to locate. Taxation authorities take a *very dim view* of organizations outsourcing to get around paying Social Security and other employment taxes.

In referring to independent contractors, which probably applies to any outsourcer, the Internal Revenue Service says clients have *the right to control or direct only the result of the work,* but *not the means and methods of accomplishing the result.* More about that later when I discuss contracted home agents.

* Outsourcing is versatile. The agents are there to handle contacts during the day, after hours, during demand peaks, and in emergencies, for whatever you want to pay for.
* Ultimately outsourcing is more expensive than the best-run call center. Why? Because outsourcers have to make money at your expense.

○ FUNCTIONS

Outsourcers can provide a wide range of call center services and functions. They include:

* Inbound and outbound sales, including customer acquisition, order desk, and reservations
* Lead generation and qualification
* Market surveys and polling
* Fundraising
* Customer service, including information, dealer and office locators, and corporate hotlines
* Customer support
* Billing
* Collections
* Employee support, including HR, sign-in, sign-out, tech support, and hotlines
* Disaster response and emergency contact handling
* Telemessaging (telephone answering, call and email screening and forwarding, voicemail, paging, and dispatch)
* IVR, web, and fax self-service
* Insourcing (managing in-house call centers)

○ ABOUT OUTSOURCERS

Outsourcers come in many different sizes, capabilities, and locations. They range from locally owned companies that provide telemessaging to large multinational bureaus with locations containing thousands of seats spread around the world. The companies are privately owned and publicly owned, based in the U.S. or headquartered abroad.

Few bureaus do all tasks equally well. Some companies are best known for outbound, others in help desk, in B2B, and in the non-profit sector. There are also bureaus that offer other services such as fulfillment.

One of the key differentiators between bureaus is the size and scope of the outsourcing contract they are willing to take on. Hence the discussion in Chapter 3 on determining inbound and outbound call and contact volume. There will be more later in this chapter on selecting outsourcers.

As a general rule, big bureaus work with large programs, small bureaus with small projects.

There are economies of scale in the service bureau business such as in setup and training costs. Each bureau has different profitability levels that contracts must meet.

If an outsourcing contract proposal doesn't reach the outsourcer's profit level on the quoted rates but if both parties are interested in making it happen—like if you were a growing or large organization that has a small program—then the bureau could offer and you accept, say higher setup fees or rates. Also the bureau you approach may have the work subcontracted to another outsourcer who specializes in smaller projects.

○ PROGRAM ACCOMMODATION

Outsourcers accommodate your program based on size, value, duration, and skillsets. They will pick or suggest locations that have the room to handle the volume and agents who have distinct qualities such as computer training, technical certifications, or language skills.

Outsourcers will house your program on one floor or center. They may also offer to spread your program across multiple centers. That is a sound strategy because it provides you with business continuity. If a disaster knocks out one center, there will be staff trained and experienced in the other center who can be brought in or stay longer to handle your contacts, or train other agents how to do so quickly.

For very large and long-term programs of more than three years, outsourcers will build completely or partially dedicated centers where you are the only or anchor client. The bureau will often design it to your specifications.

The call centers can be onshore or offshore, or a combination of the two, known as "blendshoring." Offshoring saves money and provides better-educated agents at the risk of creating cultural issues that could debilitate programs (more about onshoring and offshoring in Chapter 8).

Alternately outsourcers can have some or all of your programs handled by employees or independent contractors working at home (discussed later), which cuts costs and improves flexibility while having contacts handled by Americans. There are several outsourcers and programs including Alpine Access, ARO, IntelliCare, O'Currance, West at Home, Willow CSN, and Working Solutions, which have home agents either as employees or as contractors.

○ OUTSOURCING PROGRAM TYPES

Outsourcing programs fall under several rough categories: timing, channel, ancillary services, and agent sourcing. You can mix and match them to meet your needs.

Timing

You can pick the occasion when to outsource, when it best suits your needs.

Weekday

In this scenario *usually* the outsourcer is acting as your partner during your workweek, either supplementing or replacing your call center.

Supplementing can take the form of splitting the load with your in-house centers or handling the overflow contacts—volume over and above your maximum capacity at a given service level, measured by hold times. Supplementing can also be triaging calls and opening trouble tickets for in house or other expert outsourced professionals, including field dispatch and paging. Replacing can mean just that: doing away with your in-house call center altogether or part of its functions.

After Hours

This is a common time period for outsourcing. All contacts switch over to the out-sourcers' centers, enabling you to shutter your facilities, saving you money. You also improve your overall productivity, because for many functions contact volume drops off after hours. You still provide service while avoiding keeping the lights on for rela-tively few and idle staff.

Demand Peaks

As noted earlier, outsourcing can share the load. One of the best applications is for han-dling demand peaks, day or night. You size your call center and staff it to handle the maximum volume as per your service standards (see Chapter 3) then contacts that come in above that threshold are routed to your service bureau partner.

Emergencies

You can contract with outsourcers to provide contact handling during disasters. You need to set this up some time in advance, unless you have an existing relationship with that vendor.

Seasonal

If the demand for your products or services is seasonal, such as ski resorts and beach gear, then outsourcing for those busy periods lets you handle the load while keeping only small call centers year-round.

Channel

You can choose to outsource based on customer interaction channels. Working with service bureaus avoids making costly investments in supporting technologies that you may rarely need. Here are the choices:

Online email

As noted in an earlier chapter, online technology is not cheap. It can cost $100,000 or more for a typical call center and requires agents who are more literate than the others. Outsourcing lets you tap these tools without the big investments.

Voice

Your organization may communicate predominantly online, such as with email, chat, automated email, and web self-service. So why have voice switches and phones? Out-source that to a service bureau.

Special needs

Outsourcing can help meet customers with special needs, such as those who speak lan-guages other than English and the hearing-impaired.

Languages

Countries like the U.S. are becoming multilingual. Immigrants are holding onto the native languages longer. A Yankelovich Partners study states that Spanish dominates

ANCILLARY SERVICES

Outsourcers often provide a range of ancillary services. Leading examples include:

- Field support and dispatch
- IVR response transcription
- Online appointment setting
- Credit card processing
- List acquisition and management
- Billing
- Data processing
- Payroll
- Product fulfillment, shipping, storage, and returns handling
- Postal mail response
- Staffing and training
- Consulting

over two-thirds of Hispanic households. Use of Spanish has been rising while English has been dropping.

This trend is being abetted by cheap communications and worldwide media that enable people to stay in constant touch with family and relatives back home. Airfare is cheap compared to a generation ago. Many of their home countries have become economic powerhouses in their own right, like China.

But not every community has sufficient availability of non-English speakers to meet market needs at the right price. Your call centers may be in one of them.

For example, Hispanics are the largest American ethnic minority, accounting for over 38 million residents. But can you find enough smart, qualified Hispanics willing to work in your centers? Chapter 9 explores how to serve the Hispanic market.

Hearing-impaired

There are over 2 million American consumers who are deaf or hard of hearing. They rely on TTY, which converts text to speech over phone lines. But the technology is expensive to deploy in call centers.

There are service bureaus that have made that investment. One of them, CSD, has the TTY functionality built into its switch, enabling their agents to treat these customers like the others.

Technology

Outsourcers can also host your call center technology, including IVR, web self-service, switching, routing, and CRM software. You avoid the up-front capital costs of buying, installing, and upgrading these tools. You also obtain disaster protection because the hardware and software are offsite and hopefully—if you select the vendor right—backed up somewhere else.

Agent Sourcing

There are different ways to source agents for these programs.

Shared Agent

Shared agent is where your contacts are handled by the outsourcers' general agent pool along with contacts from other clients' customers. All agents receive standard training in how to respond or make those contacts, including specific instructions for several clients.

Shared agent upsides

Shared agent is the lowest-cost and most versatile of the outsourcing program types. That's because employees are continually busy taking care of contacts instead of waiting for those from one or two clients.

Shared agents are ideal for high-volume, minimal-depth interactions such as dealer locators, office hours, order taking, general information and inquiries, customer acquisition, collections, and market surveys. Telemessaging is virtually all shared agent.

Shared agent is the best program type to take care of sudden demand changes like demand peaks and emergencies. The bureaus do not have to provide much if any additional training to other agents to prepare them to handle those calls.

Shared agent downsides

Shared agent programs do not provide outsourcer agents with the capability to provide and build on expertise and knowledge of a client's product, service, or procedure. Customers cannot easily go back to the agent who dealt with their order or issue.

Dedicated Agent

Dedicated agent programs are those where contacts for a specific client go only to agents who have been assigned to them. These agents are trained thoroughly on that client's program. For all intents and purposes these bureau employees perform just like the client's in-house agents.

Dedicated agent upsides

Dedicated agents give greater customer service depth and build true long-term customer relationships. Agents stay longer and develop their expertise in your products and services.

Dedicated agent downsides

Dedicated agents are much more costly to provide than shared agents, with costs approaching those of your in-house agents. You pay for idle time, or as my paramedic son would say, describing the on-call wait periods in their ambulances, "for their potential."

Because inbound calls arrive randomly, managers have difficulty using dedicated agents, says Amit Shankardass, solution planning officer of outsourcer ClientLogic. Consequently, these agents remain unproductive when no calls or contacts arrive for them.

Modified Dedicated Agent

Modified dedicated agent is where the outsourcers offer to add services or clients to defray costs to you and them. Examples include:

* Email, chat, fax, and mail handling in addition to making and taking calls
* Non-call center tasks such as data processing and IVR transcription
* Shared agent overflow (directing other clients' calls to dedicated agents during slack times)
* Semi-dedicated agents (assigning one or two other noncompeting clients to the same agents)

The advantage of this approach is lowered expenses. But it adds to program complexity, such as additional agent training, and the savings cannot always be assured.

◎ OUTSOURCING SERVICE OPTIONS

Outsourcers can deliver your service and sales through several services options.

Outsourced Bureau Home Agents

You can also outsource your home agents and programs to service bureaus to substitute or support premises call centers.

Boomers and retirees who would never step foot in a call center would be more willing to handle the same functions from home. You can also indirectly access the mobility-impaired and those with childcare and eldercare responsibilities.

As pointed out in Chapter 5, these gains are real. Consultant Jack Heacock estimates that a call center can save over $10 million in five years. ARO, a Kansas City outsourcer, saves over $1 million annually compared with having agents work in a traditional call center.

More importantly, you can match agent availability more closely to volume with outsourced home agents than you can with traditional call center agents, because there's no commuting involved.

Outsourced home agent benefits

The benefits are similar yet unique compared with having in-house home agents. Depending on the outsourcer and the project, you can save money because the bureau gains from real estate savings and higher productivity from better-quality employees.

Home agent outsourcers offer agents for time increments as low as 15–30 minutes. You minimize paying for idle agents' time. You can also respond to call spikes far quicker with outsourced home agents. Traditional call center outsourcers will need to bring in and pay employees for at least an hour or so to make it worth their while (and the employees' while) to commute in.

Moreover, by outsourcing you do not get involved with sometimes complex home office setup and administration issues, including IT support and ergonomics. You leave that up to the outsourcers.

Outsourced home agent challenges

By outsourcing you will not experience all the gains you would get from an in-house home agent program. Moreover, like traditional call center outsourcing, you lose direct control and touch with your customers.

Outsourcer agents work for you. When outsourcing, make sure that your partner educates and instills in their agents that they are working as your representatives. Many do by methods such as hanging banners with your name, such as what Convergys did for Hallmark.

Contracted Home Agents

Contracted home agent out-sourcing is different from out-sourced home agent programs because it entails doing business with specialty outsourcers or contractors who arrange, train, and monitor self-employed home workers for you. The contractors seek the business and put out the contracted work to see which qualified agents want to do it through a computerized scheduling system. The home workers pay for their training and access to the switch and call-handling equipment.

Contracted agent firms and agents work like car service and limousine firms in many large cities. Self-employed drivers own their vehicles, pay base fees to the contractors, let them know their availability, and put in for offered trips.

Agent contractors have developed some very sophisticated bidding applications. Clients generally post schedules a week to a month in advance. And, as long as the self-employed agents have completed training to answer calls for a client, they can indicate when they're available to take these calls.

To ensure that all workers have a fair shot, there are limits to the number of hours they can devote to a client per week. Conversely, contractors' clients look up the number of self-employed home workers they've requested for a given hour and the number on call during that hour.

Some contractors charge their self-employed home workers a service fee for their participation in the contractors' service pool and access to the bidding system. The contractors then bill the clients and pay the self-employed workers. At tax time, the contractors issue miscellaneous income statements (IRS form 1099s) for those workers not incorporated.

Self-employed home workers are not limited to providing their services as a sole proprietor or to use the IRS term "independent contractor." Willow says that should the self-employed worker desire, the individual can establish a corporation to provide the services and be an employee of that corporation. The corporation, as the vendor, invoices the organizations that contracted it to provide services.

Contracted agent benefits

Contracted home agents are more flexible than outsourced or in-house home agents. You can bring these agents on for one campaign, program, project, or story. When the work ends, so does your involvement and financial commitment.

You can easily hire these independents to do the work and fire them if they do not do it well. Because they work from home, they are out of sight and out of mind. There are no personal attachments, unlike with consultants and temporary premises office personnel who work as independents or for contractors on your premises and which your staff gets to know.

Self-employed agents are often but not always more motivated to perform and succeed than employees. They made a financial commitment to get equipped and to sign on, so they will work hard to get the most from their investments.

With self-employed home agents you get the productivity, quality, and disaster-recovery benefits of home working without the up-front costs and time of a creating home worker program. These independent workers own, lease, pay for, and write off, their computers, phones, and lines. Because they are self-employed you avoid paying for their voice, data, equipment, and furniture.

The agent contractors handle the screening, training, and technology setup. They train the self-employed workers following your internal training program—either instructor-led or technology-based training such as through the web, CDs, or videos. The contractors stay in touch with these individuals through email, newsletters, voicemail alerts, instant messaging, and secure chat rooms.

There are also fewer regulatory and jurisdictional hassles. You don't have to comply with OSHA and workers' comp or be involved with cross-state tax issues. It is up to the self-employeds to be responsible for occupational safety, building codes, and local regulations like zoning. It is clear where they pay their taxes: where they do business.

Contracted agent downsides

Agent contractors have less control over their agents compared with in-house or bureau call centers because these call center workers are not employees.

The taxation authorities are very clear who's self-employed and who's not. You can't ask self-employed people, even through contractors, to do any more than what you and they have agreed to.

A contracted agent can tell you to go jump in the lake if you rang them at midnight and demand that they fix some problem with the web site. If it isn't in the deal then you're sunk.

There's also the flip side of flexibility. Self-employeds are not loyal to your contractor or to you. Many of them have other clients, as I did when I freelanced. If you don't have someone else covering you with their medical benefits—perhaps the major employment issue in the US—the pressure is high to be employed by someone else who can offer them. When I got a staff writing job at DM News, after 2 years of being self-employed it was sayonara to my clients.

Chances are self-employed home workers will have less knowledge about your particular field and less incentive to obtain that knowledge than your employees. Self-employed people tend to work for more than one client at the same time and have to be up to speed on all of those clients, whereas employees work for one employer at any one particular time and can focus on that employer.

The exception is when those self-employed have worked in your field before; in those cases they may know more than your employees know. The same goes for any contractor or outsourcer. There also is less incentive for you to train self-employed workers to give them that knowledge. Why pour those resources into people that are only available temporarily, at best?

You risk some of that information going to competitors, just as you do when employees leave your firm. For that reason, and to avoid the appearance that these contractors are employees, you cannot supply equipment to self-employed people, which means they should not gain access behind your firewall to your network.

Prison Labor

There is another outsourcing service option: having inmates handle the contacts in state and federal prisons.

Federal Prison Industries/UNICOR, which operates factories under a U.S. Department of Justice mandate to provide convicts with productive work, offers outsourced directory assistance and outbound B2B. The centers can also take on customer service, business support, help desk, and fulfillment.

UNICOR's programs promise lower costs; they are intended to compete only with offshore call centers. It is not permitted to go after higher-cost domestic outsourcing business. Only work that would have gone offshore goes to UNICOR.

The firm, a self-sustaining public corporation, provides skilled agents, native English and Spanish speakers, scalability, accessibility by clients, and high security. UNICOR screens projects to make sure inmates do not have access to personal data. Inmates are not permitted to call private citizens.

UNICOR's and similar programs perform another important benefit: They give inmates experience and skills that they can apply on the outside. That way they don't return back to the inside. Former inmate agents have been hired by call centers following their release.

On the other hand, you have to balance how your customers and senior management feel about having convicts taking care of some of your work. There is a risk, no matter how small or manageable. For example, Utah scrapped its prison telemarketing program in 2000, reported the Associated Press, after an inmate sent three inappropriate letters, two of them to teenage girls.

There is also the moral issue. Yes, inmates must be given skills so that they obtain work on the outside. But every job that goes to a prisoner that takes away a job for a law-abiding citizen somewhere.

Would it not be fairer for governments to sponsor training and apprenticeship programs for inmates once they have been released? Or give their employers, like call centers, wage subsidies similar to welfare-to-work programs? If convicts know that this is waiting for them, then they will have an incentive to stay straight, so that they will be given another chance to become productive members of society.

Insourcing

Outsourcers provide what is known as insourcing, where they hire, train, and manage your center for you, using your facilities and technology. You can have bureaus run existing or new centers. Many organizations are insourcing similar back office functions such as mail room and shipping/receiving.

The chief benefit of insourcing is that you leave the staffing, training, and HR hassles to the outsourcers while maintaining a greater level of control compared with traditional outsourcing. The bureaus will quite likely know better than your own HR department what makes a good agent and how to get the most out of them.

The principal downside of insourcing is that the financial benefits may not be as high as traditional outsourcing. You still have to do all the site selection, real estate investment, and work and negotiate that with other departments.

Interpreting

Except for Spanish (and in Canada, French), most organizations do not experience enough demand for agents who speak other languages to require separate teams or call centers. That goes for both in-house and outsourced call centers.

Hence over-the-phone interpreting (OPI). There are specialist outsourcers who provide OPI. OPI firms hire or contract with trained individuals to interpret both sides of the conversations.

How does OPI work? Daniel Trevor, president of Tele-Interpreters, which provides translators in 150 languages, says that if the agent taking the call has difficulty understanding the caller, he or she asks the caller which language they speak.

When the caller replies, the agent can dial up the translation service toll-free on a separate line, supply account information and then hook in the customer for a three-way conference call. Interpreting connections take only seconds to set up. You brief the interpreters what you want to have done.

Many OPI firms also train and test your interpreters. Other services they provide include document translation from English into other languages, for forwarding by email and fax.

Language Line, founded in 1984, is one of the most experienced and well-known OPI firms. It has a handy "Hold Please" training kit for trainers to teach agents key phrases in several languages.

OPI *advantages*

A key advantage is flexibility. You bring in the interpreters only when you need them.

The other major benefit is quality. Interpreters tend to have much higher education levels than typical call center agents and are often native speakers of those languages. They typically have bachelors' or sometimes masters' degrees and are working on the side while going to school or pursuing other careers like acting. That knowledge and skill helps to ensure that you and the customer are on the same page, even if the words are different.

OPI *downsides*

For high volumes OPI can cost more than recruiting and hiring your own agents or specifying outsourcers that have agents who speak those languages. You are adding time, complexity, and expenses per transaction with interpreters.

One challenge of OPI is how to keep callers on the line from the time they dial in to when they are connected to an interpreter. These can be awkward moments for agents, who may be lost for words. Another issue is that some end-customers may see OPI as patronizing, treating non-English-speakers like second-class citizens, especially if the organization using it is based or has operations in communities with large ethnic populations

Compare closely OPI and in-house or outsourced call centers located in metro areas with a diverse language base. Leading examples include New York; northern New Jersey; Boston; Washington, D.C.; San Francisco (including Silicon Valley); Los Angeles; Seattle; Toronto, Montreal; and Vancouver.

✪ ADVANTAGES

Outsourcing has key benefits.

Cost savings

The principal reason you should consider outsourcing is cost savings, direct and indirect. PriceWaterhouseCoopers estimates that outsourcing can slice your call handling costs by up to 30 percent at least in the short term. You also avoid the cost-riddled hassle and drain on your staff, facilities and IT resources, especially if your alternative is building another call center.

Service bureaus can help you reduce or eliminate a call center, depending on how many of these functions you can outsource. You may want to have a few top-level agents in house to handle tough questions or to patch in engineers and developers, or to have senior account managers give personalized service to top customers who appreciate dealing directly with your company.

If you are with a growing new or capital-husbanding company, your best bet is to outsource your customer service and sales. You obtain agents as needed, without investing any resources in the hardware, software, facilities, screening/training and people.

How Outsourcers Do It

Outsourcers obtain their financial edge in several ways. They have workstations to fill and people to fill them. Unlike your own call center, these contractors can keep their agents busy with shared agents, modified dedicated agents, and ancillary services.

The other key way that outsourcers can cut costs is that they are more flexible than many organizations. Too often a firm or a department is locked into long building leases in high-wage locations. To move into lower-wage communities requires multiple meetings and approvals that occur at glacial speeds. Senior staff or unions may put up roadblocks.

There may also be good reasons for the outfit as a whole to stay in a particular city even if it is not beneficial for the call center. Senior management may need to meet often with key customers and financiers. The metro areas may have important engineering and research schools to draw R&D talent from. That's why you will still see high-end call centers in high-cost locales like Boston, New York, New Jersey, and Silicon Valley.

Outsourcers have few such limitations. They put programs into communities with the right mix of quality and cost, and transportation access if need be. It is easier for many organizations simply to contract out to an outsourcer than to move their contact center.

Service bureaus also have the expertise and contacts to set up quicker with less cost, even though they face the same cost pressures of in-house centers. They have economies of scale in buying hardware, software, and staffing and training services. They can often afford the latest, most effective tools because they can spread those costs over many clients.

How can service bureaus achieve all of these results? Because as contractors they have to be efficient and effective. That is their culture. They are competing with other bureaus and with in-house call centers. Setting up and managing call centers and running programs *are* their core competencies.

"The outsourcer call center culture is more productive than many in-house call center cultures," consultant Paul Kowal points out. "With an in-house call center when agents come to work, they expect to get paid from the time they show up. There may be a time lag between the time they punch in and when they get on the phones. With an outsourcer, because the client is paying by the minutes its agents get on the phones right away."

Flexibility

Service bureaus are there when you need them. You can contract with them just to handle telemessaging, overflow, or after-hours calls, or you can turn over your entire multimedia customer service management. While outsourcers handle medium to large contracts they also take on small ones for the right price.

That saves you money by avoiding constantly resizing your call center and hiring people to handle daily, weekly, or seasonal spikes and then laying them off. With outsourcing you won't have workstations gathering dust and employees waiting for the screen-pops to flash. You keep your in-house centers small and busy.

If you make ski equipment, why have a big call center if it will be mostly idle in spring and summer, unless you have customers in Australia or South America? If you run promotions sporadically, like direct response ads for plastic pink flamingoes and other similar tasteless ornaments during a John Waters film festival on cable, why have vacant seats over and above your normal requirements when you're not trying to offload the blessed creations?

Outsourcing is your best bet if you're launching a new product or service or testing the market. Outbound telemarketing or driving inbound calls from radio and TV ads, web banners, email, and direct mail are the most common applications for which companies use outsourcers. Once the promotion or sale is finished, you don't need the agents or the call center.

Core Competency Focus

Outsourcing enables you to focus on your core competencies and on your best customers, however you define them. Those core competencies are typically higher-level, more involved customer service and sales, where agents really need to know your organization, what it offers, how to find information, and who to contact to meet customers' needs.

Following are a few popular examples.

Outbound Business-to-Consumer Sales

This has been the classic example of outsourcing, where many outsourcers got their start. Agents call prospects and customers from lists with heavily scripted pitches.

The agents introduce the product or service, make the windup and throws, hear the responses and record the results: sales, no sales, or more information. There is no relationship-building, no need to be knowledgeable about the company and no time for chit-chat. Pitch, hear and record the hit or miss, and go.

As noted in Chapter 2, outbound is dying from consumer hostility, backed by laws that restrict it through do-not-call lists. Instead, organizations and outsourcers are having customer service agents cross-sell and upsell customers and prospects, such as when they call in to activate credit cards.

Inbound Direct Response

To drive business, organizations run continuous or targeted new marketing campaigns, and they need live agents to answer the calls and emails. Outsourcing is a great way of handling that demand.

The inbound scripts are tightly written. The agents' jobs are to give standard information about the product or service, find out more about the prospects or customers, identify needs and potential for sales, and conduct the sale or send more information. If the prospects or customers have inquiries that the outsourced agents could not answer, they escalate them to in-house staff.

Lead Followup and Qualification

B2B and high-level business-to-consumer sales often require outbound and sometimes inbound agents to qualify leads, such as those resulting from ads, trade shows, and web inquiries.

The questions agents ask are simple and routine. In some cases, like for lesser-ranked customers or for low-value products or services, they can close the sales. But by and large they turn over qualified prospects to in-house sales staff that know the products or services and markets intimately.

Level One Customer Support

In first-level or Level One support, particularly for lower-value consumer products, reps with minimal training diagnose customers' problems using scripts tied into knowledge bases, taking customers step by step to solve issues like finding the On button. See *The Complete Guide to Customer Support* for a thorough discussion of support levels.

If the difficulties can't be resolved, the issues are escalated to in-house technicians and engineers who have the skills to come up with solutions outside of the knowledge base box. They may be able to determine that the problem lies in a blown power supply.

"Outsourcing works fine for routine first-level customer service and sales," says Mark Schmidt, vice president of consulting, training, and outsourcing with TeleDevelopment Services. "But if there are inquiries and contacts that need to go to a second or third level, then they may need to be escalated and handled in-house by agents who have extensive product knowledge or another service agency that specializes in second-level support."

Collections

Companies usually try to bring current customers in-house by working with them to find ways to settle the accounts, to retain them as customers. People often get behind in their bills but most recover and become good customers again.

There are specialist outsourcers known as accounts recovery firms that can help your customers get their accounts "cured" such as by showing them different ways to pay bills or adjust payment dates. When a company decides it isn't worth the bother to retain overly past-due customers, then accounts recovery firms have agents skilled at obtaining promises to pay from delinquent customers, using a mix of compassion and firmness in compliance with the laws regulating debt collection.

Telemessaging

Telemessaging is usually outsourced, especially after hours, and is almost always shared-agent. Companies rarely get the volume to justify in-house telemessaging staff. The exceptions are when in-person receptionists who are also handling the telemessaging.

Hassle Avoidance

Outsourcing avoids or minimizes the costly and time-consuming business of setting up, managing, downsizing, and closing call centers. The bureaus do that for you.

As this book clearly illustrates, there is a lot of work that goes into planning and putting together call centers and making them work, including sizing, adjuncts and alternatives analysis, site selection, facilities and design, staffing, training, and management. As you will see especially in Chapter 9, these functions become much more complicated if you are having American customers handled from outside the U.S. and if you are serving foreign customers.

Another hassle outsourcing shrinks but does not eliminate is in downsizing. If a program ends, you're cutting it back or the outsourcer is not working out for you the bureau, not you is responsible for what happens to the bureau's employees. Service bureaus can shift agents or close up shop, with none of the local heartache and bad press if your company laid off or reassigned staff or closed a call center.

Ability to Serve Niche Markets

Service bureaus serve markets that are too costly for individual call centers and organizations to serve themselves. You may not have enough need for agents in a particular locale or with special skills to handle those customers.

Examples include insurance, which is state-regulated and needs Spanish-language support. But service bureaus, by combining your needs and those of other organizations, could cost-effectively serve those customers.

For example, you might not have sufficient numbers of Spanish-speaking customers to warrant hiring bilingual agents or to set up a call center in a community where there are plenty of affordable bilingual agents. But a service bureau may set up in such places in the U.S. or overseas

Speed to Market

Bureaus get your program set up far quicker, in days rather than weeks or months, with results seen much sooner than if you had undertaken or kept these operations in-house.

The key to their quickness is that of any contractor: when they make a bid, they have to have the people, the place, and the equipment lined up and ready to begin delivering on the promised date at the tendered price. They've done the site selection, design, and real estate work. The larger companies have on-staff and on-call location teams. They monitor the demand at their centers and expand or open centers if they forecast a short-fall or are bidding on major contracts.

The smart bureaus also have proven, effective call center designs that simplify and speed up set-up. Their vendors are only a call, email, or fax away. If you need services like design, direct mail, and fulfillment that they do not provide, they usually have part-nerships with those that do, giving you one-stop, worry-free service shopping.

Business Continuity

Outsourcing provides business continuity in two ways: ongoing and emergency.

Because contact handling is their business, service bureaus will likely have robust business continuity strategies, processes, and tools. These include onsite generators and automatic or manual transfer of contacts to other centers, home agents, or clients' sites.

In ongoing business continuity, you should have your program handled in bureau call centers located in different climatic or geologic zones than your call centers or offices are in. A distance of 400-500 miles is a safe bet, unless you have all your call centers on the same risk corridor, like the San Andreas Fault. For large programs, you can split them up between bureau call centers or between the bureau centers and yours.

That way if one of the bureau's centers is knocked offline, the other centers or your center or office can come online. Keep in mind that if your program is handled in those centers only in emergencies, those agents will not be as well trained to make and take contacts as those in the regular centers.

In emergency contact handling, you contract with the bureaus to handle your con-tacts if your centers have been or are about to be hit. Keep in mind, too, that the agents will not be as well trained as your staff to take the contacts.

Financial Appearance

Outsourcing makes you look good financially to Wall Street, Canada's Bay Street, and your board. Why? Because outsourcing is a liability, an expense, one that can be easily cut, whereas your own call center is capital, which is messier to dispose of, with employ-ees you have to hire and lay off.

"You can outsource a program at greater cost than keeping it in-house, and Wall Street will praise you for it because you don't have the capital costs of having that cen-ter on your books, along with those employees," Rudy Oetting president, R.H. Oetting and Company points out.

✪ DOWNSIDES

There are also some downsides to outsourcing.

Loss of Control

When you outsource you give up much of your control to your bureau. The bureau makes the hiring, firing, staffing, and training decisions, sources the equipment and almost always decides on locations. The bureau's employees, not you, directly interact with the customers. While you can monitor calls and review reports, if you see a problem you go through the bureau's client rep; you can't yank the agent off the floor.

Bureaus can do only what you and the bureaus have agreed to do at a set price. If you want them to do more or less you have to renegotiate the contracts.

Bureaus and agents can't think out of the box to solve customers' problems, or go the extra mile, unless the contracts permit it. They don't know your firm, products, and services as well as you know them. Outsourcers do not have the same responsiveness to your needs as those of an in-house center because they have other clients to satisfy.

For that reason, consultants tend to say that small companies, medium-sized enterprises, and larger firms seeking a competitive advantage by stressing customer "touch" should not outsource. Keep those customers close to you.

Substandard Performance

Outsourcers may not deliver the results that you are expecting. A major reason is that their agents are not, and in fairness cannot be, as infused with your corporate culture or committed to your firm as your employees. An in-house agent may know of a new product or service, or may know many more details or help solve problems quicker because of what they hear or know inside your organization.

Mark Seeley, director of consulting, and James Trobaugh, senior vice president of CB Richard Ellis Call Center Solutions Group, say that a few of their clients took their programs back in house because they were not pleased with outsourced agent performance.

"Some of our clients are concerned about cultural issues, such as communicating their core values to end customers," says Trobaugh. "They feel their employees can do a better job of representing their corporations in their own call centers because they are living their firms' values by working in their facilities."

Lack of Access to Customers

When you outsource you sell off direct interaction with your customers. Your customer contact and relationship-building skills could atrophy, yet their value depends on how important customer relationships are to your business. Some functions, such as outbound calling, have less impact on this factor, where the called party is being pressed to buy, rather than inbound customer service where the caller is more engaged and wants something from you.

✪ CUSTOMER DISSATISFACTION AND ANNOYANCE

Customers may become dissatisfied with you and could go elsewhere if your performance deteriorates from outsourcing and they find out through media reports that you've outsourced, especially offshore. Some people may feel that the company or government agency is trying to shuffle them off to one side.

Not surprisingly most organizations rarely like to advertise that they're outsourcing because they want their customers to feel that they are dealing directly, instead of through a third party. Many service bureaus also don't like publicizing their relationships because they don't want competitors steal their clients.

Data Risks

You increase the chances that your customers' data will be stolen by outsourcing; you entrust the outsourcers to safeguard the information. Laws such as the Health Insurance Portability and Accountability Act (HIPAA) require that you establish and maintain custody of that data. HIPAA stipulates that you must trace who handles what data at each step.

You also risk sharing your information with a competitor that is using the same outsourcer, though the bureaus do take steps to prevent this from happening. I've seen agents working on competing credit card accounts, denoted by signs at their workstations, sitting across from each other, but it caused no apparent problems for either client.

Cost and Productivity Disincentives

Threatening to counterbalance cost savings from outsourcing are some disincentives. They include turnover, which tends to be greater in an outsourced than in an in-house center, leading to higher training costs, greater error percentage, and lesser quality.

There is also less likelihood that you will detect problems until large numbers of customers and leads have been burned. That could lead to contract disputes and possible litigation to resolve them.

Profit

Outsourcing may also be more costly than opening and running your own center over the long term. Outsourcers are in business to make money. Experts such as Rudy Oetting have long pointed out that an extremely well-run in-house center is almost always less expensive than an outsourced center because part of your fee goes to the outsourcer's profits—that's money you could keep in house.

Consultants estimate the tradeoff occurs at the three-year mark. If your project length exceeds that, then you may be better off opening your own center.

Hiring, setting out tasks, and managing contracted home agents like Willow CSN's (seen here) are the responsibility of both the contractor and the client. It is similar to arranging for owner-operated taxis or limousines that go through companies. You call the firms, tell them where and when you want them to go, and set the price. You can then give reasonable specific instructions to the driver like which terminal to go to. But there are fine lines in managing the service and the provider.

✪ ENABLING OUTSOURCING

If you decide to outsource or look into it further, here are several considerations that enable the practice.

Know What You Can and Can't Outsource

Outsourcing is a great tool, but it has its limitations. James Beatty, president of NCS International, has seen companies test outsourcing and then leave some functions to the bureaus while taking other functions in house. Only the outsourcers that demonstrate superior levels of customer service and investment in technology have been securing contracts.

"Companies are outsourcing their more costly functions, like email response and basic service and sales," Beatty explains. "But they are keeping in house their high-touch, high-quality calls where agents really need to know their products and their customers."

Negotiate Fees, Minimums, and Contract Terms

Service bureaus usually charge by the hour, with a minimum numbers of hours, plus set-up fees. The rule is the fewer the hours, the higher the cost per hour. The average going price for outsourcing work in the U.S. is $24–30 per hour, less if offshore (more about that later this chapter).

Pay Per Performance and Pay Per Call

You can negotiate pay-per-performance contracts (PPP) where you pay part or all of the service based on sales, customer satisfaction, or other agreed-on service measurements, compared with standard per-minute or hourly rates. You can also seek pay-per-call contracts (PPC) on inbound and outbound customer service and on direct response. Additionally, you can seek blended PPP or PPC contracts with hourly rates.

The benefit of any PPP contract is that you spread the risk. Service bureaus bear some of the responsibility for the program's success. PPP connects it to their bottom line. Also, with both PPP and PPC, you pay only for what you get. You're not subsidizing agents' idle time.

Service bureaus generally do not like PPP or PPC, because they eat into their profits. The bureaus have to shoulder more of the risk and to minimize it they have to do more work putting these programs together.

The bureaus' concerns are legitimate. If you have a bad outbound telemarketing or inbound direct mail list, or customers or prospects don't like your product, service or pricepoint, the outsourcers pay the price for factors that you're responsible for. With inbound customer service, volumes are often unpredictable, which hurts the outsourcer.

The only way PPP or PPC is fair is if you give the bureaus some control over your outbound or inbound program. They have should have a say on lists. If they insist on testing them, do so. Make it worth their while financially. The greater the risk and involvement, the greater the rewards.

"The advantages to clients are lower costs," says Mark Russell, an associate with the merger and acquisition advisory firm Kaulkin Ginsberg Company. "They are not paying for idle agents and no sales. But that puts more pressure on outsourcers to lower overall costs and improve performance."

Contract Length and Terms

Outsourcing call center contracts have typically been two to three years with performance exit clauses. But thanks to competition outside of the U.S., the terms are becoming more client-friendly.

Geri Gantman, senior partner at R.H. Oetting, reports that offshore competition is forcing American bureaus to offer better contract terms. Three-year contracts now have annual reviews. Clients can renegotiate terms or cancel at any time if the outsourcer fails to meet service level agreements (SLAs).

Offshore contractors are much more willing to accept shorter contracts than U.S. suppliers have traditionally been, as a way to encourage users to make the offshore shift, explains Gantman. These bureaus often allow clients more direct control over staff selection, training curriculums, and compensation plans.

Their SLAs are more rigid and include metrics not traditionally accepted by U.S. suppliers, such as customer satisfaction, which is conducted in surveys outside of the service bureau contract. These "soft measurements" are in addition to the typical U.S. call center SLAs including outsourcer-measurable metrics, such as average length of calls, first-call resolution and abandon rates.

In contrast, simple programs like order entry and telemessaging are much less, sometimes as few as four weeks, renewable. Also, limited-duration programs like seasonal and direct response last for the length of the sales period and promotion.

Size of Business and Size of Outsourcer

As noted earlier, when considering outsourcing one general rule is that the larger the contract or organization offering the contract, the better the deal. Outsourcing is an economies-of-scale business. Another general rule is that the larger bureaus, because they have economies of scale, are more likely than the smaller bureaus to offer better pricing.

Size isn't everything, though. Because smaller bureaus often offer more intimate attention to clients and their customers, they may well deliver better service for the money.

These bureaus are also ideal for short, low-volume and very complex programs requiring agents to be intimately acquainted with products and services. Larger bureaus usually don't look at any project smaller than 200–500 hours unless at a premium. But they often have the technology, training, and experience that smaller firms lack, which may justify paying the higher fees.

Between these extremes are a few bureaus that offer the best of both worlds: small center scalability with large center capability. The keys to their service are highly efficient networks.

The AnswerNet Network, based in Princeton, NJ, is one example. It has more than 1,600 workstations in over 50 call centers across North America. AnswerNet is also the world's largest telemessaging provider; such firms have typically been local-only, with one or two centers at best. That capability enables AnswerNet to spread out or back up such programs across multiple centers. AnswerNet also offers fulfillment services.

My advice is to take a hard look at what you need to have done. For example do you require dedicated agents? Consider having shared agents instead. Is your game plan to grow your program considerably? Some bureaus are dropping the set-up charges if you sign up for a huge, "mega-hour" multiyear contract.

Covering Your Bases

Protect your program and give yourself more options by not outsourcing all of it to the same bureau. Keep some of your capabilities in-house or split the work among different bureaus. You keep everyone honest that way.

Also look at having your program spread out among a bureau's different sites. That gives you disaster protection and possibly price flexibility.

Selecting and Assessing Bureaus

Contracting with a bureau is a serious commitment. Your business or organization's success may depend on how well they serve your customers. Therefore you need to carefully research and examine your potential partners to make sure it lasts.

You can find service bureaus from web site searches, trade media, trade shows, brokers, and consultants, the Yellow Pages. You can also ask around in your business.

Most bureaus handle general customer service and sales. There are those that have specialties in customer and internal technical support. Others have expertise in key verticals such as carriers, high-tech, ISPs and media, and in B2B.

Draft and send out requests for proposals. Get references. Do site visits and arrange to listen in on calls (the last option is the only one available for home agents). Read articles in the general, business, and trade media. Also look at contracting with professional consultants who can develop RFP criteria and aid with your assessment. Ask about what organizations they belong to and for how long. The leading bureau associations include the American Teleservices Association, the Association of TeleServices International (for telemessaging firms) and the Direct Marketing Association. In Canada there is the Canadian Call Management Association and the Canadian Marketing Association.

Test outsourcers on small projects. These can involve evening shifts, weekend shifts, or situations when your center receives more requests for support than it can handle. Ideally, you want to try out outsourcers on established products where you have a track record to judge against.

Consider more than one outsourcer. If you prefer, several outsourcing firms can be brought on board to support your customers. Or you can split customer service and support among multiple outsourcers, as long as they are capable of providing support.

Better yet, keep some of your programs in house. This way you keep outsourcers and your in-house staff hungry and honest. You'll also provide your customers with service protection in case you're forced to end one or more of your outsourcing relationships.

Certifications

Take a hard look at certifications. Many outsourcers have certified their operations against standards for quality and performance to ensure that their service is up to par, which they can then demonstrate to their clients. The certification organizations measure, audit and grade centers against standards based on industry benchmarks. If there are multiple centers, each individual call center is usually evaluated separately.

Depending on the program, certification covers quality-management program setup, documentation, measurement, and maintenance. They check a firm's accuracy in call and contact handling, customer and employee satisfaction, problem resolution and correction, and agent staffing and training. Some certifications require that you identify, provide, and manage a quality workplace.

The certification programs may also evaluate performance measurements such as average speed of answer, service level agreements, ability to hit performance targets, and how the call center's technology performs.

The standards that call centers certify are usually open, but they don't have to be. They may have standards committees comprised of leading firms that review and update the standards, usually annually.

To become certified there are several options depending on the standard and the program. Applicants must document and self-assess their processes, and an auditor accredited by an outside firm checks to see if your call center meets the standard by examining and grading your assessment and your operations.

If you do not comply, the auditors point out where. If your call center is far out of compliance with the standard, the auditors will reject your application; if the difference between your practices and the standard are slight they may pass it but ask you to make corrections. Several certification programs offer or recommend consultants who will advise how to fix those problems.

A certification is only as good as the firms certified to it. To ensure that your call center complies with the standards, the top-quality certifying bodies will periodically review and reaudit it. They may also receive and follow up on complaints by customers and employees alleging that you are violating the standards. In the most extreme cases the certifying body pulls the certification.

There are several firms and organizations that certify call center operations to particular sets of standards. Some of the certification and standards are specifically for general call centers; others are tailored for technical external and internal support desks.

The two certifications you're most likely to encounter when outsourcing are the International Standard Organization (ISO) 9001 and the COPC-2000.

The ISO 9001 covers quality system establishment, documentation, management, infrastructure, monitoring, and measurement. This includes quality staffing and work environments. Applicants must also identify, meet, and improve customer satisfaction. They must have process controls for product and service design, development, purchasing, operations, and identifying and correcting problems.

The COPC-2000 program is administered by the Customer Operations Performance Center (COPC). The certification applies to general call centers and technical support desks.

The COPC is made up of top call centers and companies that outsource customer service, order taking, and help desk services. It audits applicant in-house and service bureau call centers and fulfillment houses to see if they meet the COPC-2000 standard. The standard, based on the U.S. Malcolm Baldrige National Quality Award and adapted to accommodate industry needs, covers 29 separate items.

AWARDS

When you do your research on service bureaus you will inevitably come across them having won some award, presented by an association, business/trade magazine or vendor. That may or may not be an indication of that potential outsourcing partner's quality.

Some awards are rigorously judged, usually by outside experts and as such may have significant value in the call center community. I've been a judge for call center awards. The call centers that apply are typically very good, which makes the decisions of which ones get cut and which others are ultimately honored often very difficult.

Other awards are fee-based competitions where a call center, usually an outsourcer, pays the awarding firm to cover the processing costs. Some observers have criticized these types of awards as "pay for recognition" advertising. Others point out that they often do provide good value for money; the awards may be based on some objective criteria and they may be well known industrywide.

From firsthand exerience I know award vetting is time-consuming. In the business publishing field there are highly respected awards that also require entry fees.

If you have any questions about the awards check them and their sponsoring organizations out. Remember, ultimately the best judges of service bureaus are references, their answers to your questions and your own instincts.

The certification procedure examines the specific service that the call center provides and all the processes that go into it, such as voice, email, fax, U.S. mail handling, agent training, credit card processing, and new program setup. Each call center must be separately certified.

While these certifications are helpful, don't rely on them as your only criteria. Certifications are good only when the organizations receive them and are audited.

"Using certifications as a way of prescreening potential suppliers makes a great deal of sense," says Gantman. "However, certification should not take the place of careful and thorough selections processes."

Program Setup and Management

Once you've found the bureaus of your dreams, now the hard work begins. You will need to have the bureau's staff trained on your program, either by training the trainers and coaches, or directly training the agents.

In smaller programs you can do this without being there by laying out your scripts and FAQs and forwarding them to your bureau contact. In larger programs you may need to have one of your staff people on site for as long as six months to oversee implementation and correct any problems.

You will need to arrange for data handling. That can be as simple as forwarding disks or tapes containing prospects' names and phone numbers for outbound, to hooking your databases to the bureaus. In the latter case, your IT people will need to speak to the bureau's IT people to work out links. Ideally you should keep your customers' records in your hands, not the bureau's, to protect the data, unless you've also contracted with that bureau to undertake data management.

Constant, open, but not meddlesome communication is a must. Service bureaus provide frequent reports and monitoring. You will also need to build a rapport with your bureau contact. But don't micromanage the program. That adds to the bureau's costs and yours, defeating some of the reasons why you outsourced to begin with.

Contract Changes

Keep in mind that like any business, service bureaus expand, contract, change names, add or delete services, merge, or go out of business. Chances are by the time you read this some of the bureaus mentioned in this book may have branched into other markets, are going under another name, or are out of business.

The company you start partnering with now may be a quite different firm two or three years later. They may or may not want your business, depending on what is best for their profitability. Therefore you need to keep a fairly close eye on your service bureau and ask what the impacts will be on your business relationship from any change.

HOW TO ENSURE A SUCCESSFUL OUTSOURCING RELATIONSHIP

TARP, a leading call center consultancy and research firm, recommends that you take the following steps to ensure a successful outsourcing relationship, making sure your customers are happy.

- Survey callers. Require the bureau to continuously and directly measure customer satisfaction by a mail survey to a random sample of callers.

- Don't emphasize speed. Limit the percentage of calls answered in more than 60 seconds. Average speed of measurement should not be a productivity standard.

- Get regular and direct feedback. Make sure the bureau's agents send input to you regularly, like weekly or monthly, not filtered by bureau management. For example, USAA uses email, while Fidelity Investments relies on voice mail.

- Analyze problem calls. Require the bureau to capture detailed data on repeat and unproductive calls, including causes and calls not resolved on first contact.

- Supply the tools. Stipulate that the bureau give its agents all the authority and tools to ensure that calls and contacts can be taken care of on first contact, and that agents be permitted to move into a soft cross-sell when appropriate.

- Provide customer information. Give your bureau the customer information files so that agents know who the customers are, their value, and enough about their circumstances to cross-sell intelligently.

- Avoid burnout. Have your bureau vendor install initiatives and career paths for agents.

- Stress training. Ensure that the price they quote you includes continuous training to your standards and reinforcement of skills, as well as satisfaction measurement and adequate analysis and reporting of calls and contacts.

Chapter 7:
Site Selection

You know what functions you want your center to carry out. You've figured out how much of your calls and contacts you want to have handled by IVR or web self-service. You've taken a hard look at alternative strategies for live agents such as home working and outsourcing, including nearshore and offshore.

Now you're ready for site selection. You face tough competition for labor and facilities. Meanwhile, senior management wants to keep your firm's clients and impatient investors happy and wants this new center up and running today.

"Without question, our call center clients tend to present us with the most demanding timelines to fulfill," reports John Boyd, principal with The Boyd Company, a site selection consultancy. "Our industrial projects and major corporate office assignments usually take four to six months, if not longer, to carry out. In contrast, our work period for call centers is half that—typically two months, sometimes less."

You may have few problems finding a floor to up to 25 or 50 workstations. But consultants say that when your need is above that number, the site and property choices get more complex. Chapter 8 explores your options for onshoring (domestic call centers), offshoring and nearshoring (foreign sites serving domestic customers), and shared-service (co-locating with other corporate functions). Chapter 10 examines international options such as serving foreign markets with home working, outsourcing, and site selection. At this point you may need to either develop in-house expertise or call on a site selection expert. Chapters 11 and 12 examine facilities and real estate.

"When you're siting a call center with 50 seats or more, finding employees and space becomes more challenging," points out Ron Cariola, senior vice president with Equis, a global site selection and real estate consultancy. "When you are this size, you begin to have an impact on local labor markets."

❂ LABOR, THE KEY SITE CRITERIA

The most important call center site selection criteria is people: cost, skills, and quality of the individuals providing call center services. To twist the old real estate adage, the three most important attributes to look for in site selection are labor, labor, and labor.

Forget real estate. *Ignore* the cool buildings. *Pay no attention* to the golf course within, ahem, "driving" distance. Labor is *why* you have a call center. *Dismiss* voice and data costs.

Labor is *king*, because labor accounts for *60 percent or more* of call centers' operating costs in North America, Europe, Australia, and New Zealand, and less in developing nations like India and the Philippines.

Moreover, labor is *why* you are building the call center to begin with. When you compare locations, you look at labor costs as the prime component.

Susan Arledge, principal at Arledge Power Real Estate, supplies this example. Say you had a choice between two facilities in two different cities. You select one of the facilities only on the basis of real estate and labor costs. You plan to provide 150 square feet of space per agent, which is an optimal amount of room that includes allowance for common space like aisles, break rooms and training facilities. You will pay each agent for 2,080 hours of work per year. Each center is to house 300 employees.

Building 1 offered a rental rate of one dollar per square foot below the rate for Building 2. Building 2 offered a labor rate of one dollar below that of Building 1. Building 1 therefore cost $150 less in real estate per employee per year, given the difference of one dollar in rental rates between the two buildings and given the 150 sf allotted to each agent.

But if you occupied Building 2, it would save *$2,239* per employee per year in labor costs. You determined its savings from the difference of one dollar in labor rates between the two buildings, from the number of work hours per year for which you would pay each agent, and from the lower employment taxes you would incur as the result of its lower labor costs.

The savings that the occupant of Building 2 would experience from the lower labor costs would be significantly greater than the additional money spent on real estate. Because each building employed 300 agents, if you bought space in Building 2, you would spend *$626,700* less than if you occupied Building 1 over the course of a year.

In the above example, a difference of a dollar per hour in labor savings in Building 2 offsets a difference of $14.93 per square foot in rent in Building 1. In other words, the rental rates in Building 2 are far above those in Building 1, yet, because if you paid lower labor rates in Building 2, you still would have had lower total operating costs than if you had bought space in Building 1.

"Usually when you think of site selection, you think real estate costs," says Arledge. "However, site selection for the call center industry is really about labor, its quality, and the quantity of the people who staff the site."

Here's another way of looking at it. The difference between a good and a bad U.S. labor market is $2 to $4 per hour, explains King White, vice president, Trammell Crow Call Center Services. A call center that must pay agents $12 per hour in a large metro area to obtain the same turnover rate of a smaller $10-per-hour market will shell out approximately $347 per agent per month. That equates to $700 per workstation per month, assuming two shifts per station.

Consider the example of a 100,000-square-foot, 700-workstation call center with two shifts and 1,400 agents. The monthly payroll would total $2.4 million for the $10 per-hour location, compared with $2.9 million for the $12 per hour location.

Given rent and utilities costs of $200,000 per month, it may pay to sublease the more expensive location and open a call center in the lower-cost community, even if it means rent on two properties. Many call centers are doing that. The savings between these two centers, minus the additional lease, is $282,000 per month or $3.38 million per year.

"The rent in this example, approximately $200,000, may seem like a lot, but it amounts to less than $25 per workstation per month," White notes. "This fact allows one to really understand the minimal impact of real estate versus labor cost."

TYPICAL SITE RELOCATION RETURN ON INVESTMENT

Annual Savings

Labor

Wages in Current Location @ $12/hr	$7,488,000
Wages in New Location @ $9/hr	5,616,000
Total Annual Wage Savings	1,872,000
Turnover Cost in Current Location @ 55%	495,000
Turnover Cost in New Location @ 20%	180,000
Total Annual Labor Savings	$2,187,000

Productivity

Current Headcount	300
Productivity Gain	10%
New Headcount due to Increased Productivity	270
Total Annual Productivity Savings	$561,600

Real Estate

Current Rent Expense (40,000 Square Feet @ $23)	$920,000
New Rent Expense (35,000 Square Feet @ $18)	630,000
Total Annual Rent Savings	$290,000
Total Annual Savings	$3,038,600
Up-Front Investment	

Up-Front Investment

Labor

Relocation Packages	$50,000
Severance Packages	500,000
Recruiting & Training	900,000
Total Labor Costs	$1,450,000

Real Estate

Subleasing Costs	$500,000
Furniture and Moving Costs	$300,000
Infrastructure Costs	$650,000
Total Real Estate Costs	$1,450,000

Economic Incentives

Training Incentive ($3,500 per employee)	$(1,050,000)
Cash Grant for Improvements	(150,000)
Total Incentive Offset	$(1,200,000)
Total Up-Front Investment	$1,700,000

Financial Impact

Annual Savings	$3,038,600
Up-Front Investment	$1,700,000
Cash Savings over 10 years	$30,386,000
NPV (Net Present Value) of 10 years of Savings at 14% per year	$14,798,792
Per Employee	$49,329.31
Payback Period	.56 Years
Annual Yield on Investment	179%

Most corporate site selection decisions are based on savings potential and return on investment (ROI). Chart courtesy Trammell Crow.

✪ THE LIFE SPAN OF YOUR CENTER

Before shopping for your new call center home, you should project its life span, including how big it will become. By doing so, you can estimate the size of your center's labor pool, which, in turn, depends on your turnover. That will help winnow down your location choices. How many employees do you go through every year? In a mature economy with low unemployment and slow population growth, there are limits to your labor force and the percentage of workers interested in jobs at call centers.

Different types of call centers consume labor at varying rates. Outbound centers have high turnover because the work involves a lot of stress and the pay is lower. A second-level customer support or high-end B2B sales center usually has low turnover and higher pay.

At the same time, the labor pool for minimally skilled workers, such as outbound sales and inbound order taking, is usually larger than that for help desks, which require people with better education. There are also regions with such sufficiently high numbers of unemployed or strong population growth that your center's life span may be over before the labor pool is drained.

Too often, firms underestimate a community's population size, labor supply, and the skills that are most in demand, Arledge points out. Although firms are well aware of their turnover rates, they are often unaware of the impact turnover has on the available workforce. "You may have many people working for you when you open up, but when you need to replace them, there may be too few skilled workers to take their place," she warns.

Your call center's life span also helps determine your real estate and facilities needs. How much room do you need with to start, and how much will you need at your center's peak? You can lease and renovate existing structures or discuss build-to-suit options with real estate developers. A renovation lease typically runs five years compared to roughly 10 years for a build-to-suit. Consultants recommend that you negotiate exit and expansion options and tenant improvement givebacks.

How long you plan to be in a particular labor market also helps you choose which type of property you need. If your company plans to locate a call center with the possibility of adding to it, you should look for room where the building can expand, preferably next to the space your center will initially occupy, and put your intentions in writing.

Your options include insisting upon a right of first refusal on adjacent spaces, renting or building more than you need and retrofitting later, or agreeing to a "must take" arrangement where you pay rent on a certain amount of additional space within a year or so after you sign your initial lease. You should also negotiate exit clauses such as subleasing rights and rights to "go dark," pay rent on a space but not occupy it (more about these in Chapter 11).

"If you are not sure about negotiating and signing what your space needs will be, then you might want the right of first refusal," advises Arledge. "If you know that you are going to need the space, then a 'must take' might be better deal."

✪ LABOR FACTORS

Here are the key factors to examine when looking at labor in prospective locations.

Cost

Intimately related to the issue of supply is cost. How many employees will you attract and keep at a certain level of wages and benefits per annum? What are the prevailing wages and other compensation expenses for call center agents and supervisors in the communities you want to locate in? How much have they grown over the past year? What are the packages offered by those employers, including other call centers you would be competing for labor with?

Local economic development agencies (EDAs) should have this cost information. You can also do your own research visiting the recruiting portion of local employers' web sites, studying the help wanted ads, visiting college and government employment offices, and dropping into the companies' recruitment offices or passing by their job boards. Good site selection consultants use such practices to gain information before letting the word get out that a company is thinking of coming to town.

If you are new in town, especially if your company is not a brand name familiar to residents, be prepared to pay somewhat more than what you would in an area where you are established. Service bureaus fall into this category; people will have likely heard of the outsourcers' clients but not of the outsourcers. This will help overcome potential employee skepticism about your value as an employer, especially if you are setting up in an area that has relatively low unemployment. Remember, neither the community nor its workers know you as a good company and employer; you do not have a track record that generates word-of-mouth employment.

Carefully Assess Supply

To staff your call center you need to find places that have the people who are willing to work for you for as long as possible at the compensation you offer. Most call center jobs are entry- or semi-entry level, requiring customer service and sales aptitudes and preferably experience, say, in retail. These general skills give you access to the largest labor pool. But increasingly you may require higher-skilled agents who can multitask, handle multiple media, and deliver either service at the level of a concierge, account manager, sales person, or engineer. That shrinks the labor pool.

When you look at supply, read unemployment data very carefully! As Benjamin Disraeli said, "There are three kinds of lies: lies, damn lies, and statistics."

Governments track labor supply by measuring the number and percentage of workers who are employed and unemployed. Don't be dazzled by high percentages. While they may be indicators of plentiful labor in some jurisdictions, the raw numbers of workers available may be less than in others with lower unemployment rates. Also be aware that many governments adjust their data to reflect seasonal fluctuations, such as agriculture and fishing, where large volumes of people work for very short periods of time.

Unemployment statistics tell only part of the labor force story. Governments draw their figures from people who are collecting unemployment insurance, who are actively seeking work, and who are working. They do not include individuals who are not officially in the workforce, such as those on long-term disability, students, those who never held a job, people who have given up seeking employment, stay-at-home spouses, and retirees. Nor do they list those who are working off the books.

Also, government statistics assume and count one person holding one job as their primary income source. They do not account for multiple jobs, such as a call center agent working for two different companies. This is common in many lines of work. My son is a paramedic working for two New York City hospitals, one in Manhattan and the other in Brooklyn. He's also done "per diem," i.e., freelancing for others.

Yet many people in the "unofficial" labor pool may wish to work or are working in a call center. For example, when male-dominated, highly paid jobs in heavy manufacturing disappear, it is often those workers' spouses, who have been supported by the high incomes, who go out to work. Another example is military spouses who live on the base or nearby.

Some of these potential employees have valued skills. For example, the military attracts recruits from a wide range of ethnicities. Chances are their spouses come from the same background. That means in the U.S. you get pockets of Spanish speakers, and in Canada you get pockets of French speakers in communities or regions that have few natives.

The military also trains many of its people to undertake technically oriented tasks. So when they retire, they have strong technical skills that are ideal for call centers.

Determine and Assess Underemployment

Underemployment occurs when people work at the first available jobs they find, even though they possess skills that would help them get work in other occupations. For example, college-educated people often work in low-paid, part-time jobs such as in retail stores or restaurants.

An underemployed workforce may exist in an area that has a high quality of life or because other regions that offer jobs better suited to an educated workforce are less accessible by car or by public transportation. I know many people I went to college with in Victoria, British Columbia, one of the most beautiful cities in the world and a top tourist destination, who've taken low-skill jobs just to stay there.

Underemployment also exists in communities where there are comparatively low living costs and where larger, more-prosperous metro areas are not in commuting distance. It often does not pay to move to the "big city" if the higher expenses outweigh the fatter paypacket. Also many such cities and towns have attracted couples where one is home working while the other is unemployed or underemployed.

"Underemployment is very difficult to analyze because you have no data to track," Arledge says. "You can't measure unhappiness. All the unemployment statistics show is the percentage of people who are not working. The only way you can find out if there is underemployment is by researching the local market, including talking to local EDAs."

One factor aiding the value of underemployment is the desirability of call center employment compared with other types of service-sector jobs. Call centers are usually the service sector employer of choice. The relatively sedentary nature of the work means people who have difficulty standing or walking can be call center agents.

"Even in a tight labor market, call centers will always draw people from other service jobs," observes King White, vice president of the site selection group for real estate firm Trammell Crow. "They pay higher and provide nice offices, unlike being on your feet in a big store or a restaurant. They are white-collar jobs that someone like a high school graduate can get."

Check Out the Competition

Be sure when you are examining local labor supplies that you check to see what other call centers are there. They probably will, but may not, compete for the same workers as you.

On the other hand, you may offer better-than-average compensation, have a reputation as being a cool place to work for, or your center may be seeking specialized skills. Also, the presence of other call centers means your applicants know what to expect. You may also be successful at attracting agents and supervisors from the other centers.

CB Richard Ellis Call Center Solutions Group devised a saturation analysis that has become the industry standard. It looks at how many people are working in call centers in a given metro area. From its experience it has found that where 3 percent or more of an area's workforce is employed by call centers, turnover and compensation costs escalate. That figure is known as the saturation level.

			Saturation Rate*							
Community	Total Labor Force	Estimated Call Center Jobs	1% Low Saturation		2% Moderately Saturated		3% Saturated		4% Very Saturated	
			Max. Employees	Available Positions	Max. Employees	Available Positions	Max. Employees	Available Positions	Max. Employees	Available Positions
City A	198,518	1,800	1,985	185	3,970	2,170	5,956	4,156	7,941	6,141
City B	545,000	11,360	5,450	-5,910	10,900	-460	16,350	4,990	21,800	10,440
City C	63,282	3,450	633	-2,817	1,266	-2,184	1,898	-1,552	2,531	-919
City D	706,000	36,699	7,060	-29,639	14,120	-22,579	21,180	-15,519	28,240	-8,459
City E	371,000	12,820	3,710	-9,110	7,420	-5,400	11,130	-1,690	14,840	2,020
City F	419,000	20,072	4,190	-15,882	8,380	-11,692	12,570	-7,502	16,760	-3,312

Community Call Center Saturation Analysis

* *Saturation Rate* attempts to model the employment potential of call centers in relation to the labor force size. To assist with a market by market evaluation, this analysis has been run whereby each percentage indicates the resulting *Saturation Rate* for a particular community. As unemployment remains low and call center positions continue to grow, markets throughout the United States have become more saturated and the acceptable saturation rate has increased. As the Saturation Rate is raised, call centers run a greater risk of future recruitment and retainment challenges. The *Max. Employees* indicates maximum number of call center employees acceptable at the particular Saturation Rate. The *Available Positions* indicates the differential between the *Estimated Call Center Jobs* and the *Max. Employees* indicators. A negative number indicates that there are more existing jobs than the market can handle (at the specified Saturation Rate). Likewise, a positive number indicates the number of additional call center jobs that the market sould be able to support (at the specified Saturation Rate).

Source: CB Richard Ellis Information Services, Bureau Labor Statistics, Community Economic Development Organizations. CB Richard Ellis considers its sources reliable, however accuracy cannot be guaranteed.

This model saturation analysis chart from CB Richard Ellis compares the impact of saturation rates of different call center locations. It also shows how many, if any, new centers a locale can support.

Follow the Benchmark Employers

Consider the benchmark employers in any community as those who hire similarly skilled and similarly paid workers as those you want for your call center. They include big box retailers, fast-food chains, and large distribution centers.

Find out the applicant-to-opening ratios, the number of applicants compared with number of jobs. Ratios of 8:1 to 10:1 show a large and eager labor supply.

See how they're recruiting. If they're not advertising in the local papers and don't have Help Wanted signs, but staff say they do hire people, that's a good sign that there is plentiful labor. These outfits don't have to advertise, and people find these jobs out by word of mouth.

Find out what these employers are paying locally. If it is higher than the industry's national average, then you may have to pay more than you planned for the skill set you need, explains James Trobaugh, senior vice president of CB Richard Ellis Call Center Solutions Group. These employers also have good turnover.

"If those businesses are having trouble hiring people, then it is a sign there may be other issues in the labor market that need to be looked into," advises Trobaugh. "Find out why the wages and the turnover is so high."

Check Skills and Quality

You need to make sure that the communities' labor forces have the skillsets you require at the level of quality you demand. Does the workforce have adequate education and training? Are the local high schools and community colleges any good? If your call center is providing highly technical services that require employees with critical thinking, or supplying social services needing supportive non-judgmental employees, do the cities and towns you're looking at have open cultures that supports both these attributes?

As the services that senior management wants call centers to provide become more sophisticated and customer-targeted, then your agents will need more specialized skills. For example, if you are serving many French- or Spanish-speaking customers, you need to have access to agents with that ability. You'll probably have more luck finding them in Quebec or south Texas than in Staten Island or northern New Hampshire.

If you provide high-level customer support with minimal escalation, you probably need to have customer-service-oriented individuals who have some technical training and aptitude, usually from a good local college that offers computer science. Communities that have attracted many hardware and software developers and manufacturers are more likely to have such individuals.

Review Education Performance

You can no longer count on a high school or college education as proof that applicants can read, write, or perform math with any level of competence.

Jeff Furst, president of staff assessment firm FurstPerson, reports that in many communities, applicants have fewer skills than in the past. Reading, writing, and comprehension abilities are diminishing along with the work ethic.

In a typical community, only 30–45 percent of candidates meet the abilities and behaviors to do the work, "which really impacts the ability to hire employees that meet the job requirements," says Furst.

Many applicants do not know how to multitask or problem-solve—key requirements in today's multichannel, multipurpose call centers. Many young applicants lack the behavior attributes to work, like accountability.

"When they get into the workforce, it's a shock to them to be accountable since they have never been before," says Furst. "Especially in call centers where they have to be on time, cannot leave when they want, and their performance is monitored, measured, and graded."

There is some disagreement on these issues. Manpower hasn't seen much change in agent quality, education, and literacy over the past few years. The new generation of workers (Gen-Y) may have many attributes and skills suitable for call centers, explains Frieda Lalli, Manpower's manager of marketing operations. These young people want results, are team players, are computer savvy, and are able to multitask.

"Hiring Gen Y-ers may prove to be very positive for call centers," says Lalli. "The challenge and the opportunity for managers are how to channel and harness that energy to help the call centers. This can be accomplished by tapping them for ideas and proposals and letting them experiment with new methods."

Therefore you need to carefully review schools' performance in your site selection to ensure that you will be setting up in a locale where your applicants have these basic skills. They are especially vital if your center is higher-end.

"You have to remember that training, including remedial training, is very costly," says John Boyd of The Boyd Company. "Therefore, it pays to locate where there are educated and quality employees."

Communities' education quality has always been a key site determinant for many of Arledge Power's clients. Her firm looks at communities' SAT scores, high school graduation rates, and the ranking of local public schools.

The challenge is translating education quality into applicant quality. SAT scores are valid only to those in high school, Equis' Cariola points out. They are not indicative of those who remain in the community, because many people with high scores go to college in other communities.

"You need to test the labor pool with job and applicant fairs," advises Cariola. "You can also pre-qualify labor pools by having the local economic development agencies or corresponding schools give tests which relate to the types of jobs to be included within the call center."

When examining education as a selection screen, make sure that the educational requirements you've set fit the positions you are hiring for. Not every call center job requires a college degree.

"You don't want a community with too educated of a workforce because they won't stay long unless there is high unemployment," says White.

Examine Labor Force Traits

Workers are not automatons, no matter how little you pay them. They have attitudes and cultural traits that affect their employment performance. You need to find out what those are and see if they mesh with your requirements.

Attitude

The first and the most important trait is attitude. Do they have the right mindset for the job? The attitude to have in customer service is markedly different from the one you need in outbound sales. Customer service needs to be friendly, intelligent and helpful; outbound sales needs to be sharp, aggressive, and goal-oriented.

If you want your call center to sell products, you must locate it in a community that has what John Boyd calls a "sales culture." A good community would be one where there are aggressive retailers—like car dealers, electronics superstores, and furniture dealers—and that employ thick-skinned salespeople who think quickly on their feet to find those emotional buttons to press to get you to say "yes."

If the purpose of your call center is to provide customer service, then you need to locate it in places where there is a customer-service culture. Such locales have hotels, restaurants, and stores where the workers have a calm, patient demeanor. They are there, but not in your face. Many firms have opened call centers in rural U.S. communities because the people there are often friendly and courteous.

Yet many communities that have the right customer service attitudes may not have those for sales, and vice versa. One company, CyberRep.com, relocated an outbound call center to Florida from rural western Maryland because the agents there hated making tele-marketing calls that would disturb other people, who may be their neighbors.

More companies are converging their attitude requirements as they adopt cross-sell-ing and up-selling practices. A good outbound agent can be trained to use sales skills to solve a customer's problems. A fine customer service agent can be taught to regard cross-selling and up-selling as ways to identify and meet customers' needs and solve their problems.

Be careful when examining communities for attitudes. For example, towns whose dominant employer is the military may not make great places for customer service cen-ters. Qualities like empathy and understanding are not the ones the military wants in its fighting forces.

But their spouses are another matter. They keep the homes fires burning. If their husbands and wives are officers, they do a fair amount of socializing to promote their loved ones' careers, whose status reflects on where they are in the pecking orders. With military pay traditionally lousy, they are often eager to work in call centers.

Attitude is also expressed in what kinds of people you can expect to work in a call cen-ter. This is a function of the local economy, demographics, and culture. Where other employment is scarce, except for at the call center, the local populace will tend to regard such jobs more highly, and you will attract higher-quality applicants.

The opposite is true if there are many other positions available or if they pay better. Then it becomes: "Oh, you got a job in a call center. How nice. When are you going to get a career?" All service sector employment suffers from that stigma; I know, I've been a "rent-a-cop" and a telemarketing fundraiser.

In some communities and countries, most notably Japan, call center customer serv-ice and sales are still considered "women's work." That cuts out potentially half your potential labor force. Also, many potential women workers who are stay-at-home spouses prefer to be that way.

Dialect

A second key trait is dialect. You need to set up in locations where the residents speak clearly. If your customers can't understand your agents, then you may have customer service, support, and sales problems.

While this has become less of an issue in recent years, with increasing ethnic diversity, greater acceptance of regional accents, and paradoxically their fading away through media and mobility that encourage vocal homogenization, some regions residents' twangs are stronger than others.

New York City is still notorious for its nasal accent. The province of Newfoundland is noted for its catchy, sing-song but fast "Newfie" dialect, which is more muted (and intelligible) in St. John's, the capital and largest city, but more pronounced in outly-ing areas.

Some people, such as those from the Indian subcontinent, speak well but too swiftly for many other people's ears (more about that in the offshoring discussion).

Dialect also affects languages other than English. Many companies reportedly prefer Mexican Spanish speakers to potential employees from other Hispanic countries or regions because they talk slower and the accent is not as harsh (see Chapter 9 on serving Hispanics).

Environment

The environment is an important trait when examining locations. For example, if you are looking to locate in a large metro area be prepared to allow for traffic snarls affecting tardiness. If you pick a rural community don't be surprised if absenteeism "shoots" up when hunting season begins. The same happens when the snow is deep enough for snowmobiling and the ice is thick and strong enough for fishing: both increasingly rare occurrences thanks to global warming. Don't be surprised to find some of your agents limping in like war veterans.

Union Presence

Many communities, especially in the larger cities and outside of the U.S., have a substantial union presence. That has several impacts. But many companies, even yours, may not want unions to represent your employees. Union representation challenges your management freedom, adds employment costs, and reduces flexibility, which cuts into profits.

Pressure to unionize

A large company or call center often but not always experiences pressure to unionize. These communities have union cultures. Being part of a union is normal and expected..

Pressure to raise wages and benefits

Even if you are and you remain non-union, you may face indirect pressure to offer higher wages, especially if there is a large and unionized call center in town, like one belonging to a telco.

With unions, the wages and benefits are usually higher than in non-union shops. To keep staff you may have to follow suit, though not to the same extent, to keep agents and supervisors.

Specter of strikes

Unions also create the specter of strikes, though with proper trust and relations, unions can provide a fast, efficient conduit of information to and from the shop floor.

Union benefits?

On the other hand, the higher wages and benefits and better working conditions reduce expensive turnover. A union call center is almost always the employer of choice. If to save your company, the union agrees to a wage reduction or looser work rules, it is done, unless the union leadership is out of touch with the members on the floor.

A union presence also keeps a check on careless, greedy, or incompetent managers and executives. You will increase your odds of unionization if you treat your workers poorly; you will decrease the chances of your employees signing up if you treat them right.

○ FACILITIES AVAILABILITY

Unless you are going to mostly self-service, home agent, or outsourced offsite options, the availability of buildings is a key issue if you have a limited time frame to get your center ready.

That's because supplying space can take from 30 days from a walk-in, just-vacated call center to a year for a customized build-to-suit center. Chapter 11 covers real estate.

Outsourcers, because they are bidding on contracts that require a certain amount of agents available to take a specific volume on a particular date, must balance both labor costs and facilities supply very carefully in their site selection. They may have to go to a slightly more expensive locale if the right building is there rather than going for the lowest priced locale that does not have suitable space on hand.

Access

Another key issue to examine is access to your center. That takes two forms.

Access by employees

For your agents to start producing at your call center, they must get there. The same goes for your senior management and clients, who want to see what they're paying for.

How easily the labor force can get to work helps determine how many of them you will tap. The area that lies within commuting distance of your proposed sites is known as a commutershed. The size of the commutershed depends on the time your potential workers in those areas want to spend commuting.

The distance to your planned location is known as the commuting radius. For low- and medium-wage jobs, the commuting radius is 20–30 minutes driving and 60 minutes on mass transit, door to door. Mass transit is typically slower than driving, but the longer commute is tolerable because the workers are not driving for much of the way.

The commuting radius is higher for higher-paid supervisors and managers. They can also afford the greater gas, toll, and transit costs.

Access by management, clients, and customers

Another consideration is access by senior management, clients, and customers. If they need to be at your centers in person, then the centers need to be located within 90 minutes of an airport.

But with ongoing cutbacks by air carriers, especially to smaller communities, there are fewer locales that make this qualification. Also, consultants such as CBRE's Jim Trobaugh say lack of easy access is an attribute. It discourages other call centers from moving in, swiping your labor.

But access to major metros by air isn't going to get better either. There are many predictions of airport gridlock within a few years. Any more terrorist attacks, and you can forget air travel.

Driving is not much of an option either, with badly congested roads. Chronic underfunding of rail coupled with a fanatical obsession of opponents to that mode of transport has limited that option, except outside the Northeast Corridor, parts of California, and the Pacific Northwest.

On the other hand, increasing use of conferencing tools has lessened the need for expensive and time-consuming business travel. That goes for call centers wherever they are.

Disaster Vulnerability

Take a hard look at the risks of disasters. There is no place you can locate a call center where nothing will happen to it. The objective is to reduce the odds that some event will knock it out.

If you place it in the Northeast, expect ice storms that crumple power and phone lines. If you site it in the Southeast, expect Hurricanes Bonnie and Clyde to rob it of its capability. If you drop your call center anywhere from the Midwest to the Southeast, there is a chance that a tornado may drop pieces of it somewhere else. And if you're on the Pacific Rim, prepare for earthquakes and mudslides in California and volcanic eruptions and pyroclastic mudflows in Oregon and Washington State.

And there's the reasonable chance an asteroid, comet or wayward star will wipe out everything or the Earth's core fizzles out before the Sun dies and the Milky Way collides with Andromeda. So why worry?

While these are real risks and inevitabilities, don't be a scaredy cat in site selection. Human beings have about 200 million to 600 million years or so, depending on what happens biologically, cosmically, geologically, and socially, before the planet becomes too uninhabitable for anything but cockroaches and bacteria.

Weigh this factor against others in site selection. Many of the most vulnerable places, like the Gulf Coast, often make the best call center locations because they have affordable labor. Consultants say damage to call centers from hurricanes has been minimal.

People will live in a place, come hell or high water, if they can make a decent living there, especially if it has attributes like warm weather or a beach. By the same token, some of the safest places, like central Nevada, have too few people to support a call center. And it still won't protect you against an asteroid hit or a core collapse.

If you are locating only one call center, your best bet is to look into disaster vulnerability after other key factors, such as labor supply, and then plan for disaster recovery. Examine methods such as home working, outsourcing, and self-service to handle contacts if your center is closed. Otherwise, spread your call centers around in key labor markets, so that if one goes down calls are automatically switched to another.

Taxation

Examine the impact of different tax rates—corporate, employment, income, property, and value-added—between jurisdictions. Between two or three locations that have roughly the same labor availability, skillsets, and costs, the one that has the best regime could make a bottom-line difference in deciding where to locate.

CD Richard Ellis's Trobaugh points out that even if a call center such as a free support desk does not make money directly for a company, it could create substantial tax liabilities. For example, a 500-workstation center could create a multimillion-dollar liability amounting to 20 percent increase in the center's labor costs for a firm that generates $2 billion to $3 billion in net income.

Trobaugh suggests that with careful planning and cooperation from local communities, a company such as an Internet firm could avoid having to collecting taxes on its customers even if they set up in a state that has a sales tax. If your company makes a profit and is eligible to receive state income tax credits, but your call center is a non-profit entity that does not pay income taxes, you may be able to gain some of those breaks.

Keep in mind that many places that would make great call center locations because of labor force availability also have high personal income and sales taxes.

Working with EDAs

At some point in the site selection process, you will encounter economic development agencies. They will try to get information about underemployment and want to help in setting up your call center.

EDA Benefits

EDAs are the channels for financial incentives that can reduce the costs of locating your call center, like cash, training allowances, tax abatements, and buildings.

These agencies can put you in touch with local resources, such as community colleges to provide training for agents, and help you through the application and zoning process. If you need extra services, such as arranging for a new bus stop close to your call center, the EDA can introduce you to the local transit authority.

If you are having trouble getting phone service in time from the incumbent local exchange carriers (ILEC) EDAs can connect you with the right people at the ILEC you do business with. In contrast to U.S. practice where the ILECs seem disinterested in call centers, carriers in many other countries, recognizing the revenue call centers bring in, work closely with local EDAs to attract and establish call centers.

EDAs can obtain important information for you that can be difficult to obtain from other sources. Examples include the number of call centers in the local area, the companies that run them, and these companies' prevailing wages. EDAs can also fill you in on signing and re-signing bonuses.

"Those numbers are not listed in an area's reported average wage, but they can make a big difference on your bottom line," says Susan Arledge of Arledge/Power Real Estate.

EDA Caveats

You need to be as smart in dealing with an EDA as you would with any salesperson. The goal of an EDA is to convince you to locate in its community. Consultants recommend that you obtain as much information as possible about communities you're interested in before you meet with EDAs. When an EDA presents you with data, find out how recent the data is and confirm the source.

According to consultants, EDAs often fund market and competitive assessment research from private firms. Such research usually reflects the conclusions and views of the sponsors. They are a good resource for community information but should never be the only one.

Be wary when an EDA presents testimonials either on paper or arranges you to visit companies that have located there. According to teleservices consultant Philip Cohen, in some instances, such firms have been told to say nice things as conditions of receiving generous grants. But it distorts your decision-making and that of the entire industry.

Also, be careful of what an EDA promises. Many are private or semi-private organizations that have no real authority. They cannot guarantee planning approval or tax breaks that come from the government.

Do you homework on EDAs. EDAs like most agencies, companies, and people, run on a spectrum from being quite good to grossly incompetent. All too many of them are patronage parks. Where you do have a good EDA it may not have the backing of local governments and will have difficulty making deals stick.

You also need to decipher the confusing array of incentive packages that come from EDAs. Governments similarly offer what can be a bewildering variety of job creation grants, loans, and tax breaks. Many incentives have strings attached such as the number and types of jobs your center creates. "You may not be able to take advantage of a tax credit if your firm makes too much money," Arledge cautions. "Therefore, you have to look for grants or training dollars."

Sometimes you will find a great location with affordable, high-quality stable labor, reasonable property availability, and good access for your senior managers and staff. But the EDA and the local government are lukewarm. Or sometimes the EDA is incompetent.

At the same time many communities are no longer rolling out the welcome mat to call centers, for good reason. Call centers are not stable employers. In good economic times there are often many other better-paying or more-secure jobs available. A big box retailer may pay the same as a call center. Yet they are less likely to lock the doors overnight compared with an outsourcer that loses a big contract or who has a dominant client in that center that wants their program offshored to India. Also, retailers run the risk of losing customers and revenues if they shut down in a community; how many will drive to the store in the next town? No such considerations apply to call centers.

Outbound customer acquisition call centers have developed a reputation for "churning and burning" through labor forces. Many civic leaders and potential workers may also object to those centers because their agents interrupt people at dinnertime. Outsourced customer support call centers have been packing up and moving offshore after demanding and receiving generous incentives. Sometimes call centers have built facilities in towns, gone to the trouble of putting on job fairs, raising expectations, only never to take a single contact when the contracts they were scaling for didn't materialize.

What do you do?

First, try to educate the EDA, local governments, and community leaders, recommends CBRE's Trobaugh. Explain the benefits the call center will provide their community.

Second, I suggest that you see how the location compares to others on your shortlist. If the differences between the reluctant community and more welcoming and cooperative ones are minor, go with the latter. Why go where you're not wanted? Also, if you had a reasonable amount of publicity in your site selection, like a job fair, and you reject the community, the resulting bad press and political opponents will make the establishment think twice the next time you're looking to locate. At that time, there may be a different and much more cooperative team in place.

Third, see if you can get pressure placed on them by a senior level of government. If they care about the employment and economic activity you'll generate and believe they're going to lose those jobs and incomes to the community next door, they may put pressure on the junior levels. Local governments are creations of the governments above them, who provide much of their financing and set their policies.

Fourth, and as a last resort, go to the public via the media. But be careful. This strategy works only if you are a stable name-brand in-house employer or if you are a well-established outsourcer with clients that are willing to be named. Get to know your target communities real well. Research and approach groups that might benefit from your employment, such as young families, parents with college students living at home, ethnic groups, and retirees. But don't assume that just because unemployment is high that they're interested in working for call centers.

Don't Tip Your Hand!

Consultants advise that you not tip your hand about your final site decision. If you do, you could risk losing many incentives (see box on incentives later in this book). Before you declare your intention to locate your center, you need to have a preliminary written commitment from local officials that details the types of incentives offered, the level at which you qualify, and the conditions that you will be required to adhere to.

✪ SITE OPTIONS

This section will not tell you the location of the rich veins of affordable people and places. They may be played out by the time you read this. What it will do is suggest some options to use when placing your center.

The next chapter looks at locations strategies: onshore, offshore, and nearshore. This section looks at location options within the context of most of these strategies.

There are two options that are stand-alone because of their unique advantages and complexity: distributed call centers and shared service centers. The discussion on them follows this section.

While the focus applies to U.S. call center site selection, most of these tips will hold true in other countries, especially Canada and Australia. There's more about foreign locations options in Chapter 10.

Labor Cost Analyses

Labor costs in the U.S. vary greatly, as much as 30–35 percent, reports John Boyd of The Boyd Company. The swing is less so in Canada, 10–20 percent.

Regional Centers

Regional centers are those small or midsized cities up to 500,000 in population that service a wide geographic area. Locating there has long been a favorite strategy of John Boyd. These communities provide employment, education, shopping, professional services, cultural amenities, and commercial air service to the entire region.

Potential workers flock there for work or go to college and work part-time in call

LOCATION FACTORS

There are many factors to consider when deciding where to locate your center.

I grouped these factors around three main categories: labor, legislation, and community.

Labor

- Availability
- Wage structures
- Local fringe benefits
- Education
- Language skills
- Labor force participation

Legislation

- Telemarketing laws
- Taxation policies and rates
- Unemployment insurance and workers' compensation rates and rules
- Right-to-work laws
- Incentives, such as training grants and tax credits

Community

- Time zones
- Climate
- Telecom and electrical infrastructure
- Transportation (roads, mass transit, and inter-city air and rail access)
- Cost of living, including housing
- Public safety
- Property and site availability
- Local business attitudes (whether the community is pro-business)

centers. Airlines provide access by management to set up and supervise the centers and show senior executives and clients around them.

Rural Communities and Smaller Cities

Consultants often recommend looking at rural communities or cities under 200,000 because such locations often have higher-than-average unemployment rates, significant underemployment, and a loyal workforce with a strong work ethic, all of which could lower your costs. They now have excellent voice and data infrastructure thanks to the great late 1990s telecom boom and the spread of high-speed wired and wireless broadband networks.

These communities may also have very willing EDAs supported by local, regional, and state governments. They could offer you incentives such as free or low-cost buildings and training subsidies.

If you are planning to locate a large call center, placing it in such communities will make your firm a big fish in a small pond. You become the dominant or top call center employer, enabling you to have more control over wages and benefits. Everyone else will have to scramble to meet your compensation levels and working conditions. You may lap up enough of the labor market to ward off other call centers.

Many smaller communities also have excellent public schools and community colleges. CBRE's Jim Trobaugh has been "blown away" by the equipment, education, teachers, and skills of these institutions.

But many other smaller cities may not have the same quality educational system that is often found in larger cities, Arledge points out. If a call center needs a highly trained, specialized skill, there is a greater chance of obtaining these agents and supervisors in larger communities.

The best bet is to find smaller communities with excellent educational systems, such as those in Kansas or Oklahoma that have four-year and two-year junior colleges. "Not only do they indicate a good local education system, but they provide part-time and summer full-time labor," says Arledge.

Small City Caveats

When examining such communities, make sure their labor forces have the skills and quality you're looking for. While you can readily find people for general customer service, sales, and first-level support, you may need to locate in a large metro for more specialized workers.

Also you may not find employees with good supervisory or managerial skills in smaller communities. You may then need to bring your managers from other larger centers, at least initially. And that might cause issues if those employees are there for any length of time. You may have to relocate them and their families, which opens issues of spouse adaptability, selling old homes, finding new homes, and moving expenses. Children, especially teenagers, may not fit well in small communities.

Also, these locales may be less tolerant of diverse ethnicities, non-Christian religious beliefs, agnosticism or atheism, single parenting, gays, and interracial marriages and relationships. It doesn't take long to find this out in a community. The stares tell all.

I've lived in small cities and towns. The people are friendly but are not friends. I knew less about my neighbors, even though I owned a house, than when I rented apartments in New York City. But at least they talk to you, on occasion. Some communities are so insular and tight-knit they don't acknowledge your existence, let alone speak to you unless you were born there.

If you plan to set up your center quickly, smaller towns may not necessarily have readily available real estate. There is not the big supply of large occupancies in office buildings in small cities as there is in big metros. Instead, you're facing options like converting former retail or industrial outlets into call centers or financing a build-to-suit.

"It takes four to six months to renovate a shopping center and nine to 12 months to develop a build-to-suit," White advises. "If speed to market is critical to you, look at the larger cities, because they usually have more suitable real estate."

Further, if part of your strategy is to back up your call center with a constellation of home workers (see Chapter 5) living in smaller towns nearby, be aware that many of these homes still lack broadband.

The U.S. has been behind countries like Canada in connecting rural residents to the information superhighway. Many people are still on dial-up, which is the electronic equivalent of dirt roads. While dial-up access is fine for simple outbound and inbound call and email handling, it cannot support receiving and transmitting large amounts of data, like customer files.

Lastly, if you have to pack up shop, you could face stronger, more negative reaction in a smaller community. Especially if the local political and business elite had arranged generous incentives to get your center. Losing a 500- or 700-workstation call center employing 1,800 agents is a pinprick to the economic life of a metro area with 1.1 million or 1.8 million people. But it is a big wound for a small city with 110,000 or 180,000 residents. When you open businesses, especially large ones in small communities, the people "adopt" you. You become part of the social family. You are expected to become involved in the public life. So when you pack up shop, the effect is traumatic, like a divorce or death. The last chapter of this book covers your exit strategies.

College Towns

College communities make for excellent call center locations. The student population is often looking for part-time jobs to help put them through school.

The work for the most part is sufficiently routine to give these agents mental breaks from their studies. Yet these potential workers train quickly and have the intelligence and quick thinking to serve and sell to your top-level customers and clients. They are very computer-savvy and reasonably literate to successfully handle email and chat.

Many college towns also attract foreign students. If you can attract them to work for your call centers, you can serve callers in languages other than English.

Follow the Tourists

Remember that great vacation you took, where the people were so friendly, the service was excellent, and the prices were low beyond belief? That may be a great place to locate a call center.

Communities that attract tourists also have the same type of people who will work well in call centers: those who are friendly, trustworthy, and willing to work flexible hours at low but fair wages. Such cities include Atlantic City; Branson, Mo.; Deadwood, S.Dak.; Las Vegas, New Orleans, Orlando, and new riverboat ports like Tunica, Miss. They also have around-the-clock services like dry cleaners to support multiple-shift operations like call centers.

Even where the tourist traffic is seasonal, there are call center opportunities. You can tap into workforces that would go often on unemployment or out of town to find work, and utilize cheap-to-rent buildings that are closed for the season. With today's smaller, flexible computer and phone hardware, you can move right in. A summer vacation area like Cape Cod could support a temporary winter call center.

Military Communities

Communities where there are military bases can make excellent locations. Wherever there is a base there are usually many former service personnel living nearby.

Their education, discipline, and toughness make them excellent, reliable call center agents, especially for outbound collections and sales. Many of them have had technical training, giving them skills for support desks.

As noted earlier, spouses of active duty and retired military personnel are also prime call center material. They have a blend of both their partner's discipline and customer service skills. Military pay is, unfortunately, woefully low, which makes working in a call center quite appealing.

Military bases that are downsizing or being closed offer superb facilities. The buildings are often large, low-rise with plenty of parking, and excellent voice and data links.

Not all military communities are suitable for call centers. For example, those that are boot camps provide few stable households to recruit from.

Population Inflows

If your call center has a medium or high turnover rate or you are planning to expand the facility once it opens, you need to look for communities that attract more residents than those who leave. They are often the archetypical big cities that draw people nationwide and from abroad, like New York, Chicago, Los Angeles, San Francisco, Atlanta, and Dallas-Fort Worth.

Some communities may also attract new residents for climate and lifestyle reasons, like those in Arizona, Colorado, and Florida. They are often a big city or regional center of influence.

Downtown Locations

Downtowns, both central business districts and adjacent areas, are slowly coming back as business locations. While in other countries downtowns never went away, many American companies that took flight to the suburbs as soon as the local Interstate opened are taking a second look at what they left behind.

Crime crackdowns, cleanups, modernization, construction of convention centers, malls and arenas, and improvements in mass transit systems are making many downtown areas once again attractive and accessible. Long-distance commuter rail, light rail transit and subway lines have spawned new developments at stations. There are very popular downtown streetcar systems in cities like Little Rock, Memphis, Portland, Ore., Seattle, Tacoma and Tampa that act as people-movers between activity centers. Light rail systems in San Diego; San Francisco; and Calgary, Alberta; and Toronto, Ontario serve the same function. Trains and trolley cars, unlike buses, are fixed assets that are likely to stay around for the 15-year repayment life of a new building.

Downtown and suburban downtown areas draw labor. They have become magnets for young tech-oriented workers, attracted by colleges, a vibrant nightlife and culture, and funky remodeled buildings or newly built lookalikes. Firms are establishing themselves in older, renovated districts like Denver's LoDo, New York City's Silicon Alley, northwest downtown Portland, Hoboken, and Jersey City. Once-derelict areas with rotting warehouses and rusty railroad tracks are buzzing with new and converted offices and homes.

Many cities inside their surrounding suburban expressway rings still have higher-than-average unemployment rates and large numbers of unemployed people. Many of those who live in urban areas often suffer from greater jobless levels than those who reside in the suburbs. Yet with training, such as job skills for those with spotty work histories, they can provide a great labor source. The Hispanics and immigrants possess vital native language skills for centers that need to serve those markets.

Aiding and abetting the revitalization are the move back of older empty-nesters who don't want to maintain sprawling homes and who want cultural amenities minutes instead of hours away on backed-up freeways. They too are labor pools for supervisory, managerial, and part-time workers.

Downtown upsides

Downtown areas usually have convertible older offices suitable for call centers, which are popular with younger workers. They also have unused low-rise warehouses or industrial buildings that also work for call centers. Making these locales happen are their evening and weekend life: galleries, gyms, movie houses, restaurants, shops, and theaters that give them life that is lacking in the typical office park.

As self-service, home working, and other strategies shrink the size of call centers, the easier they are to fit into older structures and newer top-drawer Class A buildings (see Chapter 11) found in the suburban and downtown cores. Those buildings often have high vacancy rates and may be willing to do business with you.

You can't beat downtowns for facilitating face-to-face interaction. You have superior access to cafes and restaurants. You don't have to drive 30 minutes or so to another boring chain in another lookalike strip mall.

You also save on costly amenities like cafeterias, childcare, and gyms that you may have to finance for a suburban call center to attract and keep your employees. Chances are that you will find those features in walking distance. You could reduce your healthcare costs by locating downtown. Studies are beginning to show that commuters who take mass transit are fitter than those who drive all the way to work. The reason? Part of the mass transit journey inevitably includes walking.

Downtowns also tend to have more robust infrastructure. They have fat voice and data pipes. Power lines are underground, relatively invulnerable to surface storms that rip apart the overhead lines that string the suburbs. Downtowns are easier to get to and out of, even for motorists, as witnessed by the huge traffic congestion into and out of office parks. Downtown street systems are built on grids, whereas office park traffic flow is squeezed through only one or two entrances and exits, which backs up traffic on the roads they are feeding.

Office park roads are to traffic what corsets are to women and men. The tighter you make them, the more you cut off the airflow and circulation to the point where nothing moves.

Downtown Downsides

Downtown locations can be more costly than suburban locales for people and property, especially as more offices move to or back to these areas. You are still paying top dollar for people and lease and upkeep costs.

Not all downtowns have great transit access, especially after midnight and on weekends and holidays. Some areas at night can get scary too; you should pick a building near activities like restaurants, movie houses or theaters. On the other hand, the parking is less expensive and more available.

Also, many people who live in the suburbs may not want to work there, fearing crime and parking problems (though not necessarily in that order), the exceptions being where there is good mass transit with excellent security. Check out the communities to be sure.

Follow the Sun

One locations strategy that could save you expenses is to follow the sun, locating a network of call centers spread out east to west in varying time zones. This way you can have a call center open for only one or two shifts but provide continuous live agent customer service.

This provides customers greater convenience; many call centers never close. But that doesn't mean they are open all of the time, staffed at full volume.

Look for "Big Boxes"

At the other end of the scale, the presence of big-box retailers in a community is a good indicator that there is a solid labor force. The pay rates at such employers are comparable to those of call centers, and these firms are very savvy in their site selection. But make sure, as noted above, that they are not experiencing hiring and retention difficulties.

Feed Off Other Call Centers

Want to find a locale that has well-trained agents and supervisors who won't take long to get up to speed in your new center? Go for a community where there are other but lower-paying, frequently hiring call centers. Feed off the food chain. For example, service bureaus, especially those with outbound and order-taking inbound call centers, are good sources of labor for higher-paying inbound and in-house call centers.

But make sure that the skillsets are transferable. Plan to screen the staff closely. A good outbound agent may not be a great inbound concierge. Also

These employees, who work at a WalMart in Thunder Bay, Ontario may be your call center agents if you locate there and make it worth their while to work for you. Many big-box retailers like WalMart stress customer service as well as low prices. The proof is to do business with them in the communities you're targeting. Go inside to get assistance, buy items and maybe return one to see how you're treated. Credit: Brendan B. Read.

be aware that the local EDAs may not want to assist you because you're consuming the same labor pool. But if you can show that you'll create more and more permanent jobs and pump in more money into the local economy, they should not object too loudly.

Troll for Secondhand Locations

The call center industry never stays still. There could be several vacated call centers, sometimes in locations that may work for yours. Call centers have come onto the market through downsizings, restructurings, switches to self-service, outsourcing, consolidations, relocations, mergers, acquisitions, loss of clients, loss of business and budgets, and just plain bad business and political decisions.

King White, senior vice president at Trammell Crow, points out that there are many vacant centers that were not open long, and in some cases never opened. In some cases, the workstations and chairs are intact. Yet many of these call centers are in good labor markets, waiting for the right buyer.

White notes that many of these call centers were built out less than three years ago. A number offer sound design features, such as indirect lighting, ergonomically-designed chairs and workstations, large break rooms, large training rooms, and raised access floors.

These used centers offer several advantages. Because someone has already done the sselection homework, there is a reasonable chance that the location could work for you.

Also, the centers are easily occupied, often with few renovations. Sometimes the previous owner will leave the switch, wiring, workstations, back up generator systems and software, saving you money.

For example, if they left a switch in place and you can use it, you could save up to $1 million over the cost of buying a new one. While most switch manufacturers will assess

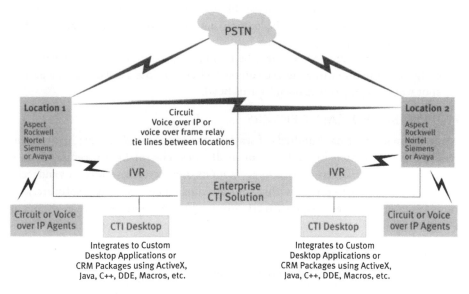

If you are planning to lease a vacant call center or are taking over existing facilities in a merger or acquisition, you can connect different switches, both PSTN and VoIP, by deploying CTI middleware. Courtesy: Norstan.

a fee for software upgrades or licensing the switch to a new user, it still costs substantially less than having to purchase a new switch.

You may be able to hire former agents and supervisors. You'll likely find the landlord, local EDA, and elected officials very helpful in assisting you. You will be putting money back into their hands that was taken out when the old center left, provided you demonstrate to them that you'll be sticking around for awhile. All of these factors can radically cut down the time it takes for you to open your new center.

"In the past, when call centers vacated space in a community, it usually meant that there were labor problems there," reports Susan Arledge of Arledge/Power Real Estate. "But many of these vacant centers are in communities with good labor markets and available labor pools."

Secondhand Downsides

The downsides are that these workforces may not have the skillsets at the compensation you offer. Facilities and equipment may not be suitable for your operation.

Roger Kingsland, design principal with architects Kingsland Scott Bauer Associates, warns that reusing former call centers risks high upgrade costs. You could be spending more money on making a secondhand center fit your needs and in running it than by having a new center designed and built

"Just because the heating systems, lighting, and furniture were used by the old call center doesn't mean they were optimized for the old center, or that they are good for your call center," he says.

Before you select a secondhand call center, find out if it can mesh into your operation and why the previous occupant moved out. Talking to previous employees and perhaps the local newspaper's business editor could give you some clues.

If the center was shut down because the company owning it retrenched or was acquired, you need to ascertain why the location drew the short straw. Also make sure that the property and equipment are sound.

"While you might find good pickings here and there, you must remember that there is usually a good reason why a merged company chose to close a call center at a particular spot while keeping others open," warns Boyd.

✪ DISTRIBUTED CALL CENTERS

If your optimal center has hundreds of seats, you may want to distribute it, especially if you locate your centers in rural areas or small cities. The smaller the center, the less daunting it is to find workers, and the more locations choices you have. Methods and technologies like network/Centrex routing, VoIP, app-hosted and web-based software, and online training tools (see Chapter 14) are shrinking the economies of scale advantage of larger centers.

Distributed Center Upsides

There are several upsides to distributed call centers.

Increased Community and Building Choices

There are many more communities and buildings that can support a 10- or 100-workstation operation compared to those that can support a 500- or 1,000-workstation call center, especially as labor supply begins to tighten up in many areas. Also, if you have a call center that has high turnover, you will burn through that labor supply far faster.

Further, should you need to downsize, disposing of a bigger center and its employees becomes much more difficult than if you had a smaller center. The negative impact on the community and the resulting bad publicity will be less.

Business Continuity

Breaking up a large center into smaller, more widely distributed facilities increases the likelihood that your call center operations can continue even if one site has an outage or has to shut down.

Cost Management

Distributed centers enable better cost control management. By having multiple locations, you can compare in real time labor prices, turnover, and quality. For example, if one site gets more expensive than others and if it is a long-term trend, you can downsize or close that center and move those operations into the other facilities.

Flexibility

Having smaller, distributed facilities gives you greater flexibility. If you get programs that require different skills, you can split them up between centers if they are smaller.

One increasingly common example is Spanish-language support. A community where you have a 500-workstation center, say in Dallas, may not have enough affordable Spanish-speaking workers. But if you have a 300-seat workstation there and open a 200-workstation facility in a border city that has such labor, like McAllen, then you could provide that service to your customers and lower your costs.

Distributed Center Downsides

There are also downsides to this strategy.

Higher Rents and Costs

The major downside is that small call centers may have to pay higher rents per square foot than their large cousins. Property owners may be less willing to fund tenant improvements such as better air conditioning to support the increased density of workers.

Management Complexity

You also need more management attention in distributed centers because you have more complex voice and data connections and technologies to look after than in single centers. Your managers, supervisors, and agents will have to be equipped with virtual communication tools like conferencing and IM technology. Service quality has to be seamless across all centers.

Multiple facilities also means more work. There are additional issues such as leases, cleaning and maintenance contracts, security, and other tenant relations to look after.

Also, senior management or clients may not be convinced that you can spread the work around more locations effectively. You have to prove it to them.

"Landlords often feel less secure with a smaller call center, unless it is being operated by a larger firm," says Mark Collmar, president of design firm DM Communications.

✪ SHARED-SERVICE CENTERS

Another site selection strategy is to co-locate your call center with other functions, such as administration, engineering, R&D, and distribution. This is most practical for very small centers with fewer than 10 seats. But planned right, it can work for larger operations. These combined facilities are often known as shared-service centers.

Shared centers can be as simple as a few seats in an office or be a separate facility housing accounting and HR. Alternatively, shared centers can be handling call center contacts in face-to-face environments, like reception, counter sales, or a customer service kiosk.

Shared centers can be in place or supplement in-house or outsourced call centers. For example, when member services staff at Kaiser Permanente's 13 Portland, Ore. area clinics are not handling in-person inquiries they take calls from the HMO's 40-agent Portland call center. Calls are received through Siemens' HiPath Teleworking rerouter, which attaches to Kaiser's HiPath ProCenter ACD.

The satellite program has dropped queue times (some clients penalize the HMO if it doesn't meet service level agreements). The initiative has also let Kaiser avoid hiring additional call center agents, says Rebecca Rowland, call center technology specialist with Kaiser Permanente.

Shared Service Advantages
Facilities and Real Estate Cost Savings

You can save from 10–30 percent in costs compared with locating these functions separately. The benefits come from shared facilities like cafeterias, on-site power backup, and parking.

But the actual amount will depend on factors such as size and function of the business units being co-located, whether the facility is a renovation of an existing building or floors, the type of facility, and other on-site assets like parking and utilities.

Companies that locate more than one function have bigger clout with landlords and developers. They can strike better deals by leasing the entire building than by renting only one or two floors. This also eliminates parking hassles with other tenants and provides better security.

If you have vacant office space and you can't get out of the lease, a shared center could fill it nicely. Or if you downsized your call center and other departments are able to move in, you can turn your call center into a shared-services center or general or regional HQ.

HR Issues

Shared-services or co-located call centers can help you keep your agents longer and maximize your corporate investment in them by giving them career opportunities. Equis's Cariola says co-locating can reduce call center turnover and enable better recruiting by allowing for greater interaction and career path development between the call center and co-located departments.

There are time-saving and profit-enhancing synergies between customer service, sales, marketing and product development departments. This can shorten new item lead times, improve design and delivery, and solve problems faster. Your customer service and sales agents are your company's front lines. The quicker it takes for the information they gather to reach the colonels and generals, the more accurate your company's response will be.

Shared Service Downsides
Labor costs Attraction, and Retention

You may end up paying more for labor and experience greater turnover in a shared services center than in a stand-alone call center. There are differences in labor markets between call centers and other business functions, such as professionals like engineers and sales and management. Call center employees are paid less, have smaller commuting distances, and are more likely to depend on mass transit than other employees.

More important, a community that may make a great locale for upper management may be the most expensive choice with the highest-priced, least-loyal labor. The cost savings by combining your call center with those other operations may be cancelled out by the higher wages and benefits and turnover expenses.

"Companies, however, need to make certain that the facility and location criteria of a center are met," says Cariola. "Being co-located within the more expensive part of an area, because the center is co-located with higher paying corporate functions, can prove unsuccessful for a center that employs lower paid workers."

Unoptimized Buildings

With modern multimedia routing (discussed later) there is little cost penalty entailed in locating a small center. There are off-premises extensions from large switches, small fully featured switches and network and Centrex routing, with IVR, CTI and skills-based routing to handle your calls and contacts.

Site not ideal for call centers

Call centers, unlike other business functions, are often 24x7, and as noted before, the workers are lower-paid and often rely on mass transit. Conventional office parks aimed at regional and main headquarters and distribution and industrial centers for shared-service centers have poor mass transit, especially in the evenings and on weekends. There are also few nearby amenities such as restaurants, dry cleaning, fitness centers, and childcare, which many call center workers depend on and which they use on their breaks.

Security issues arise, especially at night in industrial areas. After dark they can get downright creepy because they are so desolate. Businesses can get broken into. Industrial districts are often some distance away from law enforcement. If there is a truck accident or a train derailment, those areas can be blocked off, preventing workers from leaving.

Also, communities that are ideal for manufacturing or transportation may not be great for call centers or for regional offices. Distribution centers require good access to airports and major trucking lines, locations that often won't work for regional or headquarters offices, DLR Group's principal Robert Hoffman points out.

"Call centers and distribution centers are two very different operations with different geographical needs," he says. "Under some circumstances, a combined center can be effective where cities have allowed industrial, office, and commercial zones, and transportation services to commingle."

❂ ENABLING SHARED-SERVICE CENTERS

Corporate Culture

To benefit from interdepartmental synergies, observers say, a firm's corporate culture must also be more horizontal than vertical. Companies, too, must ensure there is enough land for joint operations. A distribution center, for example, needs a lot of paved space (including wide access roads) to turn trucks. "Employees at call centers, regional offices, and distribution centers are very different kinds of people," says Hoffman.

Carriers

Organizations, unless they are call center service bureaus, pick carriers for reasons other than how well they handle call centers. In the seven years I worked for *Call Center Magazine,* I did not see local or long-distance carriers begging to be advertised or written up; the publication dropped carrier coverage. My understanding was that the firms didn't see the magazine's readers as being involved in the buying decision.

But if you are going to be expanding or adding a call center to your firm, then you must take a close look at the rates and services charged by the telcos. As noted earlier, call centers consume a lot of phone time and enabling technologies. There are finally competitive local and reseller carriers such as UCN that are targeting call centers with an integrated suite of network routing services combined low-cost long-distance rates.

Well-Designed, Complementing Interiors

Modern workstation design means "equal cubes for all." If your employees work in a call center or support a call center, they have similar furniture, chairs, and phone systems. But you have to accommodate the unique needs of other occupants.

For example, if you have management, you will need higher-quality furnishings, boardrooms (often with high quality videoconferencing equipment), and private offices. You may need flashier exteriors. If the center is shared with fulfillment and shipping, you will need warehouse space, with thicker floors, loading bays, and room for trucks to enter and back up.

More Stringent Security

Unless your call center is to be a strict 9-to-5 operation along with other functions at that site, you will need to consider and plan for employee safety. That means picking the property or planning a new center accordingly.

Smart Locations

All parties need to compromise to meet mutual locations needs. The non-negotiables for call centers include:
* Viable labor markets, including mass transit access
* Nearby or onsite food service
* Needs of 24x7 centers

These requirements include locating in a low-crime area, proximity to other evening and weekend activities, short walking distances to parking and transit stops, absence of bushes or other shelters where criminals can hide, parking and walkway lighting; safe

FINDING SPACE QUICKLY

Yes, you can set up a new call center in less than 60 days. Outsourcer ClientLogic got its new Albuquerque center opened in 37 days—and over the Christmas holidays. The center, which received the go-ahead on December 8, 1998, went live January 18, 1999.

One of its keys to success is advanced planning. ClientLogic forecasted seat planning in 30-day, 60-day, and 90-day timeframes. Needs are reviewed each week. It has in-house design and human resources project teams. It contracted with site selection firm Jackson and Cooksey to find locations and space. It had preset its center plans, sizes, configurations, and facilities requirements and had agent profiles to screen staff with.

Another key is finding a good EDA. Albuquerque Economic Development helped Jackson and Cooksey find space; ClientLogic looked at eight buildings before picking a recently vacated third floor of an office building that had administrative offices and a call center. The agency introduced the outsourcer to recruitment and training firms, helped it obtain the necessary state and local permits and made contacts for the call center firm with US West (now Qwest).

A third key is constant communication. ClientLogic had cellphone-equipped team members in Albuquerque, Las Vegas, its nearest call center, and in Buffalo, N.Y.

A fourth key is serendipity. The office it took over required few modifications. For example, it already had carpeting, which saved enormous time. It reused rooms for office and training space, though it did have to install a separate computer and switch room. It also used a tie line from its Las Vegas center until its switch could be hooked in.

"We were on a tight, aggressive timeframe spanning two big holidays," explained ClientLogic senior vice president-facilities Melissa Bailey. "We were pushing US West to connect us on time. The Albuquerque Economic Development office called the right people in the company for us who then enabled us to meet our timeline."

parking lots and garages, security guards, and security cameras.

Downtowns may be win-win for shared-service call centers, especially in revived gentrified areas, or adjacent to shopping centers and busy transit terminals compared with exurb office and industrial parks that are downright creepy after dark. You're also able to hire a broader range of employees for your call center because you're near where they go after-hours—they know where you are—and because you're often on mass transit routes.

Home Agent Support

If your call center, or organization has home agents (see Chapter 5) and you require them to work in the center once a week or month, you will have to accommodate them with workstations. To limit how much space and furniture you allocate, you will need to carefully schedule these agents. There are workspace management software tools and services to assist you.

✪ ENABLING SITE SELECTION

There are several steps you need to make in your site selection. Miss one, and you risk having your project tumble, costing you time and money.

Get Everyone on Board

Site selection is a team effort. The people you should have on your team include your human resources, operations, IT, property, and finance departments.

Susan Arledge of ArledgePower Real Estate and Ron Cariola of Equis recommend that you bring on board senior management, such as marketing and business development VPs, who usually have responsibility for customer service and sales call centers. This way you have buy-in from top managers in your company on your site selection project. In some firms, such as service bureaus, final site selection reviews and decisions are made by the COO, CEO, or by the president.

"Senior management should buy-in from the start because they have to ultimately sign off on the project," explains Cariola. "You don't want a situation where they find out some details they didn't like after you started the process. They may hold up the project because they weren't provided with enough information to make a prudent decision."

Labor Attraction Strengths and Weaknesses

You also need to be aware of what you offer. You are going into geographic markets where you are competing with other employers for labor and space. You have to convince the labor force to work for you and do business with you rather than your competition.

If your company is well-known and respected and it has a reputation for being a good place to work, then you have more choice of locations open to you. Highly regarded firms that enter new regions can attract workers from other employers. Word of mouth will drive applicants to your web site long before you've made a formal announcement.

But if your firm is less visible and has a lower compensation package, consultants say you have to look at other ways to be more competitive. Examples include your firm's location, career paths, and work environment.

Service bureaus fall into this dilemma. The savvy ones obtain the permission of their clients to use their clients' names in online and print recruiting material. After all, those employees will be working on behalf of those clients.

Have a Checklist

Site selection consultants recommend that your team devise checklists (see box at end of the chapter), establish criteria such as your desired cost and labor availability, weigh each item in importance, and create a scoring system. You should gather data about each locale and run it through a matrix. Then you should rank your priorities and narrow down your choices to the three or four sites that most closely satisfy your requirements.

Picking the Site

If you have closely defined what you want your new call center to accomplish, you should have already sifted out many locations. You may have excluded others based on the views of your clients and members of senior management in your company, as well as your own opinion of a locale. You may also have eliminated some possible sites because you do not want to be too close to other call centers. If you locate your center near another

INCENTIVE TIPS

Whisper that you're going to open a call center in a region and you may end up feeling like the rich prince off to find a bride, only to discover someone in your court had leaked the word out. You'll get more offers and promises and see fatter dowries than you've ever dreamed of.

Every city and town, it seems, wants the jobs and tax revenues you will create, even at the price of going broke with grants, loans, tax breaks, free buildings. And every city and town knows about each other' goodies, just as there are no secrets in the locker and powder rooms.

It is tempting to make your decision based on which community can come up with the best deal. But don't. Like marriage, when you set up a call center you have to live with the decision, and it could be the worst one you'll ever make. You will have to put up with the consequences such as higher-than-expected costs and turnover, and not enough quality agents.

In the words of one obviously male site selection expert, who will go nameless, "The bigger the dowry, the uglier the bride."

Incentives do have a place in deciding between two or three roughly equal locations. They can reduce the costs of setting up in that community, improving your bottom line. Senior management likes such dealmaking. It shows that you're a good businessperson.

CB Richard Ellis Call Center Solutions Group offers these tips on incentives:

- Have you worked with the community to create a beneficial incentive package that will help you and them?
- Has someone explained to the community which of the offered incentives do not apply to your company? It is imperative to give specific reasons and support to the local community contacts. Remember that a community official may have to sell your incentive package to other local agencies or governments in the same way in which you have to sell your projects to your company's management. One of the best ways to help the community understand how your company will be affected is to provide a detailed analysis of their taxing structure in comparison to other communities.
- If a community offers incentives that cannot be utilized, work with the community to create alternative solutions.
- Involve your tax or finance department and check to ensure that administrative requirements for tax credit programs are reasonable. Do not assume that your tax department will have the resources to follow through with your incentive programs.

company's center, you run the risk of having to contend with more competition for labor and perhaps greater turnover.

Consultants recommend shortlisting down to three sites. Once you identify the locations that meet your criteria, you should then conduct a more intensive evaluation that includes site visits and meetings with local officials.

Then you interview comparable employers about their recruiting experiences, compensation levels, turnover and the quality of their workforce. Test a new labor market by placing blind ads and attending job fairs.

You should also visit the communities and see how you're served and sold to. Go shopping, have a meal, and stay in a local hotel or motel of mid-quality standards, or high-quality if you are looking to locate a top-end center. Check for speed of response, helpfulness, courtesy, and accuracy. Go back to those establishments two or three times to determine whether your experience was typical.

You pick the best overall community and identify the most desirable neighborhoods to open your center. You then find the most competitive property, secure any available incentives, plan your recruiting and staffing strategy, create a communications program, choose a transition team, and secure your resources to open the center.

DIY SITE SELECTION

One of the best ways to find out if a location will work for you is to find out for yourself, directly and indirectly, without letting on that you're interested in opening a call center or any other function there. It is amazing how nice people can be when there's jobs and money riding on it.

Here's some suggestions:

Research the community. Frequently visit (but don't register at) local web sites, especially those belonging to media outlets and governments. Subscribe to the local newspapers—circulation departments usually don't ask any questions, as they just want the money. But have the papers sent to your home. Pay particular attention to the business pages and the Help Wanted ads.

Find out the lay of the land. There are issues such as zoning that could affect your call center that if you went through normal channels like the planning office. Word might get out. The advice that I got when I was researching opening a bed and breakfast is to hire a well-known local lawyer. They will tell you what is going on and what you can do, with confidentiality sealing their lips.

See for yourself. Nothing beats being a tourist to see firsthand how a place ticks. Stay at a local hotel, eat in a variety of restaurants, and visit the shops. Go to some of the attractions. Strike up casual conversations with the front desk clerk or with some of the people at the bar. Bring your spouse and kids along, if the locale is attractive to them; they make great foils. Go back to some of the businesses you first patronized to look for consistency. If you had a bad experience the first time, it may be good the next; everyone has off-days. Visit again or have one of your staff travel to the community some weeks later.

What you're looking for is how well you and your family have been sold to and served by the employees, because this is the labor pool you'll be drawing from if you locate your call center there. If the salespeople and waitstaff have been attentive without being pushy then the community might work as a location. If they take their sweet time about getting to you and have an attitude (like the TV commercial where the waitress, having been told by a customer that he didn't want mayo, proceeds to scrape the piece of bread on the side of the table), leave.

Chapter 8: Onshoring, Nearshoring, and Offshoring

Whatever live agent route you choose to go, outsourced or with your own call center, perhaps the most critical decision you need to make is *where* to have your contacts answered.

The choices are onshore, nearshore and offshore. There is also blendshoring that provides the best of all worlds.

Onshore refers to the same nation as your market, which for most of you means the U.S., and Puerto Rico, a U.S. commonwealth. Offshore refers to foreign countries whose workers produce goods and services for your domestic markets. Nearshore is a subset of offshore; it applies to countries and parts of nations that are close to your domestic markets culturally and physically, such as Canada, Ireland, Australia, New Zealand, and northern Mexico. Each of these selections will determine the costs and benefits of your program.

✪ SITE SELECTION IN RELATION TO OUTSOURCING

There has been some confusion between outsourcing and site selection in this area because outsourcers have handled most offshored and nearshored call center work. Locating your call center onshore or offshore are two very different practices.

When you locate, you undertake all the site selection, design, property, HR, and equipment tasks and costs yourself. But you have complete control over what you want.

These tasks are often time-consuming outside of the U.S., especially in developing nations and in regions like the Caribbean, Africa, and Asia/Pacific. It is a different world when you are setting up an operation in a foreign country.

Cultural, economic, legal, linguistic, monetary, and political differences with the U.S. take considerable time and sophistication to understand. Even next-door Canada is a very foreign country when you begin to look at what is entailed in establishing an enterprise there, including currency, measurement, regulations, culture, and depending on the province, language.

When you outsource you entrust the bureaus with the site selection, design, real estate, staffing/training, technology and voice/data vendor selection and operations. These costs, many of which have already been incurred prior to taking on your project, are built into your fees.

Offshore and nearshore outsourcing especially incurs the benefit of relying on the bureaus doing the heavy bureaucratic, legal/political social and business networking lifting in foreign countries. They know the lay of the land, which unless you already have operations in that nation, would take years for you to acquire.

The good bureaus have partnerships with businesses and government, have figured out the laws and regulations, and know where to get staff, utilities, and other services. They're the insiders.

Outsourcing also lessens the political and security risks. The outsourcer is exposed to political instability. Their name is on the buildings, not yours. Because the outsourcers are doing nearly all the work, foreign travel, costs, disruptions, and potential threats to employees and families are minimized but not eliminated.

The downsides include loss of control but no less blame from customers and users if they are not pleased with the programs. You still need to have and trust local people on the ground to run it.

"Offshore mitigates some of the risk because you're flying fewer people abroad, for shorter periods of time," says site selection consultant John Boyd of The Boyd Company. "But outsourcing does not eliminate the risk, because you still have to have people over there to check out the outsourcers, set up the programs, and train the staff."

Yes, the distinctions are tangled. This chapter attempts to straighten them out as much as possible. Later on in this chapter I examine considerations unique to outsourcing and unique-to-in-house.

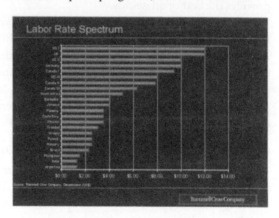

This chart shows by bar graphic the dramatic differences in labor costs between the US, nearshore and offshore locations. Courtesy: Trammell Crow Company.

✪ THE CASE FOR ONSHORE

Onshore remains the top location for call centers and outsourced programs. For the purposes of this chapter onshore refers to the U.S. There are several good reasons why onshore is still the preferred place to handle customers, employees, and the public.

Customer Preference

Your American customers may want to speak only to Americans. If your centers are all onshore, that may give you a market and certainly a marketing advantage. *Call Center Magazine* Chief Technical Editor Joseph Fleischer cites data from research firm Issues and Answers Network that found that 86% of consumer respondents said they preferred to reach American call center agents. One reason? Better service. Over 72% told the study, sponsored by Nuasis, that they expected that American agents would serve them better than offshore agents.

Cultural Affinity

Here's why. Despite the globalization of the economy and largely American culture, no one can understand and relate to Americans better than other Americans. The agents' ethnicity, where they live, when they arrived, the accent they have and when they came here matter little. Anyone who resides, or has resided in the U.S. has an idea what America is about and what matters to Americans.

No matter where they are and who they are talking to, be it a Yankee in a Maine call center helping a Mississippian out with their computer problem, a Mexican immigrant in Washington State taking an order for a new high-speed cable Internet service from an African-American in Washington, D.C., they are all Americans. They reside in the same country, live the same events, and share a bond of values no matter where they are from.

This factor, known as cultural affinity, has an impact on call center functions where there is a high degree of interaction with the contacts or contacted parties. The more the agents can relate and understand customers, the more successful the transactions are, be they service, sales, support.

Examples are high-end, high-touch, business-to-consumer and B2B service and sales and low-end, high-touch, billing, collections, catalog, order entry, and telemessaging. There are key verticals that should stay onshore, including energy, health care, insurance, securities, non-profits, and political work. This also includes region-specific call center work such as targeting the Seattle-Tacoma area for inbound sales of combination umbrellas and latte holders.

You *especially* want to keep work onshore if you are with a government agency, are government-owned or government-regulated, or your firm receives contracts or tax subsidies. Political pollsters or service bureaus specializing in serving candidates and parties should not even *think* of going offshore, unless they have a suicide wish or want to throw the game.

Flexibility and Scalability

You can have any size call center you want in the U.S., from one workstation to 10,000, part-time or full time. That goes for site selection and outsourcing. But if you go offshore, you need minimums, like 10–25 workstations in Canada and 50–100 workstations in other countries, to make it worth your while.

Small call centers and those that require frequent management and client visits should stay in the U.S. They require close, personal attention that only Americans can provide and which helps them differentiate themselves in the marketplace against competition that has nearshored or offshored their service.

Home Field Advantage Site selection/outsourcing/legal ease

You are on your home turf. You know the lay of the land, the laws, regulations, and practices. While there are often peculiarities with state rules and every city has its own zoning and traffic issues, they are easily manageable.

Legislation and Regulatory Compliance

American laws are generally enforceable only in America, unless there are effective reciprocal agreements with other nations. There are now strict laws protecting consumers' privacy.

One such regulation is the Health Insurance Portability and Accountability Act (HIPAA). HIPAA requires that organizations have custody of covered (specified) data at all times, including by outsourced vendors.

Because consumer medical data is so sensitive, you probably should not risk having agents located outside of the U.S. handle it. The political and legal firestorm that would erupt if an offshore or even a nearshore agent or service bureau let it get into the hands of criminals would be like a nuclear detonation: too much to survive.

For similar reasons consider keeping your HR and IT support on home turf. You don't want anything to happen to that information or see access get into the wrong hands.

Relocation and Travel

Having your staff set up a domestic call center can be time-consuming and disruptive, but it's nothing like going offshore. Outsourcing minimizes transportation, but for medium and large projects it does not eliminate it. Either way your travel is domestic; you can drive or in some parts take a train or ferry to visit a site.

Potential Cost Savings

A well-located and well-planned onshore call center could be less expensive than an off-shore center when all factors are considered, such as labor costs, quality, design, real estate, staffing, training, and management.

You may find that an onshore call center lowers total volume, escalation charges, and voice and data costs, because American agents better understand other Americans. You also have lower travel, training, and staff expenses. Your customer satisfaction, retention, and sales may increase.

There are call centers that have gone offshore and are coming back onshore. The total benefits did not exceed the complete list of costs.

✪ ONSHORE ISSUES

But there are costs and issues involved with keeping your call centers at home.

Higher Expenses

The U.S. is a top-dollar location. You pay a premium to be there. The average wages are $7 per hour to $12 per hour, compared with $2 per hour to $2.50 per hour in India and Malaysia, reports The Boyd Company. But if you don't have to be here, then why leave the money on the table and not put it back in your pocket?

If your competitors are offshore, you should be too. They're benefiting from the cost savings, which keeps their prices competitive. Offshoring also keeps Wall Street analysts happy too, which helps their stock prices and drives in more capital.

Shrinking Labor Supply

Young people, long the fodder for call centers, are not as plentiful as they used to be. Their forebears, liberated by birth control and changing social mores did not make as many children as *their* parents made. That means a smaller pool of workers to choose from.

There are relatively untapped labor markets, like aging baby boomers and inner-city residents. But too often call centers are too conservative, their managements petrified to move out of the box with strategies that can tap them, such as home working or locating close to downtowns and mass transit.

Some call center attitudes border on racism. I've actually heard about call centers deliberately not locating not near mass transit so they wouldn't attract "those people."

At the same time, there are people who are having more children, known as "natalists" located in the American heartland. The future labor quality remains to be seen.

Turnover and Competition

A smaller labor supply means more employer competition, higher wages, and skyrocketing turnover, hence employment expenses. Individuals who took call center jobs because there was no work available in their trained or desired fields will leave quickly when openings appear or when they have enough money for training courses to get them hired in their sought-after occupations.

Call centers do not do well in competitive markets. They are lumped in with retail and hospitality in jobs and turnover. In a tight job market, employees are willing to jump for a dollar or more in the same or similar fields.

Lower Quality

High-quality, highly motivated American agents can be difficult to find at the right price. The education system is no longer up to the standards of a generation or two ago.

A study released by the American Diploma Project said that more than 60 percent of employers rated graduates' skills in grammar, spelling, writing, and basic math as only fair to poor. The report also reveals that 53 percent of college students take at least one remedial English or math class. Employers have had to pick up the slack. Another recent study estimates that remedial training cost the employers in one state nearly $40 million a year.

What makes such data scary is that the agent positions you're looking to fill require a high level of literacy and comprehension. Agents also need to read and reply to email and do text chat, which demands superior grammar and spelling skills.

Your customers and clients have had the simple inquiries taken care of by IVR and web self-service. When they talk to an agent, they expect intelligence and service with a smile, like a five-star hotel concierge.

But Americans too often look down at service jobs. Everyone wants to be served, but no likes to do the serving. With an aging population and lower replacement rates, the labor pools to draw from are drawing down.

Moreover, some but not all staffing consultants report that too many American young people are unmotivated and lack basic skills like showing up to work on time.

"When they get into the workforce it's a shock to them to be accountable, since they have never been before," says Jeff Furst, president of staff selection firm FurstPerson. "Especially in call centers where they have to be on time, cannot leave when they want and where their performance is monitored, measured and graded."

Furst reports that in a typical community, only 30–45 percent of candidates tested meet the abilities and behaviors to do the work.

In contrast, many call centers that have moved outside of the U.S. claim that they have found better-educated and more-motivated employees. Most have college educations.

"The consequence of having a poor education system is that call centers may have no other choice except to leave the U.S., taking jobs, income, and ultimately the property and other tax revenues that go to support schools," warns Furst.

Onshore Versus Offshore Quality

Teresa Hartsaw is president of ePerformax, an outsourcer and training firm with call centers in the U.S. and in the Philippines. She has to spend 25 percent of her training budget on remedial education. Training is 10–15 percent of her firm's costs.

The money she has to spend to provide what the schools didn't has to come from other training, which hurts her ability to provide top-notch coaches and managers.

But if the school systems don't get their act together, she will have to move more work offshore. Her firm does not know how long it can afford spending money on remedial training or tolerating poor performance from underskilled agents, especially if call centers like hers want to remain competitive.

"We have to train agents wherever they are," says Hartsaw. "But the agents we hire in the Philippines—while they need training in how to communicate with Americans— they are 100 percent college graduates with excellent general communication and math skills. As a result, their aptitude for learning is much greater than our typical recruits for our contact center in the U.S. We have a much higher training certification rate in Manila than in the U.S."

✪ ENABLING ONSHORING

Here are key steps to make onshore sites and programs work.

Carefully Select the Functions

Those call center functions that require a lot of interactions and/or complex interactions with Americans need to stay onshore, either in-house or outsourced, especially transactions that involve products and services that have high price tags. Examples include concierge services, fundraising, nonprofit, government, market surveys, polling, order taking, cross-selling, upselling, B2B, and telemessaging. Keep critical functions like healthcare and social services wherer you can't afford to risk misunderstandings onshore. Make sure your high-end American customers— management, executives, and wealthy consumers—are served by other Americans.

Empower Your Agents

Get the most out of your American agents. Give them the authority to find solutions to customers needs beyond the scripts and knowledge bases.

Choose Your Locations Wisely

With looming labor shortages and quality issues, do your site selection with precision. For example, make sure that the communities you pick have good-quality public education and colleges. If you need enough Spanish-speakers (see Chapter 9) to warrant agent teams, then select areas that have large Hispanic populations. Remember that wages are 60–70 percent of a call center's operating costs. Make the right investment.

✪ THE OFFSHORE REVOLUTION

Since the mid-1990s, U.S.-serving call centers have migrated out of the country. Most of these centers belong to service bureaus, or less frequently to in-house call centers, though this split may change.

English-speaking Americans are being served by call centers with English-speakers in other countries and regions such as Africa, Canada, the Caribbean, China, Eastern India, Eastern Europe, Ireland, New Zealand, Northern Ireland, and the Philippines. Spanish-speaking Americans are having their contacts taken care of in Mexico and Central and South America.

Offshore Benefits

Cost Savings

Having contacts handled offshore can cut your costs by 10–15 percent in countries like Canada, to 20 percent in the Caribbean, and 30–50 percent in other nations.

In developed nations like Canada, organizations don't have to pay for health care, which cuts cost further. The government totally or partially pays for it, with residents funding the rest, such as through premiums and taxes.

Quality Agents

You can often hire higher-quality, more-loyal workers offshore than onshore. Most agents in offshore sites have college degrees. Canada, New Zealand, and Northern Ireland have arguably better education systems than the U.S.

Why India Still Tops Offshore Locations

India remains the classic example of quantity and quality. And unlike China, which is struggling with letting the leash off its economy because it may set wild political changes that could unseat the Communist regime, no such issues tether India.

Suresh Gupta, founder of The Paaras Group, reports that India produces almost 3 million graduates every year and has a pool of 30 million graduates available as the potential workforce. Past experience suggests that they are able and willing to perform call center duties, often at higher-quality levels.

While the country is still largely rural, there are huge population influxes into the major cities. For example, Mumbai has 18 million people and is reported to experience an influx of 100,000 migrants per day.

With what *The New York Times* reports is an unemployment rate of 20 percent or three times the reported average, the labor supply is there.

"The unemployment is not necessarily driven by lack of skills, but it is a function of low availability of appropriate jobs," says Gupta. "Hence, the availability of college grads for call center work!"

More Acceptance Offshore

There is a possibility that onshore communities may not want your call center. Instead, high-unemployment countries especially developing nations will welcome them.

The call center industry has been tainted with a sometimes-justifiable bad rep for being low-wage, high-turnover, here-today-gone-tomorrow employers. There has been bad press of firms closing up shop after receiving tax incentives even though these companies argue that the communities benefited far more in wages and investment than they got in breaks.

Many call center functions, especially business-to-consumer outbound, are distasteful to enough Americans that the claim that these jobs will go offshore with the implementation of state and national Do Not Call regulations fell on deaf ears. As labor shortages mount, communities will get picky who they want for employers.

Keep in mind that these conditions also apply offshore. The higher up the food chain countries get, especially as they become more prosperous, they less likely they will seek the lower-end call centers like telemarketing.

Market Opportunities

Offshore nations are markets for American-made goods and services By having centers in other countries you can attract and serve those customers.

India and Mexico have growing middle classes. *Business 2.0* reports that India's middle class of 487 million will spend $420 billion over the next four years, aided and abetted by investments and jobs like call centers.

Building local markets helps you ride out currency fluctuations. It's a challenge for Canada because much of the savings of that location is based on the spread between the U.S. and Canadian dollars. But Canada, with about 32 million residents, is the U.S.'s largest trading partner.

Offshoring Downsides

Risk of Lower Revenues

You may lose customers and revenues if you offshore. A BenchmarkPortal/Kelly Services study showed that 65 percent of American consumers would change their purchasing patterns if they knew or believed a company had offshored their call center. The view is consistent across all call categories, including product information, purchases, reservation, account or order status, technical support, and complaint calls.

The cultural differences between Americans and non-Americans will manifest themselves in accent, choice of words, and in casual conversation. That may or may not have an impact on your program and on your customer retention. There is even a noticeable difference between English-speaking Canadians and Americans in not only dialect but also in verbal mannerisms. Canadians are less pushy and more reserved than Americans but are more stubborn and persistent.

For those reasons you should definitely not offshore outsource any kind of political work. You will be vilified; your opponents will be laughing all the way to the ballot box. If you are a charity, it is probably not a good idea to offshore, unless your group has a presence in those countries that you can turn to your advantage.

Higher-Than-Expected Costs

Your costs may also go up from longer call lengths and added escalations to senior staff, caused by agents not understanding customers. Offshore call center workers may lack the attitudes and the skills to communicate effectively with onshore contacts. They may not have cultural affinity, leading to misunderstandings, which lengthen call time and can lead to expensive repeated calls, escalations to supervisors, and at worst, lost customers.

Cultural affinity is not an issue with low-level and highly technical interactions between knowledgeable people. But it springs to life when matters get complicated. Examples include a family trying to arrange a bereavement flight or an employee who does not understand computers calling the help desk.

Cultural affinity becomes an issue in call center offshoring because unlike with manufacturing, data processing, and to some extent programming, customers are interacting with those who are delivering the service. Therefore those agents make the service.

If customers or employees perceive a service problem and they attribute that to offshoring, then they are going to blame you for it. If that perception sets in, you've got problems if you offshore. You can pull every one of your customer satisfaction stats from your files to show otherwise and they still won't believe you. What they perceive is reality to them, and they will act accordingly. It's no different than in politics. If they have a choice when purchasing goods or services, then they may wish to exercise it.

Political and Labor Fallout

Offshoring is a hot topic especially when the economy is unsettled, because it is moving existing and potential jobs offshore. When it has occurred, there has sometimes been public outrage. Unions have threatened strike action against employers who offshore.

Elected officials have threatened to take action to restrict it; some of these threats have been carried out. The Thomas-Voinovich amendment to the federal budget prevents offshoring of new federal contracts. Tennessee requires that state contract preference be given to vendors employing workers only in the U.S. New Jersey had to retrench its electronic food stamps outsourcing contact with eFunds from India to Camden, even though it added costs to taxpayers, because of the political furor.

Outsourcing is a fig leaf; it can't hide a closed center. Employees who are about to be laid off will talk, especially to reporters.

If you received financial incentives to be in your current location, expect to get some angry phone calls and bad press. Yes, you indirectly provided jobs and pumped money into the local economy. But unless there are new clients or employers coming into the market to pick up the unemployed bureau agents, you will get tarred with the same brush. Memories of fine meals, paid by hard work, won't feed families whose household heads are now on unemployment or welfare.

If your call center is moving offshore, adding jobs in other countries but not eliminating or planning to get rid of those at home, and you are located in prosperous American communities, then you may get away with offshoring. But if your centers are in high-unemployment areas or you are letting staff go, prepare to face the flak.

Learning Others' Cultures

Going to another country is not like moving to another county or state. You have very different cultures, laws, regulations, business practices, and vendors that you must comprehend quickly. Most Americans have never traveled to another country, other than perhaps Canada or Mexico.

Even when they visit Canada, Americans' almost seemingly willful ignorance is amazingly appalling. Urban legends abound of U.S. residents bringing winter clothes across to Canada in summer, even though the climates are similar between the border states and provinces. There are always stories of Americans stopped for bringing handguns, which are highly restricted in Canada.

You need to learn the nuances of the people you're dealing with: government agencies, landlords, suppliers, and employment agencies. Sometimes "yes" does not mean yes, and there may be a greater importance of keeping one's distance and saving face.

These differences can be quite subtle. My parents are from the UK, and I grew up infused with the British culture, later reinforced by living in England and in Canada.

Britons do not like being touched by people that they don't really know. Yet Americans, especially from South, like putting their hands on others, such as on shoulders. Whenever someone does that to me I automatically stiffen up. I do not say anything—that would not be British—but I do fire a "do you bloody well mind?" glare and I immediately and unconsciously downgrade that individual to somewhere below the rank of untouchables on the cultural scale.

What may be acceptable in the U.S., like chewing out an employee in front of their colleagues, is usually a no-no offshore. *Never* resort to the argument "this is the way we do it." Anything that smacks of arrogance and heavyhandedness will get peoples and governments' backs up and invite unionization.

Unequal Education and Shallow Labor Pools

The offshore labor pools are shallower than their raw population and unemployment numbers indicate. In developed countries, working in a call center is more desirable than in the U.S.—serving people does not have as bad as a connotation—but not by much.

Few people graduate from colleges and universities in Canada or the UK with the aim of making a career in call centers. If better opportunities arise, like in computer programming or management, they will grab them.

Yes, you get college graduates working for your call centers in developing countries. But they are from the middle class and represent a small percentage of the population. Only in very large nations like India and China are the numbers big enough to provide large agent pools. Working in offshore programs requires higher-than-average English-language skills and education compared with domestic programs in their own countries.

Alton Martin, president of Customer Operations Performance Center (COPC), a performance standards, testing and consulting firm, says that comparatively few Caribbean workers have college educations. The Philippines and India, by contrast, boast agent teams made up completely of college graduates.

Site selectors say the Caribbean countries are suitable for low-end call centers such as airline reservations, credit card sales, collections, and order taking. The Caribbean's low wages and loyal workforces offer a compelling economic case to call centers in these spaces. In the U.S., these centers tend to experience high agent turnover. The tradeoffs are power facilities, telecom networks, and offices that are more expensive, or built to inferior standards, than those in the U.S., say market watchers.

Management Costs

It costs more to manage a call center in another country. A 2003 NASSCOM/Evalueserve report says it costs on average 18 percent more to manage projects offshore than onshore, no matter the size. The lower the volume, the more the impact of the additional management costs.

Diminishing Returns

Turnover is creeping up in many offshore countries and in cities. The Paaras Group's Gupta reports that heavy concentration of offshore call centers in some urban centers, such as Bangalore, Mumbai, and New Delhi is contributing to "Omaha-like turnover rates." During a recent trip to India, he found out that some call centers were reporting turnover rates as high as 70–80 percent.

Datamonitor reports speech-rec-enabled IVR is emerging as a competitor to offshoring. "The popular offshore call center markets, such as India and the Philippines, are rapidly maturing, resulting in increasing wages and higher turnover rates," says voice business analyst Daniel Hong. "This is likely to nullify labor arbitrage benefits and thus decrease the value proposition for businesses to open an offshore call center."

There are special and more prominent conditions, especially in Indian call centers, that are pushing up offshore turnover and crimping recruiting. Suresh Gupta explains that most U.S.-serving Indian agents work graveyard shifts. This is because when it is 3:45 p.m. in New York and 12:45 p.m. in Los Angeles, it is 2:15 a.m. the next day in New Delhi.

But those hours discourage women from working in call centers, inhibiting social life and causing family and relationship tensions, he says. Also, Indian agents are becoming discouraged when discovering little advancement or career growth in call centers. Only one in eight agents can hope to move up the ladder. Most Indian agents are college-educated, ambitious 20-somethings who were lured to call centers by high salaries.

These two factors have combined to knock call centers down from their status as quality jobs. Status is very important in India.

"It used to be that people were proud to say they worked in a call center," says Gupta. "Now they say they don't do 'call center work' but instead they do 'business process' or 'IT outsourcing work.' "

Long Ramp-Up Times

Locating in another country is more complex than opening up at home. These lead times can be lengthy. The Paaras Group's Gupta reports that staffing a 1,000-seat call center in India could take months. Also, in the case of developing countries, you need extra time (two to three weeks at a minimum) to train agents on customers' cultures and neutralize accents.

If English is not the agents' native tongues, they must be taught how to think in English as opposed to translating. Translation takes longer than thinking in the language. That can lead to delays, eating up costs, and to misunderstandings with Americans.

Legal Entanglements

Deity forbid that you get into a contract dispute, especially in developing nation. The courts can take decades to resolve a legal matter. And the home team almost always wins. You are the rich American, and you're there for the taking.

More Restrictive Labor Practices

Not every country has a labor market as free like in the U.S. There may be laws that tightly regulate hiring and firing practices. That goes for developed nations as well as developing countries like the Philippines.

If you are locating offshore or you are with a U.S.-based outsourcer, expect your workplace practices to come under government, labor, and media microscopes. Depending on the nation, they may not like what they see and jump all over you. That has happened in British Columbia, Canada; a scathing report on U.S.-based call centers appeared in *The Georgia Straight*, a well-respected alternative periodical.

Changes in Supervising Practices

Agents are handling American contacts and may be working for American firms. But supervision styles may not be the same as in America.

Many countries, especially in Asia, avoid the common pull-you-to-one-side-and-talk-to-you technique of coaching. Also Asian agents are not likely to contact supervisors if there are problems.

For example, ePerformax dropped critical feedback and instead relies on scorecards to rate agent performance. Coaches do not use emotion; instead they present the facts. The objective scoring is there for good reason. It is very difficult to fire employees in The Philippines. The agents and supervisors are also taught to speak up if they don't understand something.

"The Philippines' culture, like most Asian cultures, is not openly confrontational," explains Hartsaw. "Employees do not volunteer information. They would keep making the same mistakes rather than admit there is a problem. If you show annoyance or that you are upset, agents will not respond."

Questionable Infrastructure

Offshore developed countries, like Canada, Australia, New Zealand, the UK and much of Eastern Europe have or shortly will have roads, airports, voice, data, power, sewer, and water infrastructure that is equally to if not better than that available in the U.S. Their mass transit systems are generally superior. In these nations you can be reasonably assured that you and the employees will get access to the call centers and that they will function.

The same cannot be said for offshore developing countries. Except in the largest cities, you cannot depend on the infrastructure to the same extent.

Electricity is a good example. The availability and quality is often spotty. Repairs can take forever. The aluminum and copper wires that conduct the current are often stolen and sold for scrap. Consequently, most call centers that have been built in offshore developing nations like India rely on onsite generation, with the local power grid as a backup, instead of the other way around else.

Transportation is another issue. Most people in developing countries do not own cars, and the roads are terrible. That can cause problems in nations like India where because of the differences in time zones, most U.S.-serving call centers are busy at night. Employers have to pick up employees at their homes and drop them off.

Severe weather causing flooding can also create problems for call centers. Sometimes employees cannot get home. For that reason one outsourcer, ePerformax, has temporary dormitories at its Philippines call center.

On the other hand, given the severe traffic congestion, rotting roads, high pollution, financially starved mass transit systems, and sometimes-unstable power grids in the U.S., many offshore locations don't look so bad.

Public Health Risks

When the U.S. government tells you to roll up your sleeves before heading to a developing country, do so. Many of these nations have poor public health systems. Diseases such as hepatitis and yellow fever are still prevalent in many nations.

Another public health danger comes from motor vehicle accidents. Many countries' drivers have little or no respect for rules of the road. The State Department reports that Americans have been killed on Indian roads.

Yet with a sizable minority of Americans lacking any form of health insurance because it is not offered or affordable, the U.S. population is also at risk to the spread of illness. Many existing and potential employees are likely to forego medical treatment. Low wages and high bills force people to come to work when they shouldn't and in doing so infect others.

Attitudes Towards Women and Gays

Developed countries and some developing countries respect and treat women as equals. But many developing nations do not, despite the country's history of strong female leadership. Women are targets of male harassment, or worse. Gays are especially shunned and despised.

Yet some nearshore countries like Canada are more tolerant of women, especially single mothers and gays of both sexes, compared with the direction some fear the U.S. is going.

The American religious right is strongly opposed to gay marriage; some of the people in those ranks are also against single mothers, reported *New York Times* columnist Maureen Dowd on Nov. 4, 2004.

Disaster Risks and Response

Offshore countries can be hit by natural disasters, too. The Philippines is especially vulnerable to earthquakes, volcanic eruptions and typhoons.

But more important, the quality of emergency response can be spotty in the lowest-cost developing nations. They do not have the infrastructure or the resources to evacuate people quickly and effectively, protect property against looting, or to rebuild roads, telecommunications, and water and sewer infrastructure.

Security

Many offshore nations are not as safe as the U.S. Businesspeople, their families, and tourists can face increased dangers of assault, robbery, theft, rape, kidnapping, murder, and terrorism when overseas. Scams are common. Law enforcement authorities are poorly paid and corrupt. Armed militias and organized crime are not uncommon. Governments face threat of mutinies, uprisings and coups, including those of call center-friendly countries like the Philippines.

Americans are also terrorism targets. Groups affiliated with al-Qaeda have murdered U.S. citizens in the Philippines, a popular call center location. Bombs have exploded in Makati City, Manila's financial district and home to some call centers, and there have been disputes with the military.

But Americans are more secure in many nearshore nations than at home. Northern Irish liked to say, with some truth, that you were more secure in Belfast during the violence that occurred with The Troubles than you were in Boston.

The Associated Press reported that the U.S. had nearly 16,000 murders in 2001, excluding the 9/11 terrorist attacks victims. In contrast, Canada, whose population is about 11 percent that of the U.S., had just 554 murders in 2001, just 3 percent of the American 2001 total.

Staff Travel and Relocation

Whether you locate or outsource offshore, you will be committing your staff to additional travel and possibly to relocation. That has consequences both in the considerable costs entailed and in employee morale.

Travel and relocations are a dream if you are young and single. But they are nightmares when you are older and have families. The issues include culture shock, schooling, accommodating spouses, and security risks.

Foreign moves are no fun especially if you have teenagers. My family moved to the UK when I was 14. I went from jeans and T-shirts and Algebra I to itchy gray school uniforms and ties and Calculus. Less than a year later we were back in North America.

On the other hand, improving technology and growing acceptance of conferencing have made that option an increasingly viable substitute to travel. Nothing beats being there, but the costs, hassle, jetlag, productivity losses, personal issues, and security risks incurred in traveling have diminished the advantages.

BEING, NOT PARROTING!

To compensate for cultural differences, offshore call centers have gone to great lengths in teaching agents about American life, such as showing them American TV programs so they can carry on chit-chats about sports, weather, and celebrities. Many agents undergo accent neutralization and adopt American nicknames like Bill or Shirley.

But be warned, this is not the same as living in or near the U.S. Agents that don't know the cultural and day-to-day milieu of Americans are merely parroting them. If an American believes that they are talking to an American and then finds out they're talking to an Indian or Filipino, they're not going to be too thrilled about it, or about you.

Your foreign agents should be honest. Americans hate being deliberately fooled. You'll get more cooperation from customers if they know where the agents are. Being Americans, they'll be curious and want to know more about the agents, and they will likely be more tolerant of accent slips and other language mishaps.

Added Taxes

Many countries have higher income and other taxes than the U.S. Quite a few have value-added tax or VAT (called Goods and Services Tax or GST in some nations such as Australia and Canada), which is applied at each step where value is added to the product or service.

Do not let that discourage you from locating there. These taxes are paid by your employees not by your company. Also, they help finance health care and other benefits that in the U.S. would have had to come out of your pocket.

◑ THE CASE FOR NEARSHORING

Nearshoring potentially provides many of the same benefits as offshoring with less of the negatives.

Cultural Affinity

Residents of nearshore countries are going to know Americans better than those living in offshore nations. That makes for greater understanding by these agents of the day-to-day issues and realities faced by their customers.

Canadians are immersed in American culture. I get on my small city's cable service American TV stations from Boston to Bellingham, Washington. Americans visit Canada; Canadians visit the U.S. The most ardent Quebec separatists have always spoken positively about doing business with the U.S. Montreal is less than an hour from the U.S. border. My sister lives in Burlington, Vt., which is far closer to Montreal than to the nearest big U.S. city, Boston.

There is a similar cultural influence in northern Mexico, and perhaps more significantly, a continued rejuvenation of Hispanic culture into the U.S. from immigrants. There is a border tongue known as Spanglish, English words blended into Spanish.

The same goes for nearshore nations far from the U.S. geographically. Australians, the Irish, and New Zealanders and share a common language and cultural and legal ancestry with Americans.

The Irish have woven their vibrant green, white, and orange threads into America's cultural tapestry. Those Celtic-American strands have in turn brought together the warring factions in Northern Ireland to knot together what one hopes to be a lasting peace.

The more complex the service and sales issues the greater the importance cultural affinity plays in establishing and maintaining those relationships. If there is a strong understanding, then customers feel happy, and the call length is kept short. If there is weak understanding, customers are dissatisfied, which prompts repeated contacts that jack up costs.

Lowered Costs

Nearshoring enables call centers to slice 8–20 percent from their operating costs. The savings usually come from lower wages, currency differences, and government-subsidized health care in the case of developed countries.

In the case of Canada and northern Mexico, nearshoring saves on travel expenses compared to offshore sites, both in ticket prices and in lost productivity caused by being out of the office.

With many Canadian and Mexican cities located across or near the border from U.S. cities, you may be able to drive to your nearshore call center. Amtrak has a daily round trip from Seattle to Vancouver timed for business trips. Or if the center is located on Vancouver Island, where I live, you can take a quick, comfortable, convenient, and very scenic ferry ride.

Canada is a great nearshore location, but many of the best sites are a little out of the way, in former forestry and mining communities. The big benefit of having call centers there is the remoteness, which impedes an exodus to employers in higher-cost big cities. Fortunately great scheduled air service connects these cities with the major hubs. For example, Bearskin Airlines links northern Ontario communities to larger carrier connections at Winnipeg, Thunder Bay, Sudbury and Ottawa. Credit: Brendan B. Read

Management Style

Americans have a distinct and effective management style, both in contract and operations, a clear, efficient, results-oriented attitude. Americans are not afraid of confrontation. They are individualists. But other cultures are often not that way at all. They are paternalistic, collective, submissive, minimally confrontative, *but no less effective*.

There is more understanding of American style by nearshore managers than by their offshore counterparts. Where in some offshore nations "yes" means no, "yes" generally means yes nearshore. But be careful there. Listen for the subtleties in tone. "Yes" can mean yes, but....

Legal Challenges

Nearshore countries also generally have an understandable, fair, sound, and enforceable legal system, to settle civil disputes, such as contract laws. Canada, Ireland, Australia, and New Zealand have their legal and political roots in British common law, as does the U.S. There are often trade agreements between nearshore countries, such as NAFTA.

Development Standards

With the arguable exception of Mexico, nearshore nations are closer to the U.S. and to each other in living standards than to offshore countries. You can count on reliable electricity, sewer, water, and voice and data networks. You can also expect an honest and reasonably efficient civil service, quality public health, political stability, and reasonably safe streets.

The similarity in standards makes for a greater comfort level for employees assigned to work there, especially if it entails having their families with them, and for business travel to these places. Mexico and many offshore nations that have extremes between poverty and wealth can be very dangerous for Americans. They are at risk to kidnapping, robbery, and terrorism.

There is less cultural shock for most Americans in locating to Larne, Northern Ireland than there is to Legaspi City, the Philippines. Telephone calls are also cheaper. With the exceptions of Australia and New Zealand, nearshored Americans can travel home more easily and at less cost.

American women are accepted as equals to men in nearshore nations, with the exception of Mexico. There is also generally more tolerance for people of different ancestries, cultures, and sexual orientation than offshore, or in some cases onshore. For example, Canada is largely more accepting of gays and has less of a racial divide than the U.S.

It's not that these nations don't have their cultural issues. There remain strong differences between Protestants and Catholics in Northern Ireland, and disputes between the indigenous peoples in Australia, Canada, and New Zealand and the white communities. There is a vocal French-speaking minority in Quebec that still wants that province to separate from the rest of Canada.

Even so, nearshore nations surpass the U.S. in many aspects, like in public health. Nearly 40 percent of Americans have no health insurance, while the residents of these countries have some form of it.

Better Security

Nearshore nations share American concerns about terrorism and have very effective and well-equipped police and security services. The comparatively smaller differences in living standards and greater stability minimize terror threats.

British, Irish, and Canadian security services have been in the line of fire long before September 11, 2001. I lived and worked in England from 1973 to 1974 and from 1992 to 1993. On many occasions I've had to evacuate buildings, change commuting routes, and have been stuck on trains during security alerts.

Few Americans know that Canada declared martial law in October 1970 after the separatist group FLQ kidnapped British and Canadian officials and assassinated Quebec labor minister Pierre Laporte. Troops and tanks were sent into the streets of Montreal and Quebec City. Civil liberties were suspended; many people were detained without being charged.

Australia lived through its own horror with the Bali, Indonesia bombing on October 12, 2002, which claimed 202 lives, 88 of them Australian. The country is very much on its guard for potential terror activity, close as it is to nations with active terrorist cells.

A BizCosts® Analysis for Call Center Operations (1)

	Chicago IL Metro Area	Hartford CT Metro Area	Phoenix AZ Metro Area	Fort Lauderdale FL Metro Area
Nonexempt Labor Costs				
Weighted Average Hourly Earnings	$16.25	$15.90	$14.25	$13.85
Annual Base Payroll Costs	$6,188,000	$6,054,720	$5,426,400	$5,274,080
Fringe Benefits	$1,980,160	$1,937,510	$1,736,448	$1,687,706
Total Annual Labor Costs	$8,168,160	$7,992,230	$7,162,848	$6,961,786
Electric Power Costs	$46,440	$58,620	$44,400	$43,920
Office Rent Costs	$660,000	$600,000	$705,000	$630,000
Equipment Amortization Costs	$960,000	$960,000	$960,000	$960,000
Telecocmmunications Costs	$559,351	$552,863	$525,609	$569,120
Total Annual Geographically-Variable Operating Costs	$10,393,951	$10,163,713	$9,397,857	$9,164,826

(1) Costs are scaled to a representative 200-worker call center occupying 30,000 sq. ft. of suburban office space and having a monthly call volume of 2 million minutes of billable 800 service.

Source: BizCosts®, the proprietary data bank of The Boyd Company, Inc., Princeton, NJ.

Viable Foreign Markets

Nearshore nations, including Mexico, have or provide access to viable markets in their own right for American goods and services that call centers there can also support. Canada is the largest trading partner of the U.S., Mexico is rapidly growing, and Ireland and Northern Ireland are gateways to the European market.

It is almost always better to serve these foreign markets from foreign call centers than from within the U.S. You obtain access to agents who have cultural affinity with your customers and who can speak local languages other than English. Your costs will also be less.

✪ NEARSHORE CHALLENGES

Nearshore locations are not without their obstacles. For all their cultural, geographic, and political closeness, they are still foreign nations. Here is where those factors impact call centers.

Costs

Nearshore sites may not be significantly cheaper to justify the expense, management hassle, and learning curve involved. There the differences are slight, like between Canada

Spokane WA letro Area	Toronto ONT Metro Area	Wichita KS Metro Area	Montreal QUE Metro Area	Winnipeg MAN Metro Area
$12.85	$13.60	$11.25	$12.10	$10.70
$4,893,280	$5,178,880	$4,284,000	$4,607,680	$4,074,560
$1,565,850	$932,198	$1,370,880	$829,382	$733,421
$6,459,130	$6,111,078	$5,654,880	$5,437,062	$4,807,981
$41,400	$24,743	$31,762	$33,437	$41,796
$630,000	$348,600	$600,000	$311,250	$292,824
$960,000	$960,000	$960,000	$960,000	$960,000
$561,169	$555,458	$519,120	$555,458	$555,458
$8,651,699	**$7,999,879**	**$7,765,762**	**$7,297,207**	**$6,658,059**

and the U.S. Currency fluctuations play a large role in costs and value for money. A rise of the Canadian dollar in 2004 for example squelched many nearshoring projects.

Cultural Issues

Having your American customers answered nearshore is close to but not the same as having them answered in the U.S. That may make a difference in service, retention, and sales.

Foreign accents alone will turn them onto the fact that your agents are in another country. Bill Sims, which outsourced its incentives program to Canadian service bureau nTouch, had the outsourcer tone down the agents' dialects.

As immersed as they are in U.S. culture, nearshore agents do not live in it. Unless you've lived in America, you can't begin to relate or to understand Americans.

For example, it is not a good idea to nearshore a health-care program because Canadians and other nearshore citizens cannot relate to the U.S. medical system. They do not know the pain of medical bills, the hassle of changing providers, and choosing between HMOs and PPOs.

Another example: the initials I.R.A. have a very different meaning to Irish and Northern Irish call center agents. "Individual retirement account" is the *last* term that will come to their minds.

I won't explain the reference. If you don't know what I.R.A. stands for in this context then should do some cultural homework before *even thinking* of nearshoring there.

Site Selection

As close as you may be to Canadians, Mexicans, Irish, Australians, or New Zealanders, you are still setting up with countries with different cultures, laws, regulations, and practices. Unless you already have operations there, you have a learning curve to climb.

Yet despite being inundated by the U.S. media, Canada is most definitely a separate nation with a different form of government, two official languages, and a more socialistic, laid-back society. A government-managed health-care system and gun control are just two of the many facets of Canada that are as alien to Americans as Mars.

Matt Jackson, manager with Deloitte and Touche Fantus Consulting, found that cooperation between individual workers and management is not as common in Canada, even in nonunion firms, compared with the U.S.

"We interviewed a U.S.-owned call center operation and received feedback that it was having a difficult time in creating a culture whereby employees would share ideas, frustrations, and complaints with the existing management team," he says.

Consultant Bob Engel agrees. "Canadian call center agents identify with the working class, whereas American call center employees see themselves as white collar and identify with management," he points out.

Suitable Property

Attractive property may be difficult to find. Canada has a limited supply of real estate suitable for call centers, especially in small cities where there is high unemployment. New buildings are difficult to come by because call centers typically lease for no more than five to seven years; landlords and financiers need double that to break even on construction costs.

Canadians are very conservative; they rarely build on spec. But in at least once instance, in Peterborough, Ontario, the local governments underwrote a new building for Ameri-Credit's in-house call center. American developers are now building on-spec buildings near Toronto's Pearson Airport.

Intolerance

There is also a risk that your staff or customers may encounter racial, religious, and sexual intolerance from residents and employees in nearshore communities, like in the U.S. Australia passed a law that bans gay marriage; there are many Christian fundamentalists in that country. Southern Alberta, the Fraser Valley east of Vancouver, British Columbia, parts of Vancouver Island, and rural Ontario comprise Canada's "Bible Islands."

✪ ENABLING OFFSHORING AND NEARSHORING
Minimum Effective Program Sizes

There is a cost floor for nearshoring and offshoring. Martin Conboy, CEO of Australian consultancy callcentres.net, recommends looking for a minimum 30 percent cost edge over onshore programs to outweigh extra English-language skills training, remote project management, and reporting, travel, and security issues.

There are also minimum project sizes. For opening your own center, you can do it in Canada for 10–20 seats and 40 seats in other nearshore nations. But if you are planning to set up in a developing nation, you will need 300 seats or more to make opening your own center offshore worth your while.

Ron Cariola, senior vice president of real estate firm Equis warns that if you can't afford your own building or be the dominant tenant with a massive call center offshore, you could be facing a nightmare of problems. Many buildings are not constructed to proper standards for earthquakes and fires. They have washrooms on every other floor.

There may not be adequate voice, data, and power, and landlords can be notoriously difficult to deal with. In developing countries, the home team wins in any dispute.

"If you don't have the capital to build your own call center, or become the dominant tenant in a building then you should outsource," Cariola points out.

Matching Culture with Services

All cultures do not possess the same attributes in identical quantities. The Indian culture is aggressive and exact. That nation has succeeded as a location for help desks, sales, and collections call centers. Non-Indo African, Asian and Caribbean cultures are more empathetic, which is why countries like South Africa and the Philippines are succeeding for customer service. Countries like Barbados and Jamaica have long served American and foreign tourists.

Familiarity with Products and Services

If you want your agents to effectively service and sell your products or services, they should be familiar with them. Training is not enough; agents must relate to the users.

One example is bank accounts and credit cards, which Ron Cariola says Indians and others from developing countries do not use. He recommends low-cost developed countries like the Czech Republic, Hungary, and Poland, whose workers are educated and Westernized, provide good customer service, and are tenacious at sales and collections.

"These countries' workers understand Western business and consumer practices," Cariola says.

Quality

Agent quality varies by nation. Some countries have excellent education systems and job training, while others do not. South Africa's education levels are reportedly not as good as India's or the Philippines'. Barbados has a 98 percent literacy rate, one of the world's highest, and Jamaica's is 85 percent.

James Beatty, president of site selection consultancy NCS International, says that Jamaica's agent workforce consists of high school graduates, while India's agents have college educations. But Jamaica is training its agent workforces in IT for first-level support desk functions.

Existing Offices

The best way around site selection and cultural issues is to begin or locate in viable countries where you have a business presence. Your colleagues will know the lay of the land; they may have a call center already to serve those foreign markets.

Laws and Regulations

Whether you locate or outsource have your attorneys go over the laws and regulations in the host countries that could affect your program, especially contract, liability, insurance, and data privacy. For example, nearshore Canadian call centers handling U.S. customers may likely have to comply with Canada's privacy law.

Culturally Acclimation

Be aware that Americans are not universally loved, that there is both respect and resentment of Americans. Stash the ego and the flag and learn the ways of the countries where you want to locate. You'll find more doors being opened to you that way.

Site Locations

Thoroughly check out your locations. Don't assume what you're being told from the locals is correct. Work with professional site selectors who have experience in that country; get references from them and talk to the outfits they located.

If you set up in a developing country, make sure that the facility has multiple voice and data connections. Ask about transportation arrangements to make sure agents are able to get to work.

The Best Strategy for Your Organization

You may be able to deploy in-house or outsourced strategies such as home agents or distributed contact centers. Developed nearshore countries with near-universal phone and Internet connectivity and high computer ownership, such as Canada, Australia, New Zealand, and the UK, are best for home working.

OFFSHORE ADVICE

When examining whether to offshore, Dennis Smith, president of PacTac Advisors, urges that you look at the following, in order:

- Rank your most important considerations: cost, language skills, cultural affinity, business continuity, management control, and possibly serving foreign domestic markets.

- Determine your core and non-core competencies. Which of those are you most comfortable with handing off to outside contractors?

- Look at the transactions you want to handle in the U.S. and outside of the U.S. Which types are you most comfortable handling offshore? How complicated are the transactions? Is sensitive information involved?

- Will the operation be large enough to justify long-distance management?

"If a matter like servicing customers' enquiries is a core competency, yet it is not a large operation, then neither offshoring nor outsourcing might fit for you," advises Smith. "If you have significant volumes of non-core transactions, like first-level technical support or simple enquiries, then you need to consider going offshore, either on your own or by letting a specialist outsourcer go offshore for you."

Having your program spread out among several call centers in a country enables you to tap different labor markets, so if the costs climb in one, you can shift the volume to another. That strategy also protects you during disasters so your contacts will continue to be handled.

Locating or having contacts handled in smaller cities is another viable technique nearshore and offshore, even in India. The Paaras Group's Gupta says Indian call centers are getting away from 1,000- to 2,000-workstation centers in the huge cities like Bangalore, Mumbai, and New Delhi. Instead, they are opening 200- to 500-workstation centers in second-tier cities like Ahmedabad, Jodphur, and Chandigarh, where the bulk of the country's workforce is located.

"These smaller cities have excellent labor, infrastructure, lower costs of living and labor costs, and few of the transportation headaches that plague cities such as Bangalore and Mumbai," says Gupta.

Follow the Sun

You can run call centers in several countries and shift calls to these places in a practice known as "follow the sun." That enables you to open or shut call centers to match demand and volume. Customers do not even realize they are being routed elsewhere.

✪ OUTSOURCING-SPECIFIC FACTORS

There are unique factors to evaluate when outsourcing.

Onshoring Outsourcing Benefits
More choice, flexibility, and scalability

Outsourcing within the U.S. or your own country means you have a great selection of service bureaus to pick from and a greater choice in capabilities. You have much lower minimums compared with offshore. There are small, midsized, and large bureaus, many of which also have offshore centers.

If you are looking to outsource your telemessaging, stay onshore. The program size and the need for agents to completely relate to callers, with near-zero misunderstandings, usually precludes offshoring.

Management Ease

Checking out an outsourcer by a site visit, meetings, negotiations, training, and troubleshooting are relatively easy and much less expensive when the bureau's call centers are in the same country. In many cases you can drive to the offices or sites without the hassle of flying.

Staying onshore means working with bureaus that know and comply with the same laws and regulations as you comply with. Their attorneys and your attorneys speak the same language. There will not be any questions on issues like contract law and obeying statutes like the Telephone Consumer Protection Act and HIPAA. You also have ready recourse in case there is a dispute.

Onshore Outsourcing Disadvantages

Staying onshore has some downsides.

Higher Costs

Onshore is more expensive than offshore. The average rates run about $24–30 per hour compared to as low as $12–14 per hour in Africa, India, and the Philippines.

Lower Quality

The American education system isn't what it used to be. The high school diploma, for all intents and purposes, is meaningless. Sadly, all it means is that the graduates have survived what can best be described as a taxpayer-subsidized, pre-adult daycare and mating service. A majority of college students now take remedial English or math classes, reports the American Diploma Project.

✪ ENABLING ONSHORE OUTSOURCING

To make onshore outsourcing work, carefully figure out what you want outsourced. Here are the key conditions where onshore makes sense:

* Smaller programs, under 100 seats, should stay onshore, except for Canada, which can take them as low as 10 seats.
* The more complex, domestic-market-specific, and hands-on the training is, the more you should consider keeping it domestic. Examples include health care, insurance, real estate, and securities sales.
* High-end, highly skilled programs where there is a lot of interaction between customers and agents should stay onshore.
* Onshoring is preferred where cultural affinity, strong unions, public visibility, and politics can arise as issues, such as for governments, government contractors, non-profits, and utilities.
* Consider having your onshore program distributed among a bureau's multiple centers or choosing several vendors. You give the contract insurance against disasters and rising labor costs by having it distributed between centers.

American Customer Care has a financial services client who looked at offshore but decided not to pursue that avenue. Frank Fuhrman, ACC's vice president of marketing, reports that while these clients recognized that certain offshore centers have competent, educated, and well-trained agents, those employees could not relate to Americans.

ACC is also a very efficient firm. Its centers are rural communities with loyal labor forces and strong work ethics. Rather than putting money into flashy buildings, it has invested in technology and people.

"These clients believe that having their customers feel comfortable talking to an agent that they can relate to was more important to the success of their program than price alone," says Fuhrman. "They have found that what we offer them, when analyzed in conjunction with the risks of offshoring, is more than attractive enough to keep their work onshore."

✪ OFFSHORE OUTSOURCING

Offshore outsourcing means having agents working for a service bureau in another country handle your domestic customers. Offshore outsourcing is different from what

I call international outsourcing, which I explore in Chapter 10, where agents in foreign countries serve your customers in those nations. Here's why.

In offshoring:

* Agents must be trained to speak to, write to, understand, and relate to Americans. U.S. expectations and standards on call handling must be followed, such as having short hold times and giving direct answers to questions.
* Programs aimed at Americans must comply with U.S. laws and regulations.
* Voice and data networks must deliver calls and data with little or no delay.

In international outsourcing:

* Agents must relate to customers in those markets, including speaking the languages and in some cases the dialects. They must deliver service to their standards. Do not assume that what works in the U.S. works everywhere else.
* Programs must comply with the laws that cover the residents of the countries you are serving.
* Voice and data networks must meet the expectations of those markets.

Offshore outsourcing is much more intense than domestic or onshore outsourcing. All the benefits and challenges of contracting out your contact handling are greatly magnified when you leave the country.

Offshore Outsourcing Upsides
Cost Savings, Higher-Quality Agents

In outsourcing, rates roughly equal $24–30 per hour in the U.S., $21 per hour in Canada and Northern Ireland, $18 per hour in the Caribbean, Central and South America, Eastern Europe, Mexico and New Zealand, and about $12–14 per hour in Africa, India, and the Philippines. Every 150 seats moving offshore could save companies as much as $4 million a year, reports Jon Kaplan, president, TeleDevelopment Services.

Remember, these numbers can change. Also the higher up the quality chain, the more it will cost you.

Important note: The rates do not include setup, travel, and transition costs to offshore. When HP offshored a project to two locations, the move cost $100,000. Nor do these rates account for lost productivity expenses incurred by program setup, management, and troubleshooting.

Much Lower Turnover

Turnover is often much lower in offshore locations, which bolsters the bottom lines of outsourcers located there. Geri Gantman, senior partner at R.H. Oetting and Company, reports annual turnover rates are just 5–10 percent in India and the Philippines, compared with 50–100 percent in U.S.-outsourced call centers.

You can channel the savings into revenue-enhancing investments and activities. You might, for example, drill deep into outbound lists to convert more prospects to customers and increase sales, says Dale Saville, executive vice president of global outsourcer Sitel.

Testing Foreign Markets

Going offshore opens the door to serving customers in those nations (more about that in Chapter 10). That can help you get the most out of your outsourcing relationship. Who knows those markets better than the offshore bureaus?

Competitive Discipline

Offshore outsourcing, if done right, keeps your in-house call centers and your domestic bureaus on their toes. The outsourcing environment outside of the U.S. is very competitive; foreign outsourcers often offer better terms that you can get from your domestic partner.

Offshore Outsourcing Risks

End-Customer Resistance

As noted earlier, Americans may not like dealing with offshore agents; many have had poor experiences with them. That includes outsourcers. What may aggravate the public more is dealing with third parties because it adds more layers of corporate bureaucracy to go through to resolve any issues they have with products or services. The organization and the bureaus may engage in finger pointing that resolves little.

Data Security

U.S. laws such as HIPAA require that you safeguard American data no matter where it's handled. That also affects your outsourcers for which you are responsible. But it is more difficult to protect data if it leaves the country, especially to a developing offshore nation, which may not have the same safeguards or enforce them as readily like in the U.S.

The theft can be sophisticated as installing keyword sniffers that automatically look for names or keywords. Or it may be as simple as agents copying down what they see on screens.

Martin Conboy, CEO of North Sydney, Australia-based consultancy and information resource callcentres.net, is picking up anecdotal evidence that some Asian call center staff are being coerced by organized crime and industrial spies to hack into the foreign firms' computer systems.

"But that can happen anywhere," he says. "That includes one's home country. It is just another issue that organizations need to be aware of in their risk-balancing analysis when considering offshoring."

More Involved Vendor Management

You have five principal choices when you offshore outsource:

* U.S.-based bureaus with offshore call centers
* Foreign-based bureaus, with U.S. offices that have switching and routing stateside
* U.S.-based outsourced aggregators of networked, independently owned offshore centers
* Joint ventures (JVs) with offshore partner
* Foreign-based bureaus with no U.S. presence

All but the last of these options reduce the risks and raise the comfort factor, because you are dealing with large enough and reportedly reputable firms. For example, you can keep all your data stateside, which minimizes the possibility of data theft. These companies also have assets you can seize if you have to sue them for contract non-compliance.

In the case of outsourced aggregators, you are relying on the management firm to do the due diligence and ensure program consistency, which saves you from taking those steps. One U.S.-owned bureau, OverC, a subsidiary of call center hardware and software vendor Cincom, selects its offshore bureau partners based on its parent's worldwide experience selling and installing products.

A Datamonitor report, "Approaching the Indian Outsourcing Market," explains that OverC maintains control over key technologies. That includes predictive dialers, to better manage and control interactions between clients' customers and partners' agents.

On the other hand, selecting and working with an aggregator adds one more hurdle in an already-complicated option. The bureau handling your contacts is two steps removed from you, not just one as in outsourcing directly.

The last two options, joint venturing and working with foreign-only based vendors, are the trickiest options, and not for the faint of heart. For example, many emerging countries require foreign companies to enter into joint ventures or partner with local firms. Some subcontractors have more than one North American partner. For example, vCustomer has ICT, iStonish, and Hispanic Teleservices as clients.

JVs are risky. Many founder because of management conflicts between parent companies. But contracting with a JV may result in better performance than with a straight outsourcer, reports offshore consultancy Renodis (now Analysts International).

Foreign-based bureaus, especially in developing countries, may not meet your standards, or worse. During the Indian call center boom, many centers were just shells, without voice and data lines or workstations, reports Geri Gantman of R.H. Oetting, because they did not have all the licenses. Many bureaus lacked the experience to deliver quality programs.

In choosing JVs and foreign-only bureaus, you or your colleagues should go in person to assess these outsourcers and partners. You will have to bring yourselves up to speed quickly on cultural nuances, business practices, bureaucracies, laws, regulations, and political climates.

The benefit is that you can get some excellent deals for quality programs, especially in Canada and the UK. The downside is that you have to do much more due diligence on the ground in those nations.

Setup and Complexity

Outsourcing takes day-to-day program handling directly out of your hands, but you are still responsible for setup, overall management, and troubleshooting. You or your staff may have to travel to other countries, which can be time-consuming and productivity-killing. Getting to Kolkata is a much different proposition than going to Kansas City.

Cost and Benefit Fluctuations

The business case for offshore outsourcing changes depending on the spreads between the U.S. and foreign currencies, costs, and labor market conditions. For example, movement of call centers into Canada slowed down in 2004 because of the strengthening Canadian dollar. The difference was not worth the political fallout from offshoring, reported site selection experts.

Difficulty with Small Programs

Offshoring is a bigger-project play, with minimums beginning at 50 to 100 seats, one shift for a 1-year renewable contract for countries such as India, the Philippines, Caribbean, Latin America, Africa, and Eastern Europe. To go smaller is not worth the time, expense, or commitment. But if you're willing to go to Canada or Northern Ireland, which offer much smaller savings, outsourcers there can support smaller projects.

❂ ENABLING OFFSHORE OUTSOURCING

You can make offshore outsourcing work.

Match Programs with Locations

The general rule is that the more involved your customers must get with your agents, the closer to home that program should be. This also applies to your in-house or home agent call center onshore. It is especially true for high-end, B2B projects where you can't afford to alienate customers.

Take a careful look at each nation's cost and cultural factors before offshore outsourcing, just like you would when setting up your own call center. Check out Chapter 10, which looks at them in detail. For example, Indians are very smart, quick, and aggressive, but they often lack empathy to the degree Americans expect. Some Latin American countries' dialects are too fast for a majority of American Hispanic ears.

There are many low-touch, low- and high-value programs alike that can be offshored. They include basic customer service, first-level support, accounts receivable, collections, order taking, outbound customer acquisition, lead generation, and lead qualification.

Vikram Talwar, CEO of Exlservice Holdings Inc., an outsourcer with an office in New York City and operations in Noida, India, told *Computerworld* magazine that although almost any service that can be delivered remotely can be moved to India, Indian companies advise caution when deciding which aspects to outsource.

"My advice to clients is that if you have never outsourced offshore before, start with your back office rather than the call center," Talwar said. "When you give out call center work, you are putting the service provider at the India end in direct contact with the ultimate customer, and you carry a far greater risk of your business being impacted."

Suresh Gupta, founder of The Paaras Group, recommends against offshoring highly sensitive work, such as serving high-end customers. Companies can't afford to lose top-quality customers through any misunderstandings.

You should look at having your high-level work offshored only after you have gained several years' experience outsourcing and offshoring less-sensitive work and have a sufficient business case for doing so.

"There is no compelling reason to move your higher-grade work offshore," says Gupta. "There is not the same cost savings as there is with high-volume commodity calls. Also, the expenses are higher because you need to spend more money selecting the very best agents and training them more rigorously."

The closer to home the outsourcer's call centers are, the more you can offshore. Canadians and northern Mexicans understand American culture and nuances much more than Indians, Filipinos, Africans, or the Irish. Canadian and Mexican call centers are easier to get to than those in other countries; NAFTA and other longstanding trade relations provide a strong comfort level.

American companies have long had business presences in Canada. Canada was the first country to attract offshore U.S.-serving call centers back in the early 1990s, led by the efforts of former province of New Brunswick premier Frank McKenna.

Crawl Before You Walk

Outsourcing your call center, like any contracting, is a business process. It represents a binding agreement between two parties. There are a lot of factors entailed when you outsource, including terms, performance, training, management, and reporting. You legally can't do with outsourcers what you can with your own employees.

As explained earlier, offshore outsourcing magnifies all these issues and complicates them further by the fact that the work is being carried out in foreign countries with very different cultures and laws. You don't have the same access or leverage as you do with onshore outsourcing.

If you haven't outsourced, before do so at home first. Get to know the process. Be used to having someone else manage your customers or employees.

The Nearshoring Step

When you're comfortable with domestic outsourcing, have your program handled nearshore. You'll get an exposure to issues like cultural affinity, access, and laws without going out of your depth.

Nearshore contact centers and bureaus offer slightly lower costs than far offshore vendors: $18–21 per hour compared with $24–30 per hour with the U.S. and $12–14 with Africa, India, and the Philippines. But by relying on bureau agents that better relate to Americans, you may get higher productivity, better customer retention, and possibly lower total costs through shorter conversations and fewer escalations to your staff.

You can count on nearshore foreign-based outsourcers as you would with U.S.-based firms. These companies are very sophisticated, sometimes more so than their U.S. counterparts. One of the world's largest service bureaus with operations throughout the U.S., Teleperformance, is headquartered in France.

Offshoring Based on Customer Type

You can get the best of both worlds by employing a customer-management-like strategy to outsourced offshoring. For example, you can have an offshore outsourcer handle contacts from lesser-valued or lower-volume customers. If their needs can't be met, escalate them to your onshore in-house call center.

But be careful. You might end up losing those customers if they felt the service they received offshore was terrible.

Don't Fake the Location

Too many offshore outsourcers, especially Indians, think the only way to build cultural affinity is to fake the names and accents, to get Americans to think they are speaking to other Americans.

Wrong. People hate being deceived, no matter who they are. There's no quicker way to annoy and lose a customer than tricking them.

It is one thing to train agents to tone down an accent, so that others can understand them. But it is another to outright fake where they are from.

"If the offshore agents come off as being fake, you violated callers' trusts," Bryan Mekechuk, partner at Pacific Crest Consulting Group, had told *Call Center Magazine*. "If that happens, you risk losing business and make it more difficult to continue or expand offshoring."

Distributing Contacts Within Country

If your offshore bureau partner has more than one call center in the same nation, ask to see if they are willing to split contacts between them and how much additional it will cost. That way you protect your program from rising costs by having options in case one community gets too expensive. You also give it some disaster protection; chances are such events will probably not destroy both.

Outsourcing Back OfficeWork

If you are planning to offshore your contact handling, consider having the bureau's agents take care of other back office work as well. That includes accounting, payroll processing, and IVR transcribing. Also, if you decide to offshore outsource from another country like the UK, have the same center handle the work.

In either case you may get better volume deals. That's because, especially in India, you spread the work around the clock.

Most Indian agents on U.S. contracts work at night, because it is daytime in the U.S. And like in the U.S., night shifts are not popular in part because it disrupts agents' lives, especially socialization. But India is closer to the UK time zone. By offering agents business process or UK work during coveted day shifts, outsourcers can help retain them. If the bureau is more efficient, you're more efficient.

Don't Underestimate the Complexity

Be prepared to put management and IT plus some travel time into outsourcing. Just because the offshore outsourcer is savvy and sophisticated and speaks your language, that doesn't automatically mean that your program will be delivered seamlessly.

R.H. Oetting's Geri Gantman points to Lehman Brothers, which pulled back its outsourced internal help desk from India.

"Lehman underestimated the complexity of internal help desk calls, and the training and process documentation involved," says Gantman.

Double Your Due Diligence

When you outsource, you need to perform due diligence on your prospective vendors. But when you go offshore, do the same on the countries you want to handle your program.

Gupta explains that it takes several years to fully understand another nation, especially one that is as culturally complex as India.

"Just as you outsource in the U.S., you must invest in evaluating prospective vendors' experience, capabilities and customer feedback," Gupta points out. "It's doubly important in India since the call center market is immature, fragmented, and full of inexperienced vendors expecting to make a quick buck."

Unique Legal and Regulatory Issues

Countries have different laws and practices that will affect you even if you are outsourcing. That is especially true if you are dealing with foreign-based bureaus with no legal entity in the U.S. For example, if your staff is training a bureau's agents in India, often the bureau will want you to undertake the liability for insuring your staff.

R.H. Oetting's Gantman recommends that companies outsourcing to offshore bureaus get legal advice. In all cases, you should insert the master contract into RFPs that you send to outsourcers. And you must ensure that prospective vendors note exceptions in the contract.

"You never want to be in court in an emerging country because it is very difficult for an American company to win," she points out.

When outsourcing to an emerging country, see if the outsourcer's primary or backup switch lies in the U.S., Gantman also advises. While telecom networks in other nations are improving, they may not be on par with one back home.

Different Managerial Styles

You must also be prepared to work with different managerial styles. Gantman says many other cultures do not handle conflict and rejection easily. Indians and Japanese, for example, prefer to say "yes" when they mean otherwise.

Data Security

Take steps to protect privacy, in compliance with HIPAA and other legislation. Keep all data onshore, in a secure server such as a Citrix server for U.S. or for foreign-based outsourcers. Agents see only what is on their screen; they cannot download any information. Also, deploy secure transmission means such as encryption through virtual private networks.

You should record and archive calls onshore. You and your outsource partners need to consider disabling removable storage media, such as disk drives and CD burners.

Further, don't allow people to take notes, but if it is necessary as part of their jobs, provide shredders. Restrict access such as with turnstiles to prevent unauthorized visitors tailgating. Top-tier outsourcers take these measures, but lower-tier firms may not.

"No firm should compromise security, whether they outsource or locate offshore," stresses Mekechuk.

Measure the Results

Make sure you can measure the impact of offshoring through customer satisfaction surveys. If a program is working in a particular location with that outsourcer, capitalize on it by moving more contacts to it. HP recommends that you do so at the pace of improved performance, including customer response. It says you can transition a program in as little as few months.

Check with Your Attorneys

Offshore outsourcing, especially to foreign-based firms, can get tricky both on contracts and with your domestic operations. There may be legislation in place that could prohibit it.

Also, if offshoring means you are closing down an onshore center, you need to see if there isn't some nasty clause requiring you to refund or keep your facility going for a certain number of years as a condition of the incentives. With the labor cost savings in going offshore, the penalties will likely be worth it.

✪ BLENDSHORING

Just as you've gotten used to the notion of offshoring, there is a more conservative middle ground—blendshoring. This is a mix of onshore, nearshore, and offshore locations that balance cultural affinity, disaster response, currency risks, labor costs, and labor quality.

That way if a typhoon or political turmoil hits one nation or the costs climb astronomically, your calls can be rerouted to centers in others. Or if you needed more Spanish-speaking agents, you can ask your Mexican call center to add more workstations and direct more calls there.

Also, have some of the calls handled in an onshore call centers or by home agents. You can track and compare the performance of your offshore and onshore locations.

To determine the best strategy and how best to implement it, work with a consultant. Outsourcer LiveBridge, which has onshore, nearshore, and offshore centers, has a professional services division. The unit offers assistance examining operational efficiencies, technologies, vendor selection, vendor management, site selection, site design, facility management, and technology hosting.

Off-Site Data Protection Strategies

Companies can protect outsourced data with many of the same tools that they deploy at the call centers. They can stipulate these measures in outsourcer contracts and home working policies.

Call centers can supply home workers with storageless PCs, require locked cabinets and rooms, and have agents sign documents indicating their responsibility for data protection. For example, Procter & Gamble's telework agent policy says others cannot see customers' data.

Consultants and outsourcers with home working agents argue that there is less risk of data theft with such agents compared with their comrades working at conventional premises call centers. No one else other than the home-working agent is using that same computer or has access to the same files; the home worker controls access to it.

Most of outsourcer ARO's agents work from home. The firm does not allow data to reside at agents' desktops; no data is written down, reports CEO Michael Amigoni. It also inspects prospective home workers before hiring them.

SELF-SERVICE VERSUS OUTSOURCING

Check to see if you are better off having all or some of your programs handled by IVR or web self-service, whether provided by you or by your outsourcers.

The self-service option is particularly relevent for high-volume, minimal-touch programs like dealer locators, order taking, and first-level customer service. No matter where you go, in-house or outsourced, self-service will always be less expensive.

The research firm Datamonitor brought these points home in a study released in late 2004, "Voice Business in Regional Perspective: The Americas." It says a call center in an offshore location saves a U.S. company approximately 25–35 percent per transaction. But a call serviced through speech automation costs approximately 15–25 percent of the cost of a call handled by an agent in India.

"The popular offshore call center markets, such as India and the Philippines, are rapidly maturing, resulting in increasing wages and higher turnover rates," says voice business analyst Daniel Hong. "This is likely to nullify labor arbitrage benefits and thus decrease the value proposition for businesses to open an offshore call center."

The rationale for having live agents is having reassuring and helpful humans on the calls. But that benefit is diminished and quality could go down the drain if the outsourcers are able to attract only low-skilled, minimally educated, and poorly-motivated "gumsmacker" agents for their domestic call centers. With offshore labor supplies starting to shrink, turnover increases, and as offshoring moves up the skill and pay food chain, expect those bureaus to start resorting to lower-quality agents too.

The reason for having outsourced offshoring disappears if the offshore agents cannot display empathy, are brusque, or otherwise show no cultural affinity to callers or called parties. Offshore firms have spent a lot of money on training agents to understand and think like Americans, but those acts can only go on for so long before the audiences see the phoniness.

And under pressure, the masks fall off. They almost always resorts to their native dialect, language, or culture when they are annoyed or frustrated, like my wife's late Austrian mother speaking German and my talking broad Lancashire English.

Besides, with these kinds of calls, all the agents are doing are following scripts, with no deviations, pulling information from knowledge bases. Where's the human intelligence in that? Customers may be better served by talking directly to the machines, which are always available, inevitably polite, and cost pennies per interaction.

If the machines can't help the people, then escalate them to live agents who are empowered to help them, smart, mature, highly motivated, and affordable. With an aging population, one of your best bets is to find them at home, hired, trained, arranged, and managed by outsourcers.

The home working workforce is older and of higher caliber than those typically found in call centers, he points out. "They are not looking to steal data for personal gain," argues Amigoni.

Call centers can allow outsourcers to log into and out of their computers for individual transactions rather than shipping data in bulk. Even so there are risks.

OFF-PREMISES DATA PROTECTION

Data theft risks increase when the data leaves the call center, either to an outsourcer or to in-house agents working from home. Companies have less control when agents working from home handle the data, because there is no one on site to make sure rules and regulations are complied with.

With outsourcing, the data goes into third-party hands. Companies that are outsourcing have no direct control over who sees it. They must rely on the bureaus to safeguard the information.

The risks jump exponentially when the data is exported from the U.S. to in-house and outsourced call centers.

Agents are beyond American law. The centers may be in developing countries with high crime and corruption.

"An agent working in India will see one transaction at a time but after a year will see the whole database," said Jerry Brady, chief technology officer of Guardent (now Verisign).

Service bureaus interviewed say they already take strict data protection steps at their domestic and offshore centers that have met clients' stringent requirements. Convergys hosts its data at several stateside servers; voice traffic is packetized, and voice and data are encrypted through IP tunneling across their own private network.

Agents at its domestic and international U.S.-serving locations alike must log on with codes and enter their names and locations into the servers when they sit down at their workstations. They must also log out when they leave, such as when on a break. Agents' data log-outs are monitored with voice log-outs to watch for agents misusing computers.

Convergys also monitors all emails and web activities. Agents are not allowed to bring in objects, like pens, pencils, paper, phones and PDAs onto the call floor; they must store them in lockers.

Conversely, none of the tools, such as technical manual printouts, leave the floor; agents must put them in locked drawers.

The outsourcer has security cameras aimed at agents to ensure compliance. Convergys conducts thorough background checks. It checks police reports, passports and visas, verifies residences, and may talk to neighbors. The bureau may request previous employment letters and certify academic qualifications.

"We're aware of the risks in outsourcing and offshoring, and so are our financial services clients," says Dennis Ross, general manager offshore operations. "We have passed many client-specific audits; we have regular third-party audits. We strive to give clients' data the same protection worldwide as they do on their own premises."

Chapter 9:
Serving Hispanics

There are nearly 40 million Hispanics in the U.S. and growing. Yet unlike most other ethnic groups, Hispanics are retaining their culture and language. That makes serving them of special importance, requiring different strategies and methods compared with handling non-Hispanic Americans.

Heather Woodward, product manager with FGI Research, a market research firm, points to a Yankelovich Partners study which states that Spanish dominated 70 percent of households surveyed in 2002 compared with 64 percent in 2000. Households with English dominance dropped to 13 percent in 2002 from 16 percent in 2000. Woodward cites continuing immigration coupled with growing media, marketing, and services in Spanish.

Just over 40 percent, or 15 million, Hispanics are foreign-born, reports the Census Bureau; 52.1 percent entered the U.S. between 1990 and 2002 compared with 25.6 percent in the 1980s with the remaining 22.3 percent having arrived before 1980. There are over 200 million Spanish speakers in Central and South America, 104 million in Mexico alone.

"Previous immigrants have had to assimilate to obtain work and raise their families," says Woodward. "But that is not happening with Hispanics."

✪ CHARACTERISTICS AND ISSUES

There are specific characteristics and issues entailed in serving Spanish-speaking Americans that affect call center planning, operations, and costs.

Longer Calls

You may need to add handling time, typically 30 to 60 seconds on a four-minute call, compared with handling non-Hispanic calls. That means hiring additional agents. Vince Romao, director of call center services with Lexicon Marketing, explains that the Spanish language is not as precise as English, which means it takes longer to explain concepts and ideas.

Also in Hispanic households, especially households with recent immigrants, other family members may be involved with the decision-making. That may also lengthen calls.

That experience varies by organization and environment. HIP of New York, the largest HMO in the New York City area, found little difference in handle time between Spanish-speaking and English-speaking customers.

Need for Bilingual Agents

Bilingual agents' first language is usually Spanish, but they can also handle contacts in English. Learned or secondary Spanish speakers are acceptable if they are exceptionally fluent in Spanish.

The principal reason you will require bilingual agents will likely be that you will not have sufficient volume to merit employing agent who speak only Spanish, explains Miguel Ramos, vice president of strategic planning for outsourcer Precision Response Corporation (PRC). The risk is of having idle Spanish-speaking agents while your English-speaking queue backs up.

Also, many Hispanics will switch between Spanish and English. There are also English words that are tough to translate into Spanish, such as web and email (no short version; the Spanish version is electronic mail), fax, software, download, and layover. If your Spanish-speaking agents have poor command of the English language, they will be lost.

"When calling into an American company, many Hispanics may begin a call speaking in English, like my mother," says PRC vice president of marketing Alicia Miyares. "They will switch to Spanish as soon as the agent detects the accents and asks them if they prefer to speak in Spanish."

If there are family members in on a buying decision, some may not know Spanish.

"You may have situations where the customer is speaking Spanish but may switch into English, or they may solicit feedback in English from family members who know only English," says Kit Cooper, president of outsourcer Hispanic Teleservices. "Being fully bilingual enables agents to better obtain first-call resolution."

He also adds that most American firms have not localized their knowledge bases. Agents at minimum must master written English.

You need bilingual agents for outbound if the customers have not expressed a language preference.

"Just because someone has a Spanish last name, it doesn't mean they are Spanish-speaking or that they prefer to converse in Spanish," Ricky Arriola, CEO of outsourcer Inktel Direct, points out.

Hispanic Subcultures in the U.S.

Hispanic culture is not monolithic. Over 66 percent of U.S. Hispanics are Mexican in origin, followed by Central and South American (14.3 percent), Puerto Rican (8.6 percent), and Cuban (3.7 percent), reports the Census Bureau.

Mexicans are more reserved than other Hispanics in outbound cold calling, according to Cesar Romo, site director of outsourcer Telvista. But Cuban, Dominican, and Puerto Rican customers expect aggressive selling.

John Ahlman, marketing consultant with outsourcer Entel, says Mexicans place high value on families and relationships with companies and people. But Cubans focus more on understanding product or service details and substantiating benefits.

And like non-Hispanics, these national and regional groups have their own dialects. Inktel Direct's Ricky Arriola says a "lapiz" is a pencil almost everywhere, but a "lapicero" is a mechanical pencil in other places and a ballpoint pen in others.

Companies need not market to specific cultures in the U.S. And it is impossible to discern by area code or exchange the culture of a Spanish-speaking customer.

"We have agents from nearly every Latin American country, and we get along fine," says Romao. "So I don't give much credence to cultural issues between Latinos in the US."

Even so some outsourcers report that they have Northeastern clients locating their Spanish-speaking programs at contact centers near the Mexican border rather than those in or closer to their cities or in Florida. The reason: Mexican Spanish is slower-paced, neutral and more understandable than the Caribbean (Cuban, Dominican, Puerto Rican) Spanish spoken and predominant in the big Eastern metro areas and in the southeast.

Dialects and cultural issues are more of an issue when serving foreign Latin American customers rather than Latin Americans living in the U.S.

"A Central American customer living in the U.S. would be pleased to get any university-educated, Spanish-speaking agent with native language-speaking skills," Hispanic Teleservices's Cooper says. "But if that customer was calling from a Latin American country, they would have a strong expectation to speak to an agent with an in-country dialect."

Geographic Spread

Hispanics and Spanish-speakers are not as of yet spread throughout the country. The population has been concentrated in certain regions: South Florida, the Midwest, northeastern metro areas, Texas, New Mexico, Arizona, southern and central California, and central Washington state. But they are moving into other parts of the country, such western Washington state.

Access to Home Cultures

Part of the reason Spanish remains so vibrant is the proximity of the home countries and territories, aided by cheap flights, phone calls, the web, and media. With Mexico on the U.S. doorstep, entire border cities and regions become amalgams of both nations.

Site Selection Options

Hispanics are overall no different than other Americans when it comes to utilizing options such as IVR and web self-service. While the recent immigrants may not be used to these tools, their offspring become quickly used to them.

IVR systems with speech recognition have to be trained on dialects. For example, Maxxar's Natural Language Speech Recognition system adjusts to colloquialisms and dialects by storing commonly occurring regional variations and incorporating them into its recognition libraries. The system develops a library of synonyms. For example, for the word *si* (yes), the synonyms include *por su puesto*, *claro*, and *OK*.

Home agents are also a viable strategy. Home agent contractor Willow CSN has about 400 bilingual agents, or 40 percent of its workforce. The outsourcer also markets to clients in Spanish through its web site.

✪ OUTSOURCING COMPARED WITH IN-HOUSE CENTERS

There is a strong case to be made for outsourcing your Spanish-speaker-serving call center functions and programs. The key points are:

* You may not have enough demand for Spanish-speakers to justify hiring a team of agents, but an outsourcer often will.
* Outsourcers can more readily access lower-cost locations with plenty of affordable Spanish and bilingual speakers within the U.S. and offshore.
* Outsourcers who have Spanish-language programs have the language screening, testing, and training tools already in place.

On the other hand, by outsourcing your Spanish-speaking customer or client handling to a service bureau, you are giving up direct access to that market. That means you will miss out on some vital intelligence about that fast-growing consumer and business sector.

Hiring or Outsourcing Bilingual Agents

You have the option of hiring bilingual agents directly, outsourcing your program to a service bureau with bilingual agents at either U.S. or offshore call centers, and contracting with an over-the-phone interpreting (OPI) agency.

Bilingual Advantages

Having bilingual agents provides the highest touch and seamless service. It enables you to treat equally customers who converse in both Spanish and English without interruption in the conversation flow. OPI can take 20–50 percent longer to complete than an English-to-English call. The better trained the interpreter, the shorter the call.

Bilingual Downsides

Having bilingual agents requires you or your outsourcer to make the investment in time and money to recruit, screen, train, and supervise agents in their own languages. That may be difficult. Your call centers may not be located in communities with enough bilingual workers; you may have to pay premiums for them.

OPI Advantages

You also need to examine whether you have sufficient Spanish-speaking call volume to justify having an in-house or outsourced bilingual team. In contrast, OPI services are scalable; whether you have a few or a few thousand calls, those vendors can ramp the number of interpreters up or down as needed. You can hire OPI firms for call spikes and after hours. Many companies start with OPI before hiring bilingual agents or beginning a bilingual program.

Interpreters can often better understand what customers are seeking and cut down on call time by communicating on the same level as the customer. They are most often U.S. residents who are immersed in American culture and can speak fluent English.

Linda Parker, marketing director with Tele-Interpreters, reports that some of their clients dropped bilingual agents after testing with their firm.

"Trained and experienced interpreters recognize the dialects and idioms of different Hispanic cultures more readily than bilingual agents who know their own dialects and idioms," she says. "That also reduces the risk of miscommunication between customers and companies."

OPI Downsides

OPI can cost more than your own or outsourced agents for large-volume projects. The gap between answering the call and putting on an OPI can be annoying. The practice can be seen by some people as patronizing, treating them like second class citizens.

✪ ONSHORE AND OFFSHORE LOCATIONS

If you outsource or decide to open a new call center specifically to serve the U.S. Spanish-speaking market, you will need to balance cost, language skills, education, cultural affinity, security risks, and travel hassles.

Onshore Advantage

The principal advantage of locating/outsourcing within the U.S. is tapping into bilingual agents who are immersed in the American culture. The U.S. includes Puerto Rico, an American commonwealth. Also, there is the security and peace of mind being in the U.S., plus ease of access to your centers.

The traditional site for Hispanic-serving call centers has been in Florida, southern Texas, the border regions across the Southwest, central California, and Washington State. Puerto Rico is viable in that the wages tend to be at the lower end of the U.S. scale.

But with the growth and spread of the Hispanic population across the U.S., the number of potential sites and Spanish-speakers is increasing.

For example, outsourcer LiveBridge launched its Spanish-language division in November 2002 primarily from its Olympia, Wash. call center and secondarily its Portland center. CEO Patrick Hanlin reports that his firm is hiring second- and third-generation Hispanics who have retained their Spanish along with new arrivals. The outsourcer screens agents on their Spanish and English skills.

"The Hispanic population had traditionally been found in the rural crop-growing areas like central Washington," says Hanlin. "But we've seen a huge influx of Hispanics into Olympia and Portland."

Dialect differences between Hispanics are fading—though still present—as criteria for U.S. site selection just as they have for English-speaking call centers. The reasons: mass media, the spread of the Hispanic population, and the intermingling of groups and individuals.

Onshore Downsides

The U.S. is the highest-cost location for serving Spanish-speaking Americans, followed closely by Costa Rica. Some call centers find that they have to pay differentials to attract and keep Spanish speakers. Outsourcers especially are now migrating many of their contacts to offshore sites.

Offshore Benefits

Offshore locations offer much more affordable labor, principally Mexico, Dominican Republic, El Salvador, Panama, Argentina, and Chile. Their labor is 20–50 percent cheaper than that of the U.S. You also have, in the case of Mexico, Argentina, and, Chile access to large and increasingly affluent foreign markets.

There are competing advantages to the different offshore sites. Trammell Crow examined Cordoba City, Argentina, which has six major universities. Cordoba's labor costs are only $2.80 per hour, compared with Buenos Aires at $3 per hour, Mexico City at $4.25 per hour, Costa Rica at $5.25 per hour, and Chile at $5.60 per hour.

Chile has one of Latin America's highest literacy rates, exceeding Mexico and the Dominican Republic, argues Entel. Chile is also one of the most stable Latin American countries. Panama has a bilingual workforce, courtesy of the U.S. past ownership of the Panama Canal. Spherion and Dell have opened call centers there.

HTC's Cooper says his firm chose Monterrey, in northern Mexico, because it offers a highly educated and bilingual workforce. He estimates that over 90 percent of HTC's agents are bicultural. They have spent time in the U.S. and understand the nuances of the U.S. Hispanic market.

Mexican border cities are inundated with American culture; the workforce there is bilingual out of necessity. They have attracted outsourcers such as Telvista, which has a call center in Tijuana.

"Our Tijuana agents grew up next door to the U.S., know the holidays, which malls to shop in, and what's on TV," says Romo. "They adapt to English. But Mexicans living farther away from the border, like in Mexico City and Monterrey, do not."

Some outsourcers offer the choice of American border city and offshore locations. ICT Group has call centers in Nogales, Ariz., and in Mexico City. The Mexico City center also serves Mexico's growing domestic market.

Offshore Challenges

The major challenges outsourcing offshore with Hispanic Americans are a lack of fluent English speakers, followed cultural affinity to Americans, Hispanic or non-Hispanic.

The same countries that make great offshoring locales for English-speakers, such as India, the Philippines and South Africa, generally do not work for Spanish speakers, and vice-versa. Canada has a limited number of Spanish-speaking residents, mostly from Central and South America, living principally in Toronto and Vancouver.

There are limits to the labor market depth in many of these countries, which means there is risk of saturation among likely call center workers. Only Mexico, Argentina, and Chile have populations exceeding 10 million; Mexico's is above 100 million. But the number of English speakers will be much lower. Panama is the exception, thanks to the years of U.S. ownership of the Panama Canal.

There are also security risks. Many Latin American countries are notorious for their corruption and crime, including kidnapping and robbery.

For example, faced with a choice of being in a Mexican or American border town, King White, vice president of Trammell Crow, recommends the U.S. side, even at higher costs.

"On the U.S. side you have better access, security, no corruption, and better English speakers," he says.

CLIENTLOGIC'S EXPERIENCE

ClientLogic's New Latin American Center
To meet U.S. Hispanic and Latin American markets for clients, outsourcer ClientLogic opened a 150-workstation center in Monterrey, Mexico in September 2003.

The Mexican call center shaves costs by 20–25 percent compared to handling Spanish-language calls in the U.S. However, ClientLogic will keep the option of having American Spanish speakers on the calls at its U.S. centers.

Amit Shankardass, solutions planning officer with ClientLogic, said the company had selected Monterrey because it has well-educated bilingual labor and top-quality real estate, infrastructure, and air access. But it's not the lowest-cost location it could have picked.

"Many of our clients are reluctant to go to even lower-cost countries like Costa Rica because they are unfamiliar with them," says Shankardass.

The outsourcer teaches Mexican agents American culture and soft skills like empathy. It customized some of its training software in Spanish. Some knowledge bases remain in English, requiring bilingual agents.

To handle clients' customers in other Latin American countries, ClientLogic does more accent neutralization than with U.S. calls. The reason is agents are serving different countries with their own slang.

○ SELECTING SPANISH-SPEAKING AGENTS

You will need to select and screen agents for their fluency in English and Spanish. Generally, native Spanish-language speakers are better than learned-Spanish speakers because they know and understand most of the nuances. However, learned English is also fine.

Because someone is Hispanic doesn't mean they speak Spanish fluently. PRC tests bilingual applicants for their ability to think in Spanish and in English as opposed to translating words, explains Miyares. Thinking the language is more fluid and quicker; there aren't any annoying pauses as the person translates.

There are assessment tools such as Employment Technologies' eSkills multimedia simulation software that tests agents on call center skills in both languages. For example, the data entry section assesses an applicant's ability to listen in Spanish yet type in English quickly and accurately. The email composition section requires applicants to review and respond to customer emails in English and in Spanish. The typing section requires applicants to retype business correspondence first in English and then in Spanish.

"When someone hires a bilingual agent, they have an agent who already speaks Spanish interview the applicant to determine if they are proficient in both languages," says Joe LaTorre, director of product development with Employment Technologies. "That informal method is subjective and does not assess if the applicant can listen and comprehend in one language while typing in another."

You will need to decide how best to cope with accents and pronunciation. Inktel Direct's Arriola says Spaniards typically pronounce the letter "z" and the letter "c" like the "th" in "thin," while many Latin Americans pronounce the letter "z" the same as the letter "s."

PUERTO RICO AS A CALL CENTER LOCATION?

Puerto Rico is an interesting site to locate or outsource your Hispanic-speaking call center to. It is a U.S. commonwealth, which means it is neither a state nor a country.

Businesses that set up there fall under U.S. jurisdiction. Laws like minimum wage apply there, but the average pay leans more toward that bottom end of the scale than inside the U.S. Taxes are lower, there is accelerated depreciation on plant and equipment, and there are no import duties on call center gear.

The workforce is highly literate and immersed in American culture, more so than in any other Spanish-speaking locale outside of the U.S. There has long been excellent telecom and travel links with the rest of the U.S.

There are downsides to outsourcing to Puerto Rico. Site selectors question the quality of English spoken there; its only advantage over less costly locales like India is that Puerto Ricans know American slang.

Spanish speakers say the delivery is fast-paced, like Cuban Spanish. Puerto Rican and Cuban Spanish are to Hispanic ears what New York English is to American ears. It's popular only where there are many Puerto Rican Hispanics, such as in the Northeast.

"Puerto Rican Spanish is very unlike Mexican-Hispanic, which seems to be the majority of the U.S. Hispanic consumer market," says King White, vice president, Trammell Crow.

Some call centers like Inktel train to neutralize accents. Inktel conducts written and oral Spanish and English testing, backed up by monitoring and, if need be, supervisors assisting agents. They write scripts to avoid slang.

"You need to strive towards a neutral accent free of colloquialisms," Arriola points out. "So that every Spanish-speaking clients' customers can clearly understand the agents."

Spherion's Panama call center uses accent neutralization training from LDS and Associates so that agents' Spanish and English are acceptable to Americans' ears. The firm also gives generalized American Spanish as well as English-language and cultural training (for example, so agents will say "Arkansaw" rather than "Ar-kansas").

Other call centers like PRC's allow agents to speak with their dialects but train them to avoid words that have more than one meaning and could possibly give offense depending on the culture.

When preparing scripts, note that the Hispanic culture is both formal and informal. It's formal until the customer becomes conversational with the call center agent and then informal afterwards.

Raul Navarro, managing director of Hispanic/Latin America services with ICT Group, recommends that agents use the formal *usted* form and should remain as such unless a customer insists on breaking the barriers by inviting the agent to speak informally.

Once the barrier is broken, agents can become consultative and, if you want them to be, less scripted. If agents need to discuss a legal compliance issue, such as assenting to receive a credit card, they could preface the statement with "I will need to break into legal talk for a moment."

"The Spanish culture is an 'embracing' culture that mandates that formalities be discarded before business can be conducted," explains Navarro. "It is not unusual for a customer to engage in small talk about a family situation in the midst of a telesales presentation."

Because of this propensity to talk and become acquainted as they conduct business, consultative selling works with seasoned, highly trained agents who understand the usages of the formal and informal forms.

If you outsource or locate outside of the U.S., you may have to train agents how to best communicate with a broad range of American Hispanics.

Telvista teaches its Mexican agents how to overcome objections and handle rebuttals when calling into south Florida, New York, and Connecticut, where Cubans, Dominicans, and Puerto Rican customers live. HTC trains its Mexican agents to be more direct with Americans.

"Americans will take the bad with the good to get the matter resolved," says Cooper.

Chapter 10:
Foreign Markets

Creating and planning your call center to serve foreign markets entails similar yet different considerations than doing the same for domestic call centers.

Note that there is a difference between foreign and offshore (or nearshore), even though these strategies may entail setting up or outsourcing to call centers in the same country, and often the same call centers and agents. "Foreign" refers to meeting the needs of that nation's market, whereas "offshore" and "nearshore" applies to the home domestic market.

✪ FOREIGN CHARACTERISTICS

There are several characteristics of serving foreign markets with call centers that you must take into account.

Cultural Differences

If you are after business in another country then you should know the cultural and language nuances of those markets. You can have the best call center in the world, but if you're not listening to your prospects, customers, and employees, you're wasting your money.

What applies in one country does not necessarily go in another. For example, Europeans like to have beer and wine with their meals. Disney had to change its dry policies at its EuroDisney theme park near Paris. Disney also learned the hard way that European schooling and vacationing habits are different, which also affects park attendance.

Do not assume American customer service and sales techniques will be instantly accepted in other countries. These methods are still new and unfamiliar to non-Americans; many see them (and perhaps rightly so) as phony.

Then again, so do many Americans, New Yorkers especially. There is an old joke about a New Yorker bumping into a Midwesterner in a bar in Manhattan. He says, "How you doin'," and the Midwesterner tells him his life story, not realizing that in New York the phrase is an acknowledgement of your existence, not an invite to a chit-chat.

Dennis Smith, president of PacTac Advisors, points to differences between Japanese and American phone preferences. A caller in Japan has no problem in waiting for many rings before the phone is answered but does not want to be put on hold once the phone is answered. On the other hand, a caller in America will hang up long before the Japanese caller but is much more tolerant of being put on hold.

"It's not just the issue of understanding the cultures of the parties on the other end of the phone," says Smith. "It's also understanding the cultures of dealing with different management and their expectations."

Management Differences

Management techniques that are proven in other countries do not necessarily apply elsewhere. In many markets, you also have to be patient, such as in Asia.

Jon Kaplan, president of consultancy TeleDevelopment Services, says call centers in the Asia-Pacific region, such as Sitel and Teleperformance, closed because they failed to consider local practices, or they did not give local management teams sufficient time and resources to be successful.

The two outsourcers apparently confirm Kaplan's story. Both firms sold their Japanese call centers to local outsourcers and are partnering with them instead to support each other's clients.

Call centers need two to three years, and a strong local partnership, to become established in most markets, says Kaplan.

"You can't take what works in the U.S. and Europe and expect it to work [in Asia]," he says. "You also can't take what you know in one Asian country, like Japan, and expect it to work in another, like Korea or China. And you have to have patience. These are not quick-results markets."

The same goes for vendor relations. If you want to do business in another country you must learn to listen what those managers, executives, and officials say, overtly and covertly. After all, you're seeking to make money there. And the customers are always right.

Many cultures avoid direct face-to-face conflict. In Asian nations like India and Japan when someone says "yes," it may not mean yes.

Britons and Canadians do not get as excited about a deal as Americans get. They will tell you politely that they like it. To them it isn't a neo-orgasmic experience as it apparently appears to some U.S. businesspeople.

But if they do not like your offer, leave it at that. Take no for an answer. Don't play with them like you would with another American. Canadians will freeze you out, or if really annoyed, give you the verbal ticket to Hades. Britons will use their seemingly innate sarcasm to slice you verbally into ribbons.

Political and Social Issues

Just as residents in other countries think differently than Americans, they also act differently. The political and social systems are not the same. Unions and left-wing ideologies are more accepted and often stronger in other nations. The social safety nets in developed countries are much more tightly woven than they are in the U.S., as shown by government-supported healthcare.

It is a often a shock to Americans who have just moved to a foreign country to read and hear others comment about them in general, not as individuals. Especially when the words are not kind. Americans may get the cold shoulder because residents resent U.S. foreign policy. There has been a huge gulf between the U.S. and most of its ostensive allies over issues such as the Mideast and trade. Foreign nations and residents have been irked with what they say is the US talking about free trade on the one hand and enacting and maintaining allegedly protectionist trade barriers against their products on the other.

Then Americans who live abroad long enough and become part of foreign countries begin to see the world and America from foreign eyes. The revelations are akin to seeing yourself as others see you. The portraits are a mix of good and bad, sometimes overblown in admiration, other times unfair.

But that's the reality. If you're going to do business in another country you have to know, accept, and work with it.

Language

When serving customers in other countries, chances are you will be serving them in languages other than English. You will need to have agents who speak those tongues. There are also dialect differences in the same language that may require separate call centers or programs.

Native Versus Learned Language

There is a difference between learning languages when growing up and being taught them in schools. That could have an impact on where your call center programs are handled.

Having a supply of native language speakers is preferable because they will know all the nuances that help to convey meaning and establish a bond of understanding with customers. There is a cultural connection that is difficult to match. Jon Kaplan points out that there are three to four levels of politeness in Japanese. A person who has learned Japanese as a second language would not know or appreciate the nuances.

Yet learned language speakers are often acceptable on matters such as technical support, where the topics are more straightforward. You are also less limited by location. You can set up call centers in university towns that are far away from your target countries and hire students who were taught those languages.

"Fluency depends on the expectancy of the person calling," says Philip Cohen, a teleservices consultant based in Skelleftea, Sweden. "If the parties calling in are consumers, they will likely want to talk to an agent who is fluent in their language and culture. If a Portuguese housewife is annoyed when her vacuum cleaner breaks down and your agent is not conversant in that language, she will get even angrier at your firm."

Dialect Differences

Like in the U.S. there are often strong regional and national dialects. Northern Germans speak differently than southern Germans. There is formal (Mandarin) Chinese and regional dialects like Cantonese and Fujianese. Cuban, Castilian (Spain), and Mexican Spanish are not quite the same language.

I grew up speaking northern English *eee by gum* as my family is from Blackburn, a small city north of Manchester. Yet there is no mistaking the Liverpool accent for mine, even though that city lies an hour away from both cities.

These differences could affect where you locate or outsource your call centers. Kit

Cooper, president of Hispanic Teleservices, points out that Latin Americans expect to speak to agents who speak the same dialects as they speak.

Insularity

Some cultures are tough to be accepted in no matter how sensitive and savvy you are or how well you know the language or dialect. You're not from there, period. Examples include Japan, Korea, and Thailand.

The Japanese, it is said, reputedly prefer that you speak Japanese poorly; if you are fluent, they are not comfortable with you because it breaks down that barrier between them and you.

Localization

When serving foreign customers you will have to localize your scripting, training, web sites, products, and services in their languages and alphabets. Localization is typically required for non-English wording. But it may also be needed for other English-language customers. Australians, Britons, Canadians, and Irish often use different words and expressions. In Canada when someone talks about "the hydro" they're not talking about tap water. Instead they're referring to the electric company. Much of Canada's electricity comes from hydroelectric dams.

Also take into account the currency and addresses used by customers. Many data fields do not include the extra lines that European addresses require. Canada's "Zip Code" is the Postal Code and is alphanumeric.

Date formats are also different. For example, Europeans and Canadians prefer to put dates as day/month/year, whereas Americans list them as month/day/year.

"A mistake that too many American companies setting up in Europe make is insisting on the U.S. date format," says Matthew Johnson, Akibia's vice president of consulting. "That rankles Europeans because it rubs the American ownership in their faces. You resolve this by bringing overseas staff into the implementation process early on, so the American managers are aware of the importance of these issues."

Support and Documentation

With serving foreign customers or handling foreign-serving or offshore operations, companies must record trouble tickets and other time-sensitive documentation in Universal Time (formerly known as Greenwich Mean Time). That helps to make sure matters get resolved when needed.

Call centers don't have to translate every comment on a customer record into a common language. That is an expensive and time-consuming task that often leaves inaccuracies, because it loses idioms in the translation. But there must be uniform terminology, such as what trouble tickets are called and defined as, to make sense of call center performance reports.

Telecultures/Web Cultures

When considering a call center in another country, find out the size and depth of their "telecultures" and "web cultures." Do your prospective customers use phones or the

Internet to obtain customer service, and to what degree compared with face-to-face? That will help determine the size of your call center or outsourced program.

The rapid growth of cellular networks and the Internet has enabled many countries with obsolete legacy phone systems to leapfrog over the U.S. and other nations in their wireless voice and web use. This is especially true in developing nations like China and India.

Credit Card Culture

It is not coincidence that the rise of call centers has paralleled an explosion in credit cards. The more foreign customers use credit cards, the more likely you will need call centers to serve them, including cross-selling and upselling them on services like protection.

Laws and Regulations

You will have to know and comply with all laws and regulations in the countries you are locating in and those where your customers are doing business from. These can be the same or different nations, depending where your call centers are.

Many nations are more restrictive on outbound selling and on personal data than the U.S. is. Canada has a national privacy law and provincial amendments. The Personal Information Protection and Electronic Documents Act (PIPEDA), administered by Canada's Privacy Commissioner, requires businesses to obtain consent from consumers before collecting personal information that can identify them. Examples include credit card numbers; financial, health, and income data; and identification numbers.

The law requires different types of consent depending on the sensitivity of the information. For example, you can have implied consent to obtain the individual's address for the purposes of delivering a product. But you need their expressed consent to obtain sensitive information such as financial and health care data.

PIPEDA allows businesses to collect some publicly available information, such as published phone numbers. Businesses will also need consumer consent before transferring the data to other users, with the exception of debt collection, government or police investigations, and emergencies. If businesses have no use for the data, they must dispose of it in such a way as to prevent improper access.

Another example: The European Union's Data Protection Directive gives European consumers the right to opt out of corporate access to their personal data. This includes names, addresses, and phone numbers. The standard is to opt in for sensitive data such as ethnicity, political preferences, health information, philosophies, religious beliefs, lifestyle, and union activities. There are limited exceptions to the opt-in requirement; check out the U.S. Commerce Department's web site for more information (www.export.gov/safeharbor). Also, consumers must be told why the data is being collected and have the right to object, and to access and rectify incorrect information.

Many of these rules apply to the U.S. and other countries outside of where your customers are based. For example, the Data Protection Directive imposes restrictions on moving personal data to nations outside of the EU member states. This led to potential disputes with the U.S., which has a much different, more private and self-regulatory privacy protection system.

To get around this issue, the EU and the U.S. Commerce Department have a "Safe Harbor" agreement that allows U.S. firms to voluntarily adhere to the directive.

Labor laws are often far stricter than what U.S. firms are used to. Many countries require call centers to sign annual contracts with full-time employees instead of hiring and firing full-timers and part-timers at will. Unions are often much stronger. You can't dismiss them as easily as you can in the U.S.

Investment Regulations

Most countries are far more open to foreign investors like call centers than they were in the past. But there can be strict stated and unstated regulations.

Many nations, including India and Indonesia, require that you have a joint venture with a local firm. The benefit is that the local company knows how that country works. The risk is that you have to trust that partner with your customer and corporate information. Therefore, you have to research and interview potential partners thoroughly.

Some countries have deliberately written investment rules to protect local outsourcers and telcos entering the outsourcing market. For example, China requires that foreign outsourcers hold no more than 30 percent of a joint venture in the first year, rising to 49 percent in the second year.

"Naturally, under such conditions, foreign outsourcers would be discouraged from entering a market that has restrictive investment rules," observes Alex Yung, general manager of new business development at outsourcer PCCW Teleservices. "That's because they may lose their control over the company and the quality of services to be delivered, along with the cultural differences they must manage."

Also, your biggest customers may be governments. China's large and centralized government, for example, owns many industries that control purchases down to the local level. Where the state ownership of industry is disproportionately large, as in China, PacTac Advisors' Smith strongly advises that you keep abreast of political developments.

"Governments like China's have big purchasing power, and they know it," Smith points out. "They can make it clear where they would like to see you set up operations and who they would like to see you work with, but they won't make demands. It's your responsibility to read the signs and understand what will be important when contracts are awarded."

✪ SITE SELECTION OPTIONS

The viability of other tools to provide call center services depends on the development and sophistication of the markets. IVR and web self-service is accepted in developed markets, less so in developing markets. But that is changing with the spread of web-enabled cellphones. Home agents are slowly increasing in acceptance in countries with widespread Internet connections such as Canada, the UK, Sweden, Ireland, Netherlands, Belgium, France, Germany, and Australia.

✪ OUTSOURCING

But the most popular alternative to setting up your own call centers is outsourcing to a service bureau. The contacts can be handled in country, in the same regional market, or offshore, which I will explore later. The service bureaus can be U.S.-based or foreign-based.

Foreign Outsourcing Benefits

Like in the U.S., outsourcing avoids making capital expenditures on facilities and technology, provides quicker ramp-up time, and greater flexibility. Chances are you will cut total costs.

But there are additional benefits conferred by outsourcing when serving foreign customers.

Avoiding Complex and Expensive Site Selection

Locating in foreign countries takes time and money. It includes dealing with bureaucracies and learning different customs, languages, laws, regulations, measurements, currencies, and labor practices. Time-consuming and hassle-ridden relocation is minimized.

Testing the Market and Gaining Knowledge

Outsourcing is an excellent way to learn the ropes on how the system works and about cultural practices. In the meantime you determine the people, processes, and technologies you need. Outsourcing lets you obtain the knowledge and experience before spending millions of dollars and months to set up the call center.

"If it's your first time into those foreign markets, you probably are better off outsourcing unless you have experience in setting up international operations," advises Geri Gantman, senior partner at R.H. Oetting. "One would expect the outsourcer to be more savvy internally and externally. Internally when it comes to staffing and externally in such matters as the marketplace. "

Better Security

Some markets, like parts of Latin America, are dangerous, especially for foreigners who are seen to be wealthy. American companies and executives are also potential terrorist targets. Outsourcing to established and locally savvy bureaus is probably the most viable and safest strategy to serve busy but strife-ridden markets like Colombia.

Whether or not you go with a multinational or a local firm in another country depends on your comfort level.

"If companies are more comfortable working with an outsourcer with U.S. offices where they can easily visit and monitor outsourcing activity rather than going to Latin America, then they might consider using a U.S.-based rather than a Latin American-based outsourcer," advises David Spindel, Datamonitor's U.S.-based managing analyst.

On the other hand, local or locally-managed Latin American outsourcers can help you get an edge on local markets.

"Local firms know the cultural nuances of the local markets, which can result in higher performance rates from a relatively flat learning curve," says Spindel.

Fewer Headaches

In developing countries, outsourcing enables you to tap these markets without the extra hassles in setting up, hiring, and managing operations in nations with less-than-ideal infrastructure, rule of law, and political stability.

Serving Niche Language Markets

One example of this is Canada. You can have a low-cost center, say, in interior British Columbia or Newfoundland and Labrador (that could also serve your U.S. customers) and contract with an outsourcer in Quebec to serve the French-speaking market. Another example is Sweden. If you are looking at expanding there, it makes sense to outsourcing a trial inbound response program. If it proves successful but you cannot justify building your own center, build and enhance your outsourcing relationship.

Minimizing Cultural Insularity

You need to partner with outsourcers based there or have strong local presences. These firms will know the nuances, have the contacts, and can assist you in marketing and serving customers correctly.

Bertlemann Online had contracted with Japanese-owned outsourcer Prestige International to answer calls and emails from its web site. It soon learned how demanding Japanese consumers could be. Dissatisfied Japanese consumers would call companies demanding apologies and in some instances insist that the firms bring them replacement items. In a few cases, company presidents have called customers back and apologized.

"If a customer has an issue with any of our products or services, we would ask the Prestige manager the best way to respond," explained Dietmar Hering, Bertlemann Online's Asia Pacific customer service manager. "We almost always leave it to their judgement. Their advice on how to approach Japanese customers is one of the more important values they add to our product. An unhappy customer can do harm to the brand, but no advertisement is more powerful than hearing from your friend how great BOL is."

Foreign Outsourcing Challenges

Outsourcing to serve foreign countries blends the risks of doing the same for the U.S. market, both onshore and offshore. You are entrusting another company to handle your customers and employees, a firm whose agents are doing the interacting are beyond your direct control and management. You are also letting them look at customer data, extending the risk of that information being stolen.

In developed markets you will have to analyze the costs and the benefits because ultimately it is less extensive to run a program in-house. The payoffs are greater in developing countries compared with opening in-house call centers, namely avoiding site selection complexity and personal security, but so are the risks of poor performance and data theft.

There are additional disadvantages.

Due Diligence

The call centers are in other nations; contracts are subject to their laws. The bureau quality is equivalent to the U.S. if not greater in developed countries.

The challenges are the same as outsourcing in the U.S. but exacerbated because you're dealing with foreign cultures, practices, and laws. Unless you're dealing with a reputable, U.S., Canadian, European, or Asian vendor, ideally one that has offices or consultants in the U.S., finding the right bureau and working with it can be difficult. Unresolvable issues such as contract terms could arise. Remember, the home team always wins, especially in developing countries.

There are call center and direct marketing associations in many other countries which represent outsourcers (see Chapter 16). But unless you are working with a multinational, your odds of problems increase if you signed up for bureaus based in developing nations.

Diminished Market Information

If you outsource you may be missing out on valuable information about your customers in foreign markets compared with serving and selling them directly with in-house call centers. You can gather a lot about attitudes, preferences and trends through monitoring calls, examining results and by talking to and getting to know your employees and listening to them talk about the customers—data and insights that you may miss by outsourcing.

Says Oetting's Gantman; "You're gaining little knowledge of that market, its needs and trends—data and analyses that are critical for you to maintain and grow your business there."

✪ OFFSHORING

Like in the U.S., companies overseas are moving contacts from developed, high-cost, foreign nations to lower-cost locales. And like in the U.S., most of the offshoring programs have come from outsourcers, with the notable exception of large global high-tech firms.

Europe

Options for UK customers include India, the Philippines, South Africa, Egypt, Ghana, Uganda, Nigeria, Barbados, and Jamaica. Also, in Eastern Europe, English is taught widely in the Czech Republic, Hungary, and Poland.

The options for French, French-speaking Belgian, and Swiss customers are Morocco, Senegal, Togo, and Tunisia. The locations choices for Spanish customers are Argentina, Chile, Costa Rica, Mexico, and northern Morocco. Portuguese customers can be served from Brazil.

Swedish, Finnish, Danish, and Norwegian customers can be answered in the Baltic States: Estonia, Latvia, and Lithuania. German, Austrian, and German-speaking Swiss customers can be handled in the Czech Republic, Hungary, and Poland. Dutch and German calls can be handled in South Africa.

Turkey is an also an option for Dutch and German customers. Robin Goad, managing analyst with Datamonitor, and Alp Koren, managing partner of call center management consultancy Sistema Managing and Information Solutions, say many Turks went to Germany and Holland as guest workers in the booming 1960s. Many of these workers and, more importantly, their German- and Dutch-born offspring, returned to Turkey.

Asia-Pacific Region

Offshoring is happening in the Asia-Pacific region. Australian calls are being handled in India and Malaysia. Parts of China, such as the northeast, are becoming low-cost alternatives to Japan. These areas have long had ties to that nation.

"Getting large volumes of Japanese language call center work out of Japan is every regional controller's dream, and it has worked well for those that have done it," says PacTac Advisors Smith. "But moving jobs is nightmarish from the costs of letting staff go in Japan and from the pressure put on companies to keep all of their operations there."

Americas

There is some offshoring in the Americas. Some Canadian work is being handled in India, which is likely to grow if the Canadian dollar is strong against the U.S. dollar. Putting the brakes on that trend is a small domestic market, low costs and turnover, and the need to have agents who speak French.

There are unconfirmed reports that Mexican companies are outsourcing to bureaus in lower-cost Central and South American countries, even at the risk of poor quality.

Joint U.S. and Foreign Offshore Locations

Most of the English- and Spanish-speaking offshore locations are suitable for U.S.-serving call centers. Dell's and Teleperformance's Indian centers handle the U.S. and UK contacts; Teleperformance's Buenos Aires, Argentina center serves clients' U.S. Hispanic and Spanish customers. Convergys handles UK and U.S. calls and possibly future Australian end-customer contacts from its Indian call centers.

That allows Indian call centers to split the contact handling between both markets from the same workstations. When it is 2 p.m. in the UK, it is 7 p.m. in India. But when it is 2 p.m. in California, it is 5 a.m. in India.

"There isn't any English-language work, except politically sensitive government contracts, that cannot be handled offshore," says Jean-Marc Hauducoeur, senior vice president of international operations with Convergys.

Offshoring Benefits

The benefits of offshoring from foreign countries are very similar as those from the US.

Cost Savings

Cost savings run from 35–55 percent in Eastern Europe, North Africa and Turkey, to 50–75 percent in southern Africa, Asia, the Caribbean, and Latin America.

Higher-Quality Workers

Call center work is generally looked on more favorably in foreign countries than in the U.S. Up until about 2001, European countries, Australia, and New Zealand were pitching hard for call centers as means of reducing high unemployment.

Offshore workers are more likely to see employment in a call center an elite job than those in developed nations. And call centers attract university students and graduates who stay longer. The abject poverty in such nations drives people to succeed when given an opportunity to work in clean, comfortable, well-furnished high-tech and comparatively very well paying call centers.

As well, developed countries' call centers often to be located in low-cost, high-unemployment communities like former factory and mining cities that typically have lower education levels than higher-cost capital cities and regional business hubs. Or they set up in university towns with transient workforces that do not stay long enough to develop customer service skills.

Mike Havard, managing director of CM Insight, explains that many outsourced call centers located in British university towns like Bristol and Manchester. While this was fine for simple transaction-based calling like outbound collections and sales, where turnover is high, these workers are not as suitable for high-end customer service, support, and cross-selling and upselling.

"You get disinterested this-is-a-job-while-I'm-waiting-for-a-better-one agents who can just as readily turn your customers off as offshore agents who are intelligent and responsive if a little lacking on empathy," says Havard.

Poor Reputation

You call center may not be wanted onshore especially if your plan calls for it to be short-length, undertake routine tasks, or carry out controversial functions like telemarketing. Call centers, especially in Europe, have been tarred with the reputation of being low-paid sweatshops and as unstable employers that pack up and go even after obtaining generous subsidies.

These conditions and reports have reportedly turned some people off from working in call centers. Some economic development agencies (EDAs) are no longer interested in attracting call centers.

"Already there are some places that have had enough of call centers," reports Philip Cohen. "I know of a Welsh site selection consultant who was told by an economic development agency director from a poor coal mining area that 'We'll take any sort of job, but don't send us any more call centers.'"

Michael Allen, a director with Chester, UK-based Mitial Research, points out that governments want long-term and high-quality employment from investors. They do not want here-today-and-gone-tomorrow operations.

"A few years ago, local EDAs were bending over backwards to provide freebies demanded by call centers that were considering locating in their communities," says Allen. "Now they're telling such call centers to get lost. Their main concern now is for the quality of the jobs, not the quantity of jobs."

Access to Foreign Markets

Many offshore countries, especially Eastern Europe, China, and India are becoming strong markets for U.S. products and services. Offshoring to an in-house or outsourced call center to serve both developed and local markets increases your utilization and spreads out the costs.

Offshoring Downsides
Lack of Cultural Affinity

Cultural affinity is likely an issue in other countries as it is in the U.S. A May 17, 2003 *Guardian* (Manchester, UK) story reported one journalist's support nightmare from Dell after the computer maker had brought support in-house to a Bangalore, India center from an unnamed outsourcer out of UK and Irish call centers.

There are similar concerns in Australia about cultural affinity, plus added communications costs, which callcentres.net's Conboy reports has led some unnamed firms to pull back from India. He advises you balance the savings benefits with those risks.

"It's all very well to have a cheap hourly rate, but if that means you burn a whole bunch of customers through agents' inabilities to communicate effectively, what have you really achieved?" he asks.

Public Backlash

There is also the risk of public backlash from foreign offshoring. This is coming from individuals upset about seeing jobs lost and from union agitation.

Language Limitations

This is the biggest restriction in offshoring from most foreign countries. India may be a great locale for English-speakers, but it cannot support any other European or Asian language. A third of Canadians speak French, which means if you want to offshore from Canada you need a second center with French-language capabilities.

There is a limited supply of educated, fluent agents in these other nations. That goes for Eastern Europe, Africa, and Latin America. There is no equivalent of an India or the Philippines, with large populations, fluency, and customer service skills in any other language.

Serving to offset offshoring in the Asia-Pacific region is limited language skills. There are only so many fluent Japanese and Korean speakers in other nations. Also, immigrants there rapidly lose touch with what is going on and being spoken in their home countries.

South Africa's telecom costs are reportedly too high. And Caribbean country alternatives such as Barbados and Jamaica are too small to support large, efficient, UK-serving centers.

Political Instability

Many foreign-customer offshorable countries suffer from political stability and threats on their borders.

Corruption

Corruption is a major issue outside of developed countries. Public servants are often badly paid; they depend on bribes to survive.

Legal Hassles

There are practices and regulations in other countries that can impede offshoring to them. Mike Havard points out that Indian contract law cases can take up to 15 years to settle.

He recommends contracting with bureaus with U.S. or UK subsidiaries where you can then have effective legal recourse. For example, you can seek seizure rights on the outsourcer's assets should their performance fail.

Regulatory Restrictions

Look closely at the laws in your foreign market. You may find rules that govern off-shoring, especially offshore outsourcing.

For example, under the European Transfer of Undertakings for the Protection of Employees (TUPE) regulations, companies that outsource existing in-house agents, or outsourcers that decide to move some or all of their operations outside of Europe, must negotiate with the affected agents to transfer their individual work contracts.

Settlements under TUPE include buying agents out, or in more extreme cases, offering to pay them to work offshore at UK wages. Once settlements are reached, that still may not be the end of the matter. Some employees are now suing to get their jobs back.

But this law does not affect companies that plan to outsource offshore to supplement existing call centers or outsourcers adding capacity offshore.

The Data Protection Directive prohibits data on individual Europeans from leaving the EU unless it goes to countries where the laws are as strict as Europe's or where there has been a safe harbor agreement negotiated to protect that data.

Low-cost countries like India do not comply with the directive. But UK companies have been getting around it by stipulating in their contracts with Indian outsourcers that those service bureaus will guarantee that they will comply with the directive.

Security Risks

There are potential crime and terrorism dangers in lower-cost countries. Leading Brazilian cities have notoriously high crime rates, as does downtown Johannesburg, South Africa. There have been terrorist attacks in Morocco. Ongoing terrorism has knocked neighboring Algeria off many site selectors' maps.

These nations' public health systems may not always be good. If there is an epidemic these countries are first in line to get hit and will suffer the worst. Witness how AIDS has ravaged much of Africa.

John Boyd Jr., consultant with The Boyd Company, reports the pace of offshoring is slackening. "Corporations are finding that they can't get middle and senior management to travel to these locations because of security and health fears," he points out.

Diminishing Benefits

Check closely to see if offshoring from foreign countries makes sense. For example, there are arguably minimal benefits for the cost and hassle of having Spanish and Portuguese calls handled in-house or outsourced to call centers in Latin America.

"Spain is a comparatively low-cost country in Europe, with high unemployment," says Datamonitor's Goad. "With Morocco nearby there is less incentive for Spanish companies to locate or outsource to countries like Chile. Portugal also has low costs and high unemployment, so there is little need for them to go to Brazil."

Eastern Europe is a popular lower-cost offshore location for European and U.S. contacts. But there is some question how long that advantage will continue as call centers move in and their economies improve.

There is scant offshore outsourcing from Canada. The reasons are Canada's low costs, smaller demand, and its French population. There are no French speakers in India, reports Brian Bingham, program manager for CRM and Customer Care Services at IDC.

CONSIDER STAFFING AGENCIES

When locating in Europe as the Netherlands, UK and Sweden, you should consider contracting with a staffing agency such as Adecco, Manpower, or Randstad.

There are two-way benefits to working with staffing agencies there. They are effective at attracting and screening workers, and they are well known to potential employees. Because these agencies know the labor market and because they recruit and pay the employees that you contract for, they'll handle complex labor legislations.

"Staffing agencies are becoming popular," says Michael Allen, managing director of The Mitial Group. "They have already got their act together, are respected, and they offer job security even if the agency's client lays people off."

Mike Morrison, vice president-sales with ICT Canada (part of ICT Group), knows of one client who returned their program to Canada after finding poor lead-to-sales conversion rates offshore.

"The cost savings proposition is greater from the U.S. to offshore than from Canada to offshore," he explains.

More Training

Like offshoring from the U.S., the same practice from other countries will require additional training, such as accent neutralization, culture, and recognizing (but not speaking) idioms and slang. You may need to have your supervisory styles changed to meet those cultures.

✪ NEARSHORING ALTERNATIVES

To get around many of these issues, there may be nearshoring alternatives to call centers in developed countries. While the cost savings may not be as great as with centers in developing nations, that may be offset by greater cultural affinity and less opposition.

Examples include Canada and Australia as a choice for UK calls. eBay's contact center in Burnaby, British Columbia, handles customers in those two countries.

New Zealand is a potential nearshore option to Australia as it is less expensive. But while the amount of savings (10–15 percent) is large enough to warrant adding a center there from new growth, it is too small to merit relocating a center.

✪ FOREIGN MARKET STRATEGIES

There are several locations strategies for both in-housed and outsourced call centers to consider.

Individual Country

"Individual-country" means having an in-house or outsourced call center in each nation that you are serving. Those call centers serve only those country's customers. You need to have only agents who speak local languages; you cope only with the laws that apply there.

The individual-country strategy makes sense where you have large populations or strong insular cultures and unique languages that not readily available in other nations. Examples include Brazil, China, France, Germany, Italy, Japan, Korea, Thailand, and Russia. But this is gradually breaking down with the rise of offshoring in the case of France, Germany, and Japan.

Canada is difficult to serve with one center from any other nation because you need agents who speak both French and English. Accent and dialect differences coupled with high telecom costs between most Latin American countries make in-house or outsourced in-country call centers a necessity. The exceptions are higher-end and more-technical calls, where you can use a pan-Latin approach.

Kit Cooper, president of Hispanic Teleservices, says customers calling or being called from one Latin American country expect to hear that country's dialect. "But if that customer was calling from a Latin American country, they would have a strong expectation to speak to an agent with an in-country dialect."

PacTac's Smith points out that China has so many distinct regional dialects that if your objective is to serve the entire country, you need more than one center to serve them.

The main downside of the individual-country strategy is high costs. You have to replicate call centers in each nation you're selling/providing service to. Not every market is the same size. You could be spending more per customer in some markets than in others. If that is the case, consider outsourcing to serve them.

Where In-House Call Centers Are Best

There are individual-country markets where you may need to have your own call center, if only to tightly manage the operations. The political, economic, and social environment is such where that you have to be very careful in whom you trust.

One of the best examples of this is Russia. With more than 200 million residents, low costs, and a distinct culture. Russia is a huge developing market that merits its own call centers. But the country continues to be riddled with crime and corruption that can prove daunting for many businesses.

To serve Russian customers, Lawrence Moretti, a director at Deloitte and Touche Fantus Location and Facilities Strategic Practice, suggests setting up and stringently managing a Russian call center subsidiary or buying a Russian outsourcer, teaming with experienced local businesspeople.

He advises against joint venturing with a Russian partner because the relationship can divert attention from the core business.

"You need to buy the best people and stay in control," he recommends. "But you can't import proven methods and business practices from the U.S."

Where Outsourced Call Centers Are Best

There are in-country markets like China where you need a strong outsourcing partner. The world's largest nation has a rich, complex culture, is highly suspicious of foreigners, has a difficult to penetrate bureaucracy, and has an authoritarian and sometimes capricious government.

Call centers locating in China must also comply with numerous licenses and regulations that can take months to obtain.

Outsourcing to an established Chinese service bureau that knows the ropes may be the best bet, suggests James Tan, associate vice president of site selection consultancy Equis Asia. A company can gradually gain knowledge of the market by working with such a partner before opening its own call center.

* Pan-Regional

Pan-regional call centers are those that service several nations in an entire region or continent, such as Europe, Latin America, and Asia. The facilities are located in large metropolitan areas that attract labor forces, typically young people and immigrants from surrounding countries, who are native speakers of their local languages.

Such cities, especially in Europe, are either national capitals, head office and regional HQ hubs, cultural cities, or have large universities that draw in students from many countries. These cities also have international airports and sometimes high-speed rail links.

Examples of pan-regional call center sites include Amsterdam, London, Manchester, Glasgow, Cardiff, Belfast, Dublin, Barcelona, Lisbon, Prague, Sydney, Brisbane, Melbourne, Singapore, Santiago, New York, Los Angeles, Boston, San Francisco/Silicon Valley, Montreal, Toronto, and Vancouver.

The upsides are economies of scale, by avoiding opening and maintaining many centers. Cities like London are major financial centers and make up for their higher price-etags by having access to very well-qualified personnel.

The downsides are higher costs and shallow foreign language pools. By placing all your calls into one center you diminish your flexibility to quickly shift volume to less expensive sites if costs rise and to recover from disasters.

Also, the supply of immigrants and students fluctuates. You may not have enough of those speakers. If there are jobs back home or the costs of living in their new home rise dramatically, they may not arrive or stay in adequate numbers to support call centers, or they may quit to work in other higher-paying employment in their new countries.

* Sub-Regional

Sub-regional call centers are individual sites serving one or more nations that are networked together to serve an entire region. Examples would be centers in Manchester, Barcelona, and Stockholm to cover Europe; or Seoul, Hiroshima, Shanghai, and Kuala Lumpur to serve the Asia-Pacific region.

The sub-regional strategy is best of all worlds. You avoid excessive expenses of individual-country centers while shrink the risks of rising shortages of multilingual labor in pan-regional centers. You also obtain disaster protection—essential in Japan with its numerous earthquakes and typhoons—by having calls rerouted from closed to open centers. The downsides are greater complexities and somewhat higher costs from managing several centers in various countries.

You can apply a variation of the sub-regional strategy inside individual countries to avoid tapping out labor markets, for example, by spreading out your US call center capacity amongst several sites spread throughout the nation. Foreign examples include the UK (Glasgow, Birmingham, and Belfast), France (Lyons, Lille, and Mosel), Germany (Berlin, Bremen, Leipzig, and Saarbrucken), China (Harbin, Beijing, Shanghai, and Guangzhou), and Japan (Nagoya, Osaka, Sapporo, and Yokohama).

One example of a network made up of sub-regional call centers is Avis's call center system. The car-rental firm has a 170-seat center in Barcelona that serves customers in southern Europe and a 200-seat center in Manchester that serves customers in Austria, Germany, Switzerland, and the UK. The two facilities replaced eight European centers.

"The churn of people is the major cost in running a call center," says Philip Cohen. "If you haven't got that supply then you lack the prerequisite to open a call center. It is much easier to find Spanish speakers in Barcelona than in Amsterdam and vice versa. The shortage of language skills among workers is a real threat to the multilingual call center. It pushes the odds in favor of regional, language-based call centers."

Adds TeleDevelopment Services' Kaplan: "People would rather be served by agents in a call center in their own countries, speaking their own languages, and sharing the same culture, than those in a pan-Asian call center. But you must have a critical mass of customers and that strong culture to support it."

One way of applying the sub-regional strategy is locating in border cities, where many residents are fluent in more than one language. Examples include Liege, Belgium; Sonderborg, Denmark; Lille, Perpignan, and Strasbourg, France; Saarbrucken, Germany; Groningen, in northeast Netherlands; and Maastricht in the southeastern part of the Netherlands.

The cross-border workforce is often substantial. For example, arriving in Saarland, which encompasses Saarbrucken, are 21,000 commuters from the adjacent Lorraine region of France. The border cities are helping firms reach out to potential agents and train them. The Saarland Call Center Academy, which opened in June 2000, provides customized training programs.

In Canada small eastern Ontario cities like Cornwall, Sudbury, and Timmins that are near Quebec or have a large French-speaking population have attracted Canadian–serving call centers. Ottawa, the nation's capital, is viable for high-end tech-support centers. The city lies across the Ottawa River from Hull-Gatineau, Quebec.

* Canada's French-speaking population is concentrated in a region from New Brunswick to eastern Ontario, with pockets on Cape Breton Island, in Nova Scotia and in Winnipeg, Manitoba. But call centers wishing to access this work force can also set up in small, midsized, and large communities that have active military bases to tap spouses of uniformed personnel. Examples include Halifax; the Belleville, Ontario area; Kingston, Ontario; Edmonton, Alberta; Victoria, British Columbia; and the Comox Valley, which lies 140 miles north of Victoria.

Another sub-regional strategy is satellite, or hub-and-spoke. In it there is one principal call center covering most of the region supplemented by smaller local satellite centers in other countries, either in-house or outsourced. This strategy is best deployed when companies cannot find enough workers who speak other customers' languages. You tap current residents and immigrant native-language speakers.

This is an excellent method to serve Asia-Pacific customers. You tap into these small but growing markets. You also provide disaster protection.

For example, you can have a principal call center in Brisbane, Australia, and Singapore, and satellite centers in Seoul, Korea; Sapporo, Japan; Guangzhou, China; Hong Kong; and Taipei, Taiwan. The contacts would go in preference to those centers, but if the volume is high or there is an issue that can't be resolved, they would be directed to the Australian or Singapore center.

"If your business plan shows that you need less than one dozen agents to serve Japan, then you can get away with having those seats in a regional call center, like Australia," says Kaplan. "But if you require more than that—that you have a critical mass for a large business in Japan—then you should set up a call center there."

* Pan-Cultural

You can make contacts and take calls and contacts, using the same language, handled in the same center, in the country that has the best combination of costs and access. One example of that is eBay's center in Burnaby, British Columbia, which handles contacts from Australia, the UK, the U.S., and Canada.

* Global

In some instances you can go beyond pan-regional and serve the planet from one call center. There are a handful of cities that qualify. These include: Toronto, Vancouver, New York, and Orlando. They have a rich multicultural populations, strong native language retention and/or many learned language speakers, and excellent communications and transportation from most parts of the world.

Alternatively you can serve specific regions, culture or the world by follow-the-sun; locating a network of call centers spread out east to west in varying time zones, including in different countries. This way you can have a call center open for only one or two shifts but provide continuous or near continuous live agent customer service.

An international follow-the-sun implementation lets you take advantage of different labor costs in other countries. While the British Empire is no more, the sun hasn't yet set on the English-speaking "empire", its most important legacy. You can, for example, have a call center in Wellington, New Zealand handle evening and night calls and contacts from California, permitting you to close down your Fresno facility in that high-cost state. When New Zealand is ready to turn in, you can have them routed to your center in Perth, in Western Australia; then to Mumbai, India; Sheffield, England; and back to Scranton, Pa.

Following the sun requires hard work and investment to network your call centers and to train agents to be on the same program to provide consistent customer service. While long-distance voice and data costs are dropping, additional capacity is coming online, and lower-cost technologies like IP telephony are becoming reality, such networking is still not inexpensive.

○ SERVING FOREIGN MARKETS

Market Size

Take a hard look at market size, characteristics, and the cost of serving the market.

There are some countries that are economically and culturally compatible with an individual-country in-house or outsourced program. They include France, Germany, Italy, Spain, China, Japan, Korea, Thailand, Argentina, Brazil, Mexico, and Canada. The UK usually merits its own call center because Britons are big buyers of U.S. products and services, and they speak the same language.

Many nations, including India and Indonesia, require that you have a joint venture with a local firm. The benefit is that the local company knows how that country works. The risk is that you have to trust that partner with your information. Therefore, you have to research and interview potential partners thoroughly.

Some countries have deliberately written investment rules to protect local outsourcers and telcos entering the outsourcing market. For example, China requires that foreign outsourcers hold no more than 30 percent of a joint venture in the first year, rising to 49 percent in the second year.

"Naturally, under such conditions, foreign outsourcers would be discouraged from entering a market that has restrictive investment rules," observes PCCW's Yung. "That's because they may lose their control over the company and the quality of services to be delivered, along with the cultural differences they must manage."

But oftentimes a foreign market is too small to effectively service with a call center based there. Therefore either outsource or serve that country from another market that has a similar culture.

For example, if you have Australian customers but perhaps not enough to merit a separate higher-end center, it can make business sense to serve them from a center that also handles other English-speaking customers. eBay's center in Burnaby, British Columbia, handles Australian as well as American and Canadian customers.

"If you're getting a toehold in Europe then you should look at setting up a small pan-European call center," advises Philip Cohen. "But if you already have a presence there and you need to expand, then a regional European strategy is your best choice. You will have much better access to workers who have the language skills you need."

Alternately, examine home working if the designated country has high Internet usage and computer ownership. You avoid paying directly or indirectly for facilities.

* Carefully Examine Locations Options

Just as you would in the U.S., look closely at the locations choices in other nations. Like in the U.S., small cities and rural areas with high unemployment also provide value for call centers in other countries. The Galicia region in northwestern Spain offers 13 percent lower costs than the Spanish average with the added benefit of serving neighboring Portugal. In Japan, less populated and farther-away islands Hokkaido and Okinawa can cut your labor costs.

In Europe and Canada especially there are many especially former industrial cities that have high unemployment and loyal staff. Some communities have special skills, such as multiple language abilities. Examples of the latter include Liege, Belgium; Metz and Nancy, France; Saarbrucken, Germany; Cornwall, Sudbury; and Timmins, Ontario.

There are lower-cost options to Tokyo. They include Nagoya, Osaka, Sapporo, Yokohama, and Okinawa. Nagoya and Yokohama are within an hour from Tokyo, whose real costs are one-third to nearly one-half less than in the Japanese capital.

"You have just as good labor supply in these communities as you do in Tokyo," Kaplan points out. "But like in the U.S., you have to offer good working conditions and wages and benefits to attract and keep them, because you're competing against other service industry employers."

Dialect Issues

Many countries have national languages and distinct accents. The best example is China. While China's official language is Mandarin, there are strong regional dialects in cities like Guangzhou and Shanghai.

PCCW's Yung says you can find staff that can speak those accents in Beijing as well as in the other two cities. All three cities are also home to China's main colleges and universities, which can provide high-quality agents.

While home cities offer many more native dialect speakers, their telecom infrastructures are not yet robust. Networked call centers are still in their infancy.

But this is changing as local banks and telcos become more aware of the benefits of networked call centers. You could, for example, direct calls from Shanghai to Shanghai-dialect speakers in Beijing or in Shanghai, depending on call volume.

"Everyone can understand Mandarin," says Yung. "But if you want to provide competitive, more customer-friendly service, you will need agents who can speak those dialects. For the time being, you will have to find them in Beijing and other major coastal cities or set up call centers only to serve those regional cities and not link them together."

Political Differences

There may be serious political rationales that you have to weigh into your site selection. For example, there are the Chinese who live in the Hong Kong, which for more than a century had been a British colony, and Taiwan, which the Chinese government regards as a wayward province. You *do not* want a call center in Taiwan taking calls from the mainland.

PCCW sidesteps these touchy dialect and political issues with five separate call centers: in Beijing, Guangzhou, Hong Kong, Taiwan, and Shanghai. The Beijing center mainly serves northern China, while the Guangzhou and Shanghai centers service their cities and surrounding regions. Hong Kong's and Taiwan's call centers serve their markets; they do not contact the rest of China.

Cost and Benefits

Some countries may offer cost savings as foreign offshore sites. But as with U.S. offshoring, the benefits must be high enough to make the move or outsourcing partnership worthwhile.

As a nearshore location, New Zealand is less expensive than Australia. The amount of savings, 10–15 percent, is large enough to warrant adding a center there from new growth but too small to merit relocating a center.

"Most companies won't go through the disruption for such small savings," says PacTac's Smith. "Also, New Zealand doesn't have the breadth of languages that Australia has."

Acceptable Locations

Some countries are more acceptable than others to base or outsource your operations in. One example is South Africa compared to India, reports Michael Allen of The Mitial Group. The country has a strong customer service culture and a sophisticated domestic call center industry.

Another benefit: You won't get as harsh a backlash from British customers and unions as you would by serving the UK from India. British consumer resistance to offshoring is twice as high to India compared to South Africa.

"There is more of an affinity between Britons and South Africans because many Britons have moved there, vacation there, and there are longstanding business relationships between both countries," says Allen.

Look at Co-offshoring American and Foreign Programs

With "co-offshoring" you could cut your total costs and spread them out amongst more work. Most of the English- and Spanish-speaking offshore locations are suitable for U.S.-serving call centers. Clients indirectly save money on seats, terminals, and cabling, the costs for which are passed on.

Minimize Accents

Stay clear of accents, pronunciations, and idioms associated with any one country. Require bureaus to train agents to speak with neutral tones and non-offensive words.

Compliance with Foreign Laws

Whether you serve foreign customers with in-house or outsourced call centers, make sure you comply with foreign laws. For example, if you have European customers, you need to adhere to the Data Protection Directive. As noted earlier there is a Safe Harbor agreement between the U.S. and the EU to enable compliance by U.S. firms.

To assist that process, the Direct Marketing Association (DMA) provides a EU Safe Harbor Enforcement Program free to its members. This program ensures online and offline data privacy through a third-party dispute-resolution mechanism for European consumers. This service would otherwise cost upwards of $7,000 per year for non-members.

Because many Safe Harbor principles will require companies to adopt new practices, the DMA says companies may find it beneficial to join an organization for verification and possible enforcement action.

Consider Insourcing

Look at insourcing, that is, having a service bureau manage your in-house call center. You retain the bureau's call centers for overflow, after hours, and disaster recovery.

"Insourcing is a common practice in Asia-Pacific countries because it enables the call center owners to familiarize themselves with their customers," says Kaplan. "[They also learn about] labor and management markets and practices in their host countries, while maintaining control of their data and operations."

Start First with Existing Offices

If you are serving foreign customers, chances are you will have a presence in those markets. That can be a sales office or distribution center. Build out that facility first as a call center and look at outsourcing variable-volume or after-hours contacts. Get to know the lay of the land. Gain experience with cultural, language, and labor issues. As volume builds, examine your options: in-house, self-service, outsourced, offshore, or home agents.

If outsourcing, R.H. Oetting's Gantman recommends that you have a local presence, such as a sales office, in the country or market you're targeting. That makes outsourcing much easier to manage than if you attempted to do it from the U.S.

The only exception is Canada, where there are great cultural similarities with the U.S. and many small service bureaus, some offering bilingual support.

✪ SERVING FOREIGN CUSTOMERS WHILE ABROAD

There may be occasions where you will need to serve your foreign customers in their language while in other countries, including the U.S. Examples include business travelers.

Japanese-owned Prestige International made its name by providing Japanese-language concierge services for clients. For example, if a Japanese businessman is staying at New York City's Waldorf Astoria and wants theater reservations to see the ninth life of Cats, a Prestige agent thousands of miles away will handle the request.

✪ SERVING FOREIGN CUSTOMERS AT HOME

There are occasions where it makes sense to serve foreign customers at American call centers, either in-house or outsourced. These are:

* Low-volume, high-value calls where the demand is too low to merit a separate center.
* When costs of centers abroad do not outweigh cost savings and added complexity.

Pearson Education had outsourcer Connextions.net design, develop and provide web-based, multichannel customer service to international customers of its Longman e-learning division in 10 languages: English, Arabic, Chinese, French, German, Italian, Japanese, Korean, Portuguese, Castilian Spanish, and Latin American Spanish.

"Connextions.net is helping us to develop a new kind of e-learning environment," says Rick Altman, vice president of sales and operations for Pearson's Longman English Success division. "Utilizing Connextions' Orlando-based contact center is a much more cost-effective and attractive alternative to operating multiple centers in multiple overseas locations."

There are services to help American call centers serve foreign and immigrant customers. Language Line and Tele-Interpreters provide translation services.

The Great Voice Company provides foreign-language prompts and recordings in many dialects. The Telephone Doctor offers instructor-led, streaming video training on how to deal with customers' accents.

Puerto Rico, with a highly educated labor force, security and stability, is in play to attract high-end Latin American centers such as financial services. While Puerto Rico has the U.S. minimum wage, residents do not pay income tax.

John Boyd Sr. reports that some U.S. firms are looking at setting up call centers in South Florida, Tampa, and New Orleans as options to locating in Latin America.

"While the wages are nominally higher, this is compensated by lower travel costs and less complexity in call center setup and operations," he points out.

Precision Response Corporation handles Latin American, U.S. and Canadian clients' customer calls from its south Florida call centers. It has multilingual employees, including French- and Portuguese-speaking as well as Spanish-speaking agents.

LOCAL OR GLOBAL BUREAU?

When outsourcing, whether or not you go with a multinational bureau with U.S. offices or a local firm in another country depends on your comfort level and how much you're willing to perform due diligence.

With a U.S.-office bureau you have the convenience and security of working with an American firm. You can also monitor activity from there, minimizing travel. Researching and holding vendors accountable is straightforward; they have to obey U.S. contract law.

But with a locally based bureau you may get better knowledge of the markets. The downsides: due diligence, travel, and being subject to foreign contract law.

When comparing multinational to local outsourcers to serve foreign markets, Philip Cohen notes that the local bureau may offer better service if its call center is designed and managed in keeping with cultural norms.

"But there are no strong advantages to doing business with a local outsourcer compared with a U.S. or foreign-owned outsourcer," says Cohen.

Steve Ferber, PRC's executive vice president of strategic planning and international development, cites cultural, dialect, and economies of scale among the challenges for companies going with a country-by-country approach.

The savings come in vendor relations, centralized management, IT, and other consolidated infrastructure costs, plus any client travel to locations.

PRC handles the same program for those three different customer bases from the same center with the same equipment, compared with outsourcing to three different vendors and three different call centers.

"You could locate or outsource, say in Panama, and get agents for a lower labor rate than in the U.S.," says Ferber. "But you may lose money and quality if your objectives are to serve the entire Latin American market, because you will also have to locate in several other countries to properly handle those consumers."

○ FOREIGN FACILITIES AND MANAGEMENT

Using foreign and offshore centers have similar issues concerning facilities and management. In both cases you have to be aware of the practices in those nations.

Management Policies

U.S.-based outsourcers locating offshore realized that Asian agents do not like direct confrontations with supervisors. Instead companies like ePerformax rely on scorecards to rate agent performance. They also had to scrub emotion; instead, just present the facts.

Americans can be also seen as bullying in cultures where nuance and tact matter more. Canadians, who are the most acquainted with Americans, will get stubborn if they feel they are being bossed around too much. They will use "American" as an epithet.

For those reasons the smart U.S. firms hire local managers to run their call centers. They set goals and standards but do not micromanage the operations.

Many foreign nationals such as Canadians have more us-them attitude in labor and workplace relations than Americans have. They don't trust management, especially those working for foreign owners. There is a greater acceptance of and tolerance for unions and labor disputes in these nations.

Workplace Restrictions

Workers have much more protection in foreign countries than in the U.S. on matters such as dismissals, layoffs, hours worked, and overtime. That creates additional costs and paperwork. "You have to practically write out contracts with employees showing what is expected and how you are going to measure their performance," says ePerformax president Teresa Hartsaw. "Your only defense in firing someone is to show objectively in writing how they underperformed and how this is in breach of the expectations when they were hired."

This is ePerformax's call center in Manila, the Philippines. But it could be anywhere. Top-quality foreign and offshore call centers do not look any different than those onshore and nearshore. The tasks are the still the same: serving people efficiently. The key is the first part: If you are outsourcing foreign-shore or offshore, make sure you pick a quality vendor. Photo courtesy ePerformax.

Property

If you are locating a call center in another country there several property considerations such as availability, access, security, climate, infrastructure, and standards.

Limited Availability

Call centers may have challenges in locating suitable space in other nations like Canada. Canadians are conservative people when it comes to money. They avoided overbuilding commercial space, which means that finding available property can be very difficult in many locales. A generally strong economy and reasonably vibrant downtowns have meant fewer big vacant office or retail buildings.

On the upside Canadian cities are livelier, which means less risk in picking downtown locales. Mass transit is generally better, with extensive weekend service, and more popular. That enables you to lower the number of parking spaces you need. Cycling is a year-round transportation alternative in Vancouver and Victoria, British Columbia. You may also find Monday-Saturday scheduled bus service in cities as small as 30,000.

In Europe, top-quality space, especially conventional offices, may also be difficult to find. You may have to build instead.

Developers are now converting former textile mills and port warehouses, which have large floor plates and plenty of natural lighting. They're also interested in shipping sheds, which have efficient, large floors.

"Ironically the cities and regions of cities that suffered the most bombing damage during the Second World War, like Berlin, London's Docklands, and to some extent Liverpool and Southampton, have the best property," Michael Allen of The Mitial Group points out. "They have either been undeveloped until now or redeveloped with large-floor plate buildings."

Space

Offshore call centers will need more training room for cultural education and accent neutralization.

Climate

Other countries have even greater extremes in climate than found in U.S. cities. For example, in Canada many vehicles built for the Canadian market have engine block heaters. Yes, those prongs you see out of Canadian cars do not mean they are battery-powered, sad to say. If you locate to more northern cities, you have to equip parking stalls with plug-ins for those heaters.

Voice, Data, and Power

Voice, data, and power lines are reliable, even more so in developed foreign markets. The same cannot be said for developing countries (or for that matter, the U.S. electricity grid).

You will need to ensure that there are multiple voice and data connections, such as through satellite. You must have onsite power generation. In developing nations it is the electrical grid that backs up the site generators, not the other way around as in developed nations.

Transportation

More people rely on public transit in foreign countries than in the U.S. Few developing country residents own cars. In the Philippines, ePerformax has a pool of cars and drivers that will drive employees to and from major transit stops.

You will need to have your call center on bus or rail routes. This also goes for some developing offshore countries, like China and India that have extensive rail networks. You may need to provide shuttle transportation for your workers in India to and from homes or major rail stations.

With most U.S. offshoring work occurring during the night, in India shuttle companies need to provide shuttles to bring them to and from the workplaces safely and reliably.

To handle buses and shuttles, Convergys' call center in Gurgaon, India, has bus loops and shelters. Convergys also installed motorcycle racks. Many agents who get to work by themselves ride motorcycles, which are less expensive to buy and maintain. They're also more flexible to operate on India's chaotic roads than cars are.

Amenities

When you plan a call center in Europe, you will need to provide access to high-quality amenities, either on site or nearby. If there is not any suitable property that offers this in outlying business parks, you may have to locate your call center in downtown areas, which have high-quality mass transit.

"If you open up a call center in a European business park and the agents have nothing to do during their breaks except eat at your little cafe, they're going to leave your call center very quickly," warns Allen.

On-site amenities are vital in locations like India and the Philippines. Call centers must provide cafeterias, large washrooms, and sometimes dormitory and shower space.

"You don't have the culture or the outlets for fast food or restaurants off-site in India as you do in the U.S. or Canada," explains Dennis Ross, Convergys' manager of offshore operations.

ePerformax created a training and community center located two blocks from its Manila center to support employee wellness and development outside their regular work hours. The building has three floors plus a rooftop terrace, which is ideal for employees who want to smoke, as no smoking is allowed inside the building.

The building's amenities include a transition area between commuting and working. It has large-screen TVs showing popular shows to allow employees to "veg out" in front of their favorite show. There is also a nurses' station and a sick room for minor medical attention.

The center has a café that serves coffees, cappuccinos, lattes, teas, iced fruit drinks and snacks, similar to a Starbucks. There are computers with Internet access for those employees who like to surf the web; pool tables for those who need to get the competitive juices flowing.

"In the Philippines, employees travel an average of two hours to get to work and two hours home, and many of them take one or more forms of transportation to make this journey," explains Hartsaw." As an added challenge, traffic or bad weather can extend the transit time by an hour or more. Since schedule compliance is a condition of employment in our business, most reps have to give themselves additional time for travel just in case. So many end up coming in early."

The center also has two training rooms. They provide continuing educational programs that also support personal development. The continuing education programs are held before and after major shift changes and will range from wellness education to career development. There are also mens' and womens' locker rooms complete with showers, fresh towels and all of the amenities required to freshen up before work.

The training rooms also double as sleeping rooms. So, when there's no training, employees can take advantage of the space to take catnaps and even spend the night if the weather or other circumstances make traveling too and from work very difficult.

"It's a bit of the home-away-from-home concept that enables employees to feel good about the time they are transitioning from their families to their work environment and back again," explains Hartsaw.

Design Elements

Designs should reflect the country's culture and milieu. Convergys' call center uses high-quality glass and marble, because these materials cost less to purchase and install. And they help cooling. Tinted glass also reflects light, cutting heat. Marble is a "cool" material, which means it does not trap heat: an important consideration in hot climates like India's. There are a few other interior differences. Indians tend to like more vibrant colors than Americans, including shades of blue, purple, and red.

"One other difference we found when we built the center was in the building support pillars," notes Ross. "In the U.S., designers hide them with plasterboard and other coverings. But in India the agents preferred to have them uncovered and asked us to backlight [the support pillars] because they showed strength."

ePerformax had its Manila call center designed in accordance with the principles of feng shui, an ancient Chinese philosophy that believes that the positioning and physical characteristics of a given space affect the fortunes of the owner and its inhabitants. The Filipino people are a mix of Chinese and Malay extraction.

"Our space has been designed optimize the flow of positive energy or chi (and deter negative energy) to improve the health and well-being of the company and its activities, as well as our employees," says Hartsaw.

Security

Take steps to ensure security in foreign locations, especially in developing nations and if there is the risk of terrorism. Consider having security as stringent if not more than that at home when opening a call center offshore. Do so if there are other tenants in the building or you face legal requirements such as HIPAA compliance for handling healthcare data. If you're worried about being identified, do not have your name on the side of the building.

Gary Kimnach, director of facilities for outsourcer APAC Customer Services, recommends that security measures include electronic passkeys, guards, and restricted access to areas such as computer rooms.

"We apply the same level of security whether in the United States or offshore," he says. "In the Philippines we ensure that our security is very tight because our insurance program requires HIPAA compliance."

Improvement Financing

When locating in other countries be prepared to finance tenant improvements (TIs), like increased heating, ventilating, and air conditioning (HVAC) capacity, rather than having the landlord pay for them like in the U.S.

In offshore countries, these buildings are typically in shell condition with inferior or high-cost HVAC systems for supporting a 24x7 operation, reports King White, vice president of Trammell Crow Call Center Services.

"In the Philippines, almost all of the time, the tenant will be responsible for 90 percent or more of the construction costs," he explains. "The cost will be similar to the U.S., which means you should expect to pay at least $40 a square foot to build out your call center plus your furniture, fixtures, and equipment."

In India, the costs are the same; however, the financing source varies depending on the city and who the owner is. When the landlord will fund the TIs, they will typically put it into a separate agreement, which is more like a loan.

"Most companies do not realize the impact the TI issue will have on the ability to establish centers offshore," White points out.

Project Management

Site selection firms such as Trammell Crow and call centers that have located offshore recommend that you hire a local or regional architect or construction manager who has experience in these locations when considering when an offshore center.

"Local architects and managers know the regulations, bureaucracies, trade labor, markets, and subcontractors," says APAC's Kimnach. "The local experts are qualified to successfully complete the project, as they have expertise in call center facility build-out."

✪ CONNECTING FOREIGN AND OFFSHORE SITES

Many companies prefer packetized voice (IP) and run it over managed or private networks to their overseas call centers. For example, Kevin Wilzbach, Convergys' director of integrated contact centers, reports that all of his firm's voice and data traffic from the U.S. to its international centers is IP over private leased lines but that voice traffic within Europe is circuit-switched.

IP is also becoming popular for domestic lines, especially in developing countries where Internet connectivity has surpassed circuit-switched voice.

"In India, for example, it is much easier and less expensive to link up call centers with VoIP than voice circuits," explained Bernard Drost, vice president of technology for Akibia "When a person has an Internet connection, it is very easy to get VoIP than to get the local telephone company to install a phone line."

Drost says many developing nations have poor and high-cost phone networks but good Internet connections. People there tend to use wireless devices rather than wired.

Also, the quality can be quite good. Drost was talking to someone in India through VoIP and didn't realize it until he was told.

There are other issues. For example, companies must record trouble tickets and other time-sensitive documentation in Universal Time. "Trouble tickets have date and time stamps on them, and companies require them to be resolved in a given time," explains Drost. "If the time is not correct, then the issues do not get solved when they should."

Call centers don't have to translate every comment on a customer record into a common language. That is an expensive and time-consuming task that often leaves inaccuracies because it loses idioms in the translation.

But there must be uniform terminology, such as what trouble tickets are called and defined as, to make sense of call center performance reports.

PLEASE REMOVE YOUR *WHAT?*

There is one sign that was once common outside of Canadian buildings whose words would make an American think twice and scratch their heads.

They read: "Please remove your rubbers."

Canadians may be strange people who drink strong beer, hit each other with sticks, speak funny, wear snowshoes, have gun control, and practice socialized medicine. But when it comes to sex they, or shall I say "we," are pretty normal.

The signs were placed to dissuade people from tracking in snow and slush on their galoshes, mucking up the floors.

Chapter 11: Property Considerations

You have selected a community for your call center. Now you need to look at property. That can be challenging because there are many considerations.

You should project the life span of the center (including how many agents you will need initially, at your peak, and at the end. That will determine the type of building you will require. Chapter 3 explores the methods to help determine how many workstations you need based on volume, after funneling those off to self-service, home agents, and outsourcers.

Also, decide whether your call center is going to be stand-alone or share with other users. Ask yourself whether the call center is a prominent part of your operation that you want to showcase or squirrel away in a back office. These considerations also affect your property selection.

If you decide to co-locate or showcase your center, then the property must be suitable. The buildings must be in highly visible locations with corporate exteriors and interiors. If your call center is to be located with a warehouse, then the building and property must be constructed and sited to handle trucks. The different classes of office space will be discussed in this chapter.

○ PROPERTY REQUIREMENTS

Now you need to set out your requirements, just like buying a home.

Space

Whether your center will need 30 or 3,000 workstations, you will have to plan for their space and for the needs of the employees who will work in them. A good rough calculation, provided by specialist design firm Kingsland Scott Bauer Associates (KSBA) for figuring out how much total room you will need for the call center plus hallways, training, break and administrative space, is to multiply the number of productive (non-supervisory) workstations by 90–140 square feet per workstation. If a workstation is a conventional cube 6x6x1.5 feet and you include circulation, it takes up about 60 square feet, or about half of the total area required to accommodate individual employees. No more than one person can carry out their tasks at any one desk. .

This formula provides for supervisory stations. Agent-to-supervisor ratios range from 8:1 for help desks to 25:1. The mean is around 10:1 to 12:1. The actual rate you select is based on your experience.

There are cost and performance tradeoffs in planning density. Yes, you can squeeze people in as low as a total of 80 square feet, compared with 90 to 140 square feet. Think about this real carefully. In tighter labor markets do you really want to go there and see turnover and staffing costs skyrocket and performance drop? The other side is that you can waste money without seeing any improvements by supplying too much room.

Ron Cariola, senior vice president at Equis, estimates that costs for real estate and facilities associated with renovations, build-to-suits, or leases, can range between $2,500 and $5,000 per seat, assuming 100 square feet per workstation. The cost of IT infrastructure often ranges between $1,500 per seat to $2,500 per seat. Higher amounts typically apply to knowledge-intensive help desks. Call center hardware and software usually costs between $1,000 and $2,200 per seat.

"Your IT costs depend on how densely populated you want your space," Cariola points out. "If you move to a less dense [workfloor], say 125 square feet per agent, it may cost you more, but you may achieve a better working environment."

Training Needs

Take a look at how you train your agents (more about that in Chapter 14). If you train in on-site classrooms, allocate 10 percent of your total space to them. By going to alternative methods such as instructor-led conferenced or technology-based training, you can add more workstations into the same footprint or reduce the amount of space by the same amount.

Air Quality

Air quality is the most vital environmental element in your center. Good air keeps people awake and lively; poor air puts them to sleep or spreads illness.

Don't take it for granted that just because nearly all buildings have air conditioning that the air is good. It often isn't. Because call centers are higher-density occupancies than other office uses, the air systems in existing and new designs may not be optimized for your needs. Therefore you need to spell them out.

With modern buildings so airtight there has been a growing problem known as sick building syndrome: symptoms such as headaches, dizziness, muscle cramps, edema, and chronic fatigue that arise from sources like formaldehyde, cleaning solvents, and mold. These symptoms often fade away when employees leave for home.

Too often, offices are poorly ventilated, with minimal air changes, keeping airborne diseases and toxins inside. The cover story from the June 5, 2000 issue of *BusinessWeek*, "Is Your Office Killing You?", reports some buildings draw in only *five cubic feet* of fresh air per person per minute.

There are voluntary standards for how much HVAC you need. HVAC is supplied through rooftop or ground-level units. The American Society of Heating, Refrigerating and Air-Conditioning Engineers (ASHRAE), recommends that you provide 20 cubic feet of outside air per minute, per employee.

Even more critical, specify that the air intakes are not at ground level, especially where

EFFICIENT AND SECURE EQUIPMENT OPTIONS

Shaping your total space requirements is the equipment you select for them. Here are two examples:

Flat-panel monitors
Flat-panel monitors (FPMs) are becoming the standard because of their appearance, but they are also cheaper. FPMs slice the square footage by 10–20 percent without giving up functional room. They are much more energy-efficient, creating less heat to be removed, and they require less wiring than conventional cathode ray tube screens.

Storageless PCs
Storageless PCs take up less room on or under desks. They can also rest on top of existing monitors, taking up as little as 24 square inches. These PCs are small enough to mount on desk or wall brackets, which gives more space and legroom.

Storageless PCs have other benefits, including lower power consumption and less generated heat. That means lower HVAC expenses and greater employee comfort.

A study by Thin Client Computing indicates that typical annual savings based on having as few as 100 thin-client storageless PC users can range from $3,000 to nearly $6,000 per year, depending on the power costs paid in a specific region.

Storageless PCs are also more secure. They do not have internal data storage or external connections like USB ports. That prevents employees from uploading spyware and viruses and downloading data. There is no data residing in the hardware's local memory.

Some PCs go further. Neoware's Capio and Eon units include customer-controlled setup security, access to which is controlled by password. Users cannot modify the setup.

Also, storageless PCs have no street value, which virtually eliminates hardware theft. "No data, no connection to the server, no use," says Neoware product marketing manager Jim Powell.

Cascade Callworks is an outsourcer based in Vancouver, Wash. The service bureau bought 85 Capio storageless PCs in spring 2003. The units, which sit behind the screens, take up less workstation space, which gave more desk area for agents.

"We also avoided damage and scuffing to the PCs with Neoware," says Cascade CEO Shawn Suhrstedt. "They replaced tower units that employees would accidentally kick or rest their shoes on."

The appliances also enabled the bureau to reduce power consumption by 7,920 kilowatt-hours per month. That saved the bureau $546 per month plus its share of HVAC expenses—Cascade shares the building with other tenants.

The Capios provide better security, a key reason why Cascade Callworks choose them.

"It is much easier to back up data in one spot, scan for viruses, and firewall with one server," explains Suhrstedt. "With our use of the Windows 2000 Server, we are able to grant very specific user rights to eliminate the threat of users downloading, installing, browsing, and deleting non-approved files. It has lessened the burden on our operations staff greatly."

there are roads, loading docks, parking lots, smoking areas, or doorways, which employees also use for "butt pads"—i.e., smoking. Also, you should have no air intakes above parking garages. You suck in poisons that will injure your staff and cause your productivity and turnover to plummet and health costs to soar.

Placing air systems in such bad locations; and designing and allowing parking garages to vent fumes into existing air intakes is tantamount to criminal negligence. Like discharging carcinogens into the public water supply.

The upsides of such specifications are clear benefits from healthy buildings. Cleaning the air could save businesses $58 billion and improve employee productivity by $200 billion annually. The *BusinessWeek* article cites Lawrence Berkeley National Laboratory researchers William Fisk and Arthur Rosenfeld, who said the benefits are eight to 17 times greater than the costs.

When drawing up your specifications, select the HVAC capacity that matches the climate and the amount of direct sunlight you expect to have pouring through your building. A hot, humid climate requires more air conditioning capacity than a cold dry, one, which needs more heating. A call center that is bathed in direct sunlight, even one in a more temperate climate, will place more load on a cooling system.

It is a good idea to have a backup cooling unit when setting up in hot climates. The loss of air conditioning could force your call center to close.

Washrooms

Some people regard washroom facilities as amenities, like cafeterias. I beg to differ. They are essentials, unless you want everyone to wear diapers or have to sneak out behind the dumpsters or into the bushes.

Building codes and laws such as the Americans with Disabilities Act (ADA) govern issues such as how many washrooms you need and accessibility. But standards aside, there are often not enough washrooms.

Women use the restrooms more often and for longer periods of time than men. If your workforce is predominantly female, you have to accommodate that. Analyze the demands in your existing centers, including observing queues and listening to employee gripes. Then have your facilities people lay out your requirements.

David Meermans, director of Product Management at EADS Telecom, points out that adding washrooms during construction is economical compared to the cost of an operation with too few.

"The single greatest and most easily resolved source of complaints you are likely to observe is inadequate restroom accommodations," he points out. "This sounds simple, but it is amazing how easily such simple things can get out of hand."

Access

Your employees have to get to work, unless they are exclusively working from home (see Chapter 5). Therefore you need to accommodate them.

The first factor you have to consider is commuting distance. Specify property where most of your workforce lives within 30 minutes driving time or one hour by mass transit. A 50-minute drive is a headache; a 50-minute bus or train ride less so, unless they're packed in every day like sardines.

Second, in most but not all locations you need adequate parking. That not only includes ensuring the maximum number of spaces that will be occupied during the shift but also additional slots during shift changes, when employees arrive before the others leave.

Parking is typically expressed in ratios referring to the number of stalls per 1,000 square feet of space. Ratios of 5:1 to 7:1 are adequate for most call centers. But you can get away with less if you choose property on or near mass transit or where a number of your employees cycle or walk.

Carefully consider how your employees are going to reach your call center. Locating in downtowns and on new rail transit lines is an excellent strategy. Salt Lake City, Utah has, like an increasing number of North American cities, a successful and growing rail transit system. Credit: Brendan B. Read.

Driving Alternatives

Call centers typically offer low-paid employment. That often means that many workers can't afford their own vehicles. To attract and keep these workers you may need to locate on a transit route with service levels that meet your daytime; or, even better, your afternoon; shift.

But there are other compelling reasons to write transit, cycling, and pedestrian access into your requirement. You can get away with lower parking ratios, which means you have more building and property choices, such as those near rail stations, bus stations.

Also, you will be complying with emission standards. Many metro areas are failing to meet Clean Air Act requirements. Their highway systems are also badly congested. So be part of the solution and encourage your workers to not drive.

Access Evaluation

You have to determine where the bulk of your workers will be coming from and by what means. If they are mainly suburban spouses or your center is in a small city, they will come by car and will want to stay in the suburbs or that city. Your call center will need to be in a location (almost always a suburb) with good road access, and the property must have excellent parking. If they are college students, trendy young people, or people who live in downtowns, they may come by mass transit or bicycle, with some by car. Therefore your call center must be on a transit route, near a station, or downtown.

Accommodating Smokers

You have to find some way to accommodate those employees that are nicotine-addicted and who prefer consume this drug by smoking—i.e., smokers. I described smoking this way for a reason. You can consume this substance by other means that have fewer consequences to people outside of yourself.

Many people who work in call centers are smokers. You therefore need to lay out in your requirements for a firesafe area for smokers away from entrances and air vents. To determine the size of that area figure out the percentage of your employees who smoke, based on your existing ratio of smokers to non-smokers, and map that out per workstation so that you know how many people will be using it per shift.

That call center employees smoke more than other office occupants have given call centers a bad reputation among many landlords. Without a preplanned smoking area, the smokers will congregate outside of building entrances, forcing others to be poisoned as they walk by, creating fire risks and garbage. Or they will hang out in the parking lot.

The property you pick must have an area that you can create for them, preferably outdoors. These are sheltered and easy-to-clean "butt pads," away from entrances and vents, with neat-looking and easy-to-clean fireproof ashtrays. The smokers should be responsible for cleanup.

Having this in the specs up front should help open more doors with landlords and developers. It shows that you're aware of the problems and have planned for them.

Technology

When planning and seeking property for your call center, you have to take your technology requirements into account. The hardware takes up space and must be connected, backed up, and protected. Hardware also creates heat that must be removed.

This book is not about technology. There are many fine titles such as *A Practical Guide to CRM* by Janice Reynolds and *The Call Center Handbook* by Keith Dawson, also published by CMP Books, that delve into the topic. Also, because hardware and software change rapidly, read; *Call Center Magazine*, the best source on products and services tailored for the industry.

What is important to note is that call centers have intensive technology needs over and above those required by standard offices. They include automatic call distributors (ACDs), CTI and CRM software, IVRs, and predictive dialers.

Technology is not cheap. Equis's Cariola reports that call center hardware and software usually costs between $1,000 and $2,200 per seat.

Therefore you need to decide what functions you want your technology to accomplish and look at the options. They include:

Premises-Based

Premises-based technology means you lay out the capital for hardware and software.

You will have to provide dedicated space for the servers and have in-house or outsourced IT staff to support the equipment.

Premises-based technology typically includes switches, routers, IVR, auto-attendants, predictive dialers, and servers.

With premises-based technology you get what you need to match your specific requirements. You also own it and can change it to meet your needs. The downsides are that in the case of hardware you need to find room for it. Switches, servers, and IVR ports ideally need to be computer or phone rooms that are climate-controlled and access-controlled. You must also outlay the capital to acquire or license the gear.

Hosted

Hosted means another company owns the hardware or software, or both, and is responsible for maintenance. Instead of purchasing the product, you in effect lease it; you pay a setup fee and user charges. Hosted applications are common for ACDs, contact management, CRM software, IVR, and switching.

Application hosting is supplied by specialized technology firms, known as application service providers, outsourcers, systems integrators, telcos, and the hardware and software vendors themselves. Applications service providers work like car dealers. That gives you the choice of buying or leasing the gear.

Application hosting eliminates or reduces the need for the servers and switches in your call center. That means less room is needed and energy costs are lower. It also provides disaster protection because the hardware and software are offsite; with switches you can have them reroute contacts to other centers, temporary facilities, off-site recovery rooms, or to home agents. On the other hand, you may need a fatter pipe from your site.

Application hosting can save installation time. For, say, an enterprise-wide CRM application, hosting can save approximately three to nine months compared to buying and owning. This translates to 12–18 months instead of 18–24 months for a large application and to as little as 30 days instead of three to nine months for a mid-size CRM package. You also avoid buying additional hardware and arranging for more networks and applications.

Carrier Connections

You will need to know your expected voice and data volume from carriers and write that into your requirements. Some buildings, especially in smaller cities, may not have all the capacity you need on the spot. You may to wait for additional lines to be provided, which could lengthen lead times and impact your performance.

In the first edition of this book I asked a leading authority, author and consultant on telephone systems, Jane Laino, founder of Digby4 Group, to contribute information for this section. Her advice is restated below.

Connections to the outside world, for all calls, can be via individual circuits (one pair of copper wires per outside line) or by high-capacity circuits known as T1 or PRI (primary rate interface, a more sophisticated version of T1), she explains. They handle the equivalent of up to 24 outside lines. The T1 handles incoming and outgoing calls.

To figure out capacity requirements, Laino says you must remember that there must be a separate outside line for each call in progress (either incoming or outgoing). For example, if you have two T1s (24 calls each), you can have 48 simultaneous telephone calls in progress.

For an existing operation, you determine the number of calls handled in the busiest hour of the day and refer to a statistical traffic engineering table to identify the number of outside lines needed to handle that call volume. To use the tables, you must convert the call volume into a total number of minutes or seconds in the busiest hour of the day. See *The Traffic Engineer Handbook* by Jerry Harder.

"For new call centers make your best guess of the number of simultaneous calls and double it. It is better to have excess than to have callers reach a busy signal," advises Laino. "Excess capacity may be absorbed or turned back in if not used."

The data circuits needed for a call center depend upon what business is being transacted there and Internet access. As with voice communications, there are statistics that consider the amount of information that needs to be transmitted and the speed requirements. This dictates the bandwidth of any data communications circuit.

There are new technologies that affect cost, bandwidth, timing, and feasibility that are worth examining. Some firms are sourcing high-capacity wireless connections for encrypted VoIP and data from carrier points of presence to bypass the local carriers on the "last mile" from the central offices to buildings. There will be companies deploying wireless communications made to the WiMax standard, which delivers wireless broadband signals up to 30 miles. The wireless carriers promise swift installation, which often does not happen with telco T1s and at much lower prices.

If these technologies interest you, see about their current and future availability and coverage when looking at property. If you plan to be the sole tenant or owner of a building, getting those connections may be easier than if you are sharing the structure with other tenants. The antennas are small so you should have minimal troubles.

Tolerance for Competition

Call centers in the same area are like having two or more cats (my wife and I have four) sharing the same house. Each is jealous of their territory.

Therefore, you may or may not run the risk of a staffing war. Instead, you may want to locate your call center elsewhere in the same region, tapping into some but not all of the same workers as the other centers.

If the presence of other call centers troubles you, see who is out there, whether they have the same function as yours and compete for the same labor pool, and whether they are in the same immediate area as you plan to locate in. A job handling debugging software requires quite different skillsets than one selling day-glo vinyl siding. There is, however, growing sales and service skills convergence, as companies want every agent to provide both functions.

But if you are not bothered by other call centers being in the same labor market, fine. You may even benefit from them. If your company has a great reputation with customers, clients, and employees and pays well, you'll be able to pull other call centers' experienced and trained workers into your labor pool.

Security and Business Continuity

Security and business continuity have become more important as call centers become critical to organizations and in the wake of the 9/11 terrorist attacks. Also, if your call center handles health-care data, you are required under HIPAA to have some form of disaster response plan. You therefore need to write these needs into your property requirements.

Risk Minimization

Specify that the building or property (if you plan to build your own call center) is to be in a low-risk area. Unless your call center is co-located with a warehouse, avoid industrial areas, especially where there are petrochemical plants and railyards that pose fire dangers from accidents, derailments, and sometimes arson.

Also airports, defense contractors, foreign embassies, and famous buildings possess security and terrorism dangers. I once worked as a security guard at a defense contractor around the same period another similar firm was bombed.

Restaurants are also a fire and theft risk, though they are a convenience to employees. Therefore consider specifying that your call center should not be adjacent to one.

It is not a good idea to locate on a floodplain, for obvious reasons. Neither it is smart to be on an active fault or on old lava or mudflows. When Mount Rainier south of Seattle blows, expect a huge loss because homes have built on the latter.

There are also indirect risks. These include risks from bridges, tunnels, and other traffic chokepoints on the key access routes that could prevent emergency teams from arriving quickly and impede evacuation.

Suburban office parks are notoriously bad for evacuations. There is often only one way in or out, which is not good when frightened people are leaving while emergency response crew are trying to get in. The access roads and nearby expressways are jammed at the best of times.

Safe Surroundings

The call center should be in an area that is safe with relatively low crime, especially at night. You want your employees to get there and back safely, particularly from their cars and bus stops. I'll say more later on evaluating buildings and sites.

Building Security

Write electronic keyless entry and other high-security features into your requirements. If your agents are working after hours, stipulate live cameras covering entrances, walkways, and parking. Also stipulate on-site security guards, preferably two per shift so that one guard can accompany employees to and from their vehicles.

Consider demanding buildings with entrances that have security guards stationed between the entrance and inner doors. See about having biometric (fingerprint-reading or iris-scanning) pads at entrances to high-security rooms.

Voice, Data, and Power Protection

You need to determine how critical it is for your call center to remain operational during a disaster and for how long, and your tolerance for outages of voice, data, and power lines.

Voice and data outages are responsible for 35 percent of outages, Kurt Sohn, principal consultant with call center business continuity planning firm 180cc. Natural disasters account for 35 percent, while man-made disasters account for 30 percent in call centers.

But the methods to enable center functions to withstand disasters can cost hundreds of thousands of dollars. Some techniques, like directing contacts to home-working agents, require managers to change their thinking. To properly protect your contact center also demands more careful planning to minimize risks when you select your locations.

Self-service, home working, and outsourcing provide backup methods that do not affect your property design, as does rerouting calls to other in-house or outsourced centers or offsite recovery sites.

Ask yourself how vital to your enterprise is having your center continue to function in a disaster, compared with shutting it down and evacuating the staff, and what is that downtime in a disaster worth to you? How long can you afford to be offline? What is the return on investment of the various response means?

The answers to those questions will help you decide which backup systems you want the building and the site to have.

Following is a look at the tools and technology.

Battery-Powered UPSs

The most basic tool is a battery-powered uninterruptible power supply (UPSs). UPSs keep computers, switches, and other vital components energized during most outages, until the lights return.

UPS permit backing up of data offsite as long as you have voice and data links. You can also download the data onto CDs or disks.

UPSes can work with backup generators. You should not have a generator *without* a UPS. You need the UPS battery to maintain continuous power to smooth over the gap between when the outside electricity kicks off and the generator kicks in. Otherwise the resulting voltage spike can zap your equipment.

UPSes also helps prevent disasters. They protect against damaging voltage spikes and sags in power supplies. UPSes and surge protectors on your phone lines can help prevent your center from being zapped.

UPS power supplies are large and heavy, especially if they are supplying entire floors. That is an important consideration if you are locating above the ground floor, because some buildings' floors may not handle the weight.

When you hear the words "backup generator," you think of big, hulking diesel engines that had been stripped from freighters or railroad locomotives. While many such units are larger, there are also many types are not, like this Kohler set that has a twin cylinder producing 15 hp.

The battery cabinets alone can be as heavy as 400 pounds per square foot, but a typical above ground space permits 45 pounds per square foot of floor loading. Tell your architect if you're planning to have a UPS so they can come up with the specs.

Onsite Generators

Onsite generators are for truly critical applications. They can maintain your operation for days.

But these units cost several thousands of dollars. Not every building or site will allow you to have them, because there are often zoning restrictions on these units, which create air and noise pollution. Fuel leakage and explosion and fire risks are also a concern. If your generator uses natural gas and the line snaps, then you've lost that fuel source. You will also need signed contracts to guarantee fuel, such as diesel or propane that is stored on site. When a disaster hits, take a number.

Blended Generation and Feeds

Consider having both multiple feeds and onsite generation. If the local utility goes down, it will not matter how many substations you are hooked into.

"Multiple locations and feeds mitigate but do not entirely eliminate; the need for onsite generation," consultant Bob Engel points out. "Unstable power is a problem nationwide that strikes everywhere. It isn't just a California or a Northeast problem anymore."

HOW MUCH DOWNTIME COSTS YOUR CENTER

This worksheet determines the estimated dollar savings that a UPS can provide your company. simply fill in the information to calculate your cost of downtime for one hour.

1. Number of critical loads
2. Number of employees using critical loads
3. Employees average hourly earnings
4. Estimated cost of lost business per hour of downtime ($1,000, $5,000 $10,000...)
5. Cost of service calls per hour (average cost is $100 per hour)
6. Cost of replacing hardware (if applicable)
7. Cost of reinstalling software (if applicable)
8. Cost of recreating data (if applicable)
9. Lost employee time (lines 2 x 3)
10. Lost business (line 4)
11. Service (line 5)
12. Replaced hardware and software (lines 6 + 7)
13. Recreating data (line 8)
14. Estimated total cost per hour of downtime

Backup power isn't inexpensive but neither is having your call center down. This worksheet from Exide Electronics (Raleigh, NC) helps you decide whether a UPS system is justified.

Fat Pipes

As noted in Chapter 3, call centers handle a lot of data. That information, along with critical applications, should be backed up off premises before disaster strikes. Temporary power systems like UPSes should be purchased with enough capacity to permit at minimum orderly data backup and computer shutdown.

To enable that added data load, specify additional cabling into the property.

Daniel Frasca, vice president at The Alter Group, a firm that develops call centers, says the increased wiring is needed because the data transfers are occurring at the same time as normal operations data is entering and leaving the premises. But in disaster mode, the normal data flow stops and the critical data is shipped off premises.

"With this enhanced transfer program, there is less risk that call centers will lose critical information, which may happen when disaster hits," says Frasca.

ACD Failover

PBXes use AC current to operate. But you can get around power failures by having power failure transfer (PFT) equipment to place in front of the PBX or ACD. PFTs route calls to basic ACD phones, though without skills-based routing, during power outages.

Satellite Backup

You can back up your network with satellite feeds. But make sure the actual downtime is worth the added satellite hardware, installation, and transmission costs. Also, some buildings or landlords may not permit them.

Multiple and Protected Feeds

You can seek or specify property with feeds from more than one source. These include multiple lines, central offices, and grids. Also look for properties fed underground instead of by poles and wires. While the former faces the odd slice by a backhoe, the latter is more likely to be downed by bad weather and lousy drivers.

These options are less expensive than onsite generation. But they limit your property and potentially your location choices.

✪ DESIGN FEATURES

There are a host of interior design features that you need to decide on. Chapter 12 goes in depth on many of them.

Floors

You will need to decide whether you want your call center to be on one floor or multiple floors.

Call centers work best on single floors. This makes for much more efficient supervision. You don't want your supervisors running up and down stairs or playing elevator tag. A single floor provides installation economies of scale. You don't have to run wire between different floors, which can be a headache. If you need to expand you can build outward, if your center is on the ground floor.

Bob Engel cites the example of JP MorganChase in New York City. It moved one of its call centers from a Manhattan building that had a large open floor to another tower with agents on different, smaller floors. Turnover skyrocketed and performance dropped as agents no longer could get assistance from management, who did not walk around as often, hampered by waiting for elevators.

If you have heavy power backup needs, a multi-floor may not cut it if you are not on the ground floor. As noted on page 218, the upper floors may not support the weight of large UPS systems. And it is awkward and probably not allowed to have a generator on an upper floor, unless you pick the top floor and the landlord permits a rooftop unit.

On the other hand tools like email, instant messaging, and web conferencing liberate employees from being on the same floors, or for that matter, from premises call centers. Studies show that managers and colleagues are switching to emails and instant messaging for internal communications because they are less time-consuming and more efficient compared with visiting cubes or offices.

Natural Lighting

Natural lighting makes a big difference in your call center's look and mood. Many studies have shown that people feel better when they receive natural rather than artificial light. How much natural light your center needs also varies by location.

But natural light presents two problems: glare and heat load, even in winter, making your air system work overtime, adding to costs. Specifying natural lighting impacts property selection and costs. That feature will run up the tab on postwar retail or industrial conversions, popular with call centers, because windows and skylights will have to be punched into the walls.

You can make your café area a real happening place, like at ePerformax's call center in Manila, The Philippines. Photo courtesy ePerformax.

Amenities

Amenities make your center livable and productive. They give agents, space and time to re-energize. They provide room for them to become human, network with co-workers, and blow off a little steam.

Amenities also impact real estate requirements and property selection. What you supply your agents will cost you money.

Break/Meal Facility

Agents and supervisors alike need to get out of the workfloor pressure vessel. You can let them unwind, say, on video games and foosball, and by reading.

Your break space should also be where employees eat. You do not want staff to munch at desks because crumbs and spills can ruin keyboards. They also attract bugs of multilegged varieties. David Meermans says to size your eating area to accommodate the maximum number of agents who will eat at any one time.

The break space and its features depend on the building you move into and the location. If your call center will be near cafés and restaurants, a simple area with a sink, coffee machine, and small refrigerator will do. If you have agents working evenings and weekends, look at vending machines.

If you locate in an office park, then you may need to consider having a space for contract caterers. In those locales restaurants are too far away for employees to get to.

Child Care

Ask any working parent what is the toughest thing to find and they'll tell you childcare. It is a chore to run the cute if troublemaking carrier of your DNA to the center in between work trips and shopping trips. On the other hand, your call center may appeal to other employees, such as college students, which in most cases do not have children.

Childcare is viable only when your workforce is mostly made up of parents. KSBA's managing partner; Roger Kingsland says that childcare facilities are rare in call centers because they cost a lot of money to set up. It's justifiable only if you have a large call center, 700 to 1,000 seats.

Gyms

Gyms are great in concept. Fit employees are healthier and more productive. Your healthcare costs also go down. But in practice, gyms have not often worked out. Call center staff prefer to exercise at home or fitness centers, and gyms are expensive.

Typically, the most frequent users of gyms are management, not the agents, explains Rick Burkett, of burkettdesign. He has heard of a study that reported that less than 10 percent of a workforce uses the gym regularly.

Employees like and need the little things too, like drinks, snacks and personal items. If you have a large call center and the best site for you is some distance from a convenience store, consider having your own "mini-store" like this one at a Convergys call center near Orlando. Credit: Brendan B. Read

He advises dedicating a small room with cardiovascular equipment as opposed to a full-size gym, which requires you to install costly showers and lockers for both sexes. This may be sufficient to attract agents without busting your budget.

"It is more successful and cost-effective to install 10 lockers at $200 apiece than buying three cardiovascular machines at several thousand dollars each, not including the real estate devoted to them, that no one will use," Burkett recommends.

Or as with childcare, select property that is a few minutes' from a gym. You get the best of both worlds, exercise without the expense.

Cafeterias

Cafeterias can be worthwhile. They limit lateness from lunch. They are essential if you locate in office parks or industrial areas that are far from restaurants.

But if the items from a cafeteria or food service vendor are priced too high or the quality and selection isn't very good, agents won't eat in them. Also employees may in that area prefer to brown-bag it.

Cafeterias and kitchens eat up 1,000 square feet or more. They also entail equipment and staff costs.

They can be dispensed with, even in isolated office parks. Burkett recommends using the break room for eating. A caterer can come in at meal times, offering one or two hot selections.

"The break room is typically only heavily used twice per shift, at the breaks," says Burkett. "So why not also use it for [serving] lunch?"

Caveats

When you supply amenities, you are undertaking additional administration. You will have to manage contractors for childcare and cafeteria operations unless they are part of the building and the lease. The value of amenities depends on the culture and demographics of your workforce. Engel has seen call center cafeterias empty in many communities because most workers brown-bag it or go to nearby fast-food restaurants.

Demographics also change, which outmodes some amenities and requires others. You may have childcare and gyms that appeal to a young workforce, but if you start to attract an older labor pool, then those amenities gradually become abandoned.

Be prepared for employee gripes with any employer-provided amenity, no matter how good it is. The service becomes a focal point for complaints of all kinds, rightly or wrongly. Many people are not comfortable with an Army-like lack of choice doled out by the boss.

With these issues in mind, Kingsland's firm is designing centrally located break areas for all employees to enjoy. Instead of walking down a 100-foot hallway to a couple of closed doors and into a big room where the sole interesting feature is a vending machine, they enter a large open central area, like a mall, with landscaping, more open and private seating, a cafeteria, eating areas, and a conference room.

"With this design you promote a sense of community in a call center," says Kingsland. "You're also spending more money to make the call center nicer for everybody."

WHY LANDLORDS MAY DISLIKE CALL CENTERS

You did your site selection and found a great community. You saw the building of your dreams. You then contact the landlord and they squirm when you mention "call center."

There are reasons for this. Too many landlords have seen call centers as lower-paying, here-today-gone-tomorrow tenants in comparison to other office types and as being more trouble than they are worth.

The higher densities create parking problems and more demand on air systems. Also, the failure by some call centers to provide adequate smoking areas have led to excessive smoking outside of entrances, leading to complaints from other tenants. The fumes often infiltrate into other offices.

Snobbery is also why landlords sometimes say no to call centers. Many want better-dressed, fitter professionals in their building rather than the more casually garbed, lower-paid, and often overweight call center agents (full disclosure here: I'm overweight). Fortunately, as call centers become more professional for longer-term, higher-paying customer service and sales, rather than fly-by-night outbound telemarketing, as smoking declines and as others become more accepting of plus-sized people, these issues are declining.

What you can do just in case there are problems is have some references from land-lords. Have some photos of your existing facilities (including smoking areas) and nota-tions how long you were there. The more professional and stable you are, the better your chances of lining up that dream call center.

✪ PROPERTY CHOICES

There are two principal occupancy options. You can have a building constructed for you or move into an existing structure.

Who says call centers have to be in boring "bankers boxes"? The Co-Operative Bank's call center in the UK shares space inside this I.M. Pei-styled building, near Manchester.

New

New buildings provide you with attractive, hopefully efficient facilities constructed to the latest building including firesafety codes. You are the sole occupant; you don't have to worry about your employees competing for parking or complaints about smoking.

You can hire an architect to custom-design it on land you select. This is the most expensive choice available in new buildings, but you get a building in a location that is optimized to your needs.

You can own the building, but you are better off selling it once constructed and then lease it back from a professional management firm. Real estate is not productive from a corporate capital standpoint; it doesn't enable you to directly provide goods and services to your customers. You also don't want it dragging down your books.

Or you can contract with a developer who has land picked out, known as build-to-suit. Build-to-suits are similar to buying homes in subdivisions. Typically the developers have an array of proven designs that you can pick from. There are firms like The Alter Group that is very experienced at constructing call centers.

One alternative to build-to-suits are are *shells,* which are spec structures designed for several different types of occupancies and fit them out to customers' requirements. You can open your center within a new and attractive modern space in as little as 30 days.

The big downside for all new buildings are high costs; you need to locate a minimum of 100 workstations to justify a standalone as opposed to shared centers. You may face long lead times, five to seven months for custom or build to suit, sometimes longer, depending on the building and zoning approval process, the climate, and the terrain.

You have fewer location choices with build-to-suits and shells as they tend to be in higher-cost higher-saturation metro areas where developers have greater business and resale opportunities. That means they may have smaller floorplates and multiple stories that may not work well for your call center.

The biggest downside of going into a new building is that you give up flexibility. If you own the building, you must sell it, and if you're doing so in a market where there is little commercial property demand, then you lose money. If you lease after selling a custom-built structure, or if you are in a build-to-suit a shell, you're locked into leases of seven to 10 years and for good reason: the payback period on new construction is 15 to 20 years.

In an industry that is changing as fast as call centers, especially with the growth of new self-service technologies and practices like offshoring, you risk locking yourself into a structure that may be gathering dust 5 years from now. Especially if the labor market that worked for your call center, like a small city, is unattractive to your other departments or to tenants or buyers like other companies, institutions and governments.

Existing

With existing buildings, you can renovate to meet your needs. The principal advantages are shorter lead times and shorter lease terms. Lead times can be as little as 30 days, and lease times can be as little as three years. The costs may be much less with existing than with purpose-built centers, depending on how much work has to be done.

Existing structures are far more common, especially in smaller, lower-cost labor markets. You have much more to choose from.

If you move into a just-vacated call center (later in this chapter), the redesign may be minimal. If you're going into a former factory or store, the construction will be considerable and may match the cost of purpose-built. You have to work around the existing structure and lot.

Either way, you may not be in the most efficient building or be allowed to deploy disaster response means like backup generators. If you are sharing the structure with other tenants, you may face parking, elevator access, security, and fire issues.

The Oak Creek Center in Lombard, Ill. Constructed by The Alter Group, it is an excellent example of call-center-suitable spec space. It is single floor with 35,000 square feet and a 6:1 parking ratio.

Comparing Build-to-Suits with Existing

When call centers are choosing property types, they often waver between existing and build-to-suits. There are differences in lease rates. King White, vice president at Trammell Crow Call Center Services, says there will typically be a 20% premium on build-to-suits (BTS) rental rates compared to existing properties. But tenant improvement costs (capital expenses to modify the buildings to your needs) are lower per month with BTS because these capital expenses are amortized over a longer period. "Both of the projects will require 10 years without increasing the rental rates significantly," says White. "The large tenant improvements for both BTS and conversions are about $50+ per square footFifty dollars amortized over 10 years adds $5.50 to your base building shell rental rate. If you amortized in over five years, you can see the impact."

✪ EXISTING BUILDING OPTIONS

Let's look at types of buildings available for call centers. They fall into two broad categories, which I call conventional and conversion.

Conventional offices are buildings that are built to house business specific functions. Conversions spaces are those that have been converted to office spaces, such as a former warehouse or factory.

You can concider both conventional and conversion by fabrication date: whether they were built before or after World War II. There are some very important differences between the two types. The two eras are separated by the ascendancy of the automobile as the primary transportation means and by the design trend to spartan "modern," away from fancy, rich, and fascinating exterior details such as gargoyles and scrollwork.

Postwar Properties

Postwar offices or conversions are usually characterized by spartan neo-International Style exteriors of steel, glass and concrete, with few adornments. Postwar conversion spaces, like stores and factories, have little natural lighting, as originally built.

Postwar buildings are often sited outside of the downtown area, with on-site parking and little mass transit access until recently. They generally have no more than four stories with large level floorplates. The exceptions are downtown headquarters buildings in large cities; these are 10 stories and up, usually with small floors and little parking.

Postwar Advantages

Postwar conventional and conversion spaces in non-downtown locations are the most popular property types for call centers because they possess efficient large floorplates and plenty of parking. The lease and construction costs are also reasonable. Suburban areas and smaller cities also tend to have less planning bureaucracy than their big city counterparts.

These building types and their demand by businesses like call centers and corporate offices reflect the reality that most people live and work outside of the downtown area and drive to work. Suburban offices have good parking ratios of 4:1.

Workers at all levels usually don't want to drive far or take inconvenient mass transit alternatives that tend to be little or nonexistent in the sprawled out suburban areas and small cities. Many don't want to go downtown because of lack of parking and security concerns.

Postwar retail conversions are especially attractive because existing and potential candidates can be found in most cities and are easy to get to—by car. They also have amenities on their doorsteps like gyms and cafeterias.

Postwar Disadvantages

Postwar properties tend to be the most expensive of the types, especially top-drawer conventional space. They are also bland and uninteresting, especially to younger workers; the conventional offices are nicknamed "bankers' boxes." Postwar conversions require costly renovations such as punching out windows to add natural lighting. But they still look like their former uses.

The once-appealing office park locations are now enmeshed in traffic snarls, which shrinks labor pools by reducing commuting radius. Badly designed and overcrowded road networks inhibit emergency access and evacuations. With recent rail, rapid bus, and ferry systems opening across the U.S. and Canada, it is easier to get to downtown on transit than across town in a car.

With the exception of postwar retail conversions and downtown offices, postwar properties are in the middle of nowhere, in sterile office parks with no amenities nearby and no transit access. You are forced to rent space in a building with a cafeteria or provide your own, at additional cost.

There are new transit-oriented commercial properties being built at or near rail stations. They minimize the disadvantages of postwar while capitalizing on the benefits of large floorplates and accessibility.

Prewar conversions are "cool". They are funky, appealing exteriors, many windows, exposed supports and are often conveniently located to mass transit. Their downsides are high renovation costs, small floorplates and less usable floor space thanks to fat columns. Credit: Brendan B. Read.

Prewar Properties

A prewar office or conversion will typically be in or near a downtown or in older suburban communities and tend to be multi-story.

The structures are brick or stone, or have terra cotta facings, often with outside detail, like gargoyles and trim. They have plenty of windows, including retail and industrial space.

Prewar Upsides

Prewar buildings are often less expensive than postwar structures if they have been renovated. They tend to be much more attractive, with excellent natural lighting and funky appealing exteriors.

Because prewar structures have windows and trim, you can easily make a prewar factory or store look like a top-level office. You can't do that with their postwar counterparts.

Prewar buildings are easily accessible by transit, cycling, and walking. Many cities are expanding mass transit systems that are offsetting the inconvenience of driving and lack of parking. That saves you money on parking costs.

There are often excellent amenities nearby, such as childcare, gyms, clubs, restaurants, and shops. There is no driving 15 to 20 minutes on a traffic-snarled cookie-cutter street only to find an overcrowded franchise barfeteria with snotty kids playing food fight accompanied by clueless preoccupied parents and five-years-out-of-date teenagers attempting to look cool.

These factors combined will help your call center attract and retain workers. There are many young people who want to work in downtowns and older suburban centers. Baby boomers are moving back from the suburbs to take advantage of amenities. There is an underemployed and relatively untapped labor force of principally black, Latino, and Asian people who can be trained to work in call centers.

The voice, data, and power systems in prewar buildings tend to be very reliable and in plentiful supply; the telcos and electric companies started out from downtowns. The cables are usually laid in weather-and-traffic-proof underground conduits rather than on ugly, vulnerable poles. Also, many landlords renovated older structures during the late 1990s dot-com boom with fat pipes and quality wiring.

Prewar Downsides

Prewar buildings have little or no parking, unless it has been provided later on site or nearby, as most people in that era commuted to work on foot or by mass transit. If many of your agents will come to work that way, you avoid paying for on-site parking.

Many, but not all, prewar buildings have small floorplates. Also, depending on the building, they often have fat support columns, taking up room. Real estate consultant Susan Arledge, principal at Arledge/Power Real Estate Group, estimates they can soak up 25-30 percent of your space requirements.

Safety is an issue with prewar buildings because of where they are located, though this is becoming less of an issue as crime rates decline and as these neighborhoods make their comeback.

Unrenovated buildings can be expensive to upgrade with new wiring walls, HVAC systems, and windows. Many buildings are undergoing seismic upgrades. Elevators may have to be expanded or installed to meet ADA requirements. Rodent infestations can occur, especially in the older cities.

Companies who occupy such spaces will concede that they may have had to make design compromises and be much more careful where they place their workstations and wiring than in postwar buildings. Yet their officials point out that the location and building's advantages far outweigh the installation challenges.

Great design can turn nearly any building into an excellent facility. Cingular's Midland, TX call center is located inside a former Builder's Square retail home improvement outlet.

✪ CONVENTIONAL OFFICES VERSUS CONVERSIONS

The key property choice call centers typically face is between postwar conventional offices and postwar conversions.

Postwar Conventional

Postwar conventional is what everybody thinks when you say the word "office." They feature fluorescent-lit cube farms stretching as far as the eye can see, surrounded by rifle-slot-sized windows, set into boring, if cost-effective, featureless, solid-colored flat exteriors, often located in so-called "office parks." So-called because rarely do you see people enjoying them as parks, being surrounded by fumes and noise from passing vehicles. Instead, the individuals who occupy the structures are too busy driving in and out as if they were metal-clad wheeled beetles, or Beetles.

Postwar Conventional Upsides

Postwar conventional buildings are popular because they work for most business types and functions, including call centers. They have the large floorplates, optimized HVAC systems, lighting, and parking, and they look corporate. These buildings are mostly located in suburbia or exurbia, where most people live. Employees and visitors can get there in their cars.

You know what to expect when you sign a lease for or build such a structure. You don't have to spend much money and time renovating them into an acceptable call center. Your employees know what to expect when they move in. Subleasing and re-leasing them for other office uses is straightforward.

Postwar Conventional Downsides

What you may want to watch for is whether the buildings have the required density and parking and whether this will cause problems for you either directly or indirectly by other tenants complaining. There may also be hassles from adjoining communities.

Many postwar nondowntown offices are located near residential neighborhoods. Once word gets out that your call center is going in, the prospect of large bursts of traffic or vehicles coming and going in the wee hours of the morning may prompt complaints and political action to stop your center from going live. To shut them up (or more politely, to "ameliorate their concerns") may require you and the landlord to spend money to alleviate the problem, like adding a traffic light or turn lane.

Most postwar conventional offices do not have parking that meets call center needs. Their ratios are typically 3:1 or 4:1. Designers did not take into account shift changes that occur with call centers, which require more parking because employees are arriving before others are leaving.

Also, you may not find the voice, data, and power quantity and quality that your call center requires. Until relatively recently, offices usually did not consume much power, voice, or data. Bob Stinson of Stinson Design suggests that you might be better off seeking an existing building or doing a build-to-suit in an office park where office and light manufacturing and distribution are located.

"The electrical system is robust, there's room to grow, and there's more parking," Stinson points out. "And there's going to be less complaints about your traffic."

Postwar conventional offices may not be as easy for employees to get to with other property types. Your workforce will have to drive. Also, unless the buildings are on a major arterial, you can't use your presence to advertise for employees. Amenities may be some distance away, requiring investments in cafeterias or catering.

Also postwar conventional buildings may not be where your desired labor force is. Many small cities that are great locations because they have plenty of labor lack such buildings.

Former retail outlets like this ex-Beaver Lumber in Nepean, Ontario, Canada make for excellent call centers with their large floorplates and plenty of parking. Their downside is that they are not the most appealing of buildings. Even so, cost-conscious computer makers like Compaq/HP (seen here) and Dell have been taking advantage of this building type for their support centers. They are often the only property available in desired labor markets. Credit: Brendan B. Read.

Postwar Conversions

Postwar conversions are buildings renovated from their original uses for other purposes. Call centers have been made out of bowling alleys, cinemas, and department stores. Some centers are in malls. Others have been carved out of factories and warehouses.

Postwar Conversion Benefits

The big benefit of conversions is their availability. You can find in nearly every city, no matter what the size and location, a postwar conversion candidate, usually commercial, but sometimes a factory or warehouse. Wherever you want to open a call center, chances are that there will be several potentially suitable properties.

Converting industrial properties into other uses like call centers is fairly common, but tech outsourcer Safe Harbor literally "went nuclear" when it located its offices and call center onto the site of a never-finished nuclear power station in Washington State. The property is now known as the Satsop Development Park. It is located just 20 minutes away from Olympia, the state capital, and the I-5 corridor.

For example, Dell opened a call center in a former Albertson's supermarket in September 2002. The center is in the small city of Roseburg, Oregon.

Helga Conrad, director of the Umpqua Economic Development Partnership, explained that there were no other buildings with the square footage Dell needed. Also, the computer maker's time frame for securing a build-to-suit was too short: three to four months.

Conversions have or could have the same if not more floorspace and parking: 8:1 to 10:1 ratios than the 3:1 to 4:1 found in postwar conventional offices. It can give you many of the same benefits of constructing your own building without the time lag and potential zoning hassles.

Also, agents know how to get there, which is important when you are recruiting and retaining employees. Everybody knows where the old K-Mart or Ford assembly plant is.

With amusement and retail conversion candidates especially, there are amenities like restaurants, dry cleaning, gyms, and childcare nearby.

Another key advantage is excellent voice, data, and power supplies. Conversion candidates are either on main roads that had fat pipes laid for their original owners and tenants. Line crews won't get lost looking for them.

Conversion candidate buildings also have high ceilings, which help to reduce noise. This gives you the option to make some nifty additions, like adding a mezzanine housing offices and training facilities.

Postwar Conversion Downsides

The main downsides of postwar conversions are that they often require extensive structural and infrastructure renovations. There is little if any cost savings with them compared with other building types.

You will probably need to add HVAC units, toilets, plumbing, sewer lines to the street, and power. These structures weren't designed to accommodate large numbers of people sitting inside all day. You will probably need to install skylights and windows to bring in natural lighting.

Also, retail conversions, like those in strip malls, tend to look like what they were no matter how nice you make them. This may be a liability if your company places a premium on corporate appearance.

Another risk is disasters. Neighboring uses can be prone to fires, vandalism, assaults, robberies, rapes, and murders. Parking lots are often not safe in the evening. Former factory and industrial properties are often in the middle of nowhere, leaving afternoon and night shift workers vulnerable as they walk to and from their cars.

Conversion Type Pros and Cons

There are key differences amongst the types of property converted or are candidates for such work. Commercial conversions are buildings that had been originally been constructed for uses such as retail or amusement. Industrial conversions are those that been design and built for uses such as factories, repair shops or warehouses.

Commercial conversions are convenient to amenities, unlike industrial conversions. Your employees won't have to spend half their lunch hour going to and from the friendly neighborhood "Kolesterol Kate's."

Also, if some of your employees come by mass transit, commercial conversions, especially shopping centers, are often on major bus routes or near suburban rail stations. Many industrial candidates lack this.

"Remember, with shopping centers all the hard work—parking, traffic, and site selection—has already been done," says site selection consultant John Boyd of The Boyd Company.

Commercial conversions can be risky in a major metropolitan area. You have to consider what properties your call center will replace when it moves into a former mall or store.

"The numbers may say there's availability of suitable conversions, but when you take a hard look at it there's not," Equis's Cariola points out. "Many times a retail center or mall failed because it may not have been located in the best neighborhood. A strip center where there's a pawn shop, a church, and a vacant department store is not really where you want to put a call center."

Industrial distribution conversions offer one unique advantage over commercial conversions. They are bland and faceless, so they can be made to look more corporate than the other type. They are big, usually featureless boxes, though many also have clear and tinted windows for direct light and opaque glass brick, providing indirect light. You can also co-locate fulfillment operations in these structures with little hassle from the neighbors.

Moreover, these sites tend to have robust power and phone connections. You also won't have a problem installing and running a backup generator. Chances are there will be a generator pad already there, or one can be easily constructed for such units.

You may find considerable assistance from local economic development agencies and the community if you pick such a conversion candidate. Some of these buildings have become eyesores, home to vandals and the like. Some states encourage "brownfield" (reuse) redevelopment as part of efforts to limit "greenfield" sprawl. In many cases the property will not be used for its intended purposes again, especially if it is older and smaller.

Unless the property is completely demolished and the grounds unearthed and cleaned up, there is scant chance there will be other uses, such as residential, since they have to be in the right spot for many other uses.

The big problem with such conversions is that you may incur more renovation costs, especially if there is environmental cleanup. Many buildings have pits, such as for processed steel looping and vehicle maintenance, that must be filled in. The grounds have to be cleared up. And they may be in the wrong spot for your call center labor force.

"You would consider a factory or warehouse conversion if there is nothing else available in your location," advises King White. "You would need more on-site amenities such as a cafeteria or gym in a warehouse conversion, because these buildings are in more isolated industrial areas."

Comparing Prewar Conventional and Conversions

Prewar conventional and industrial conversions have much of the same appeal, which is the attractive location, enhanced by new transit systems and funky design. But they also have similar downsides: fat floorspace-eating supports and high renovation costs and minimal parking. Also, there is a limited stock of these structures.

Prewar conventional has a few unique advantages. The structures are likely very well known, centrally located with the best transit access, and will likely cost less to redo because they were designed as offices. Their major failing is having very small floorplates. The development of elevators, design revolutions like steel frameworks, and high property costs made them go up and skinny than out and fat.

Prewar industrial conversion structures have strong upsides, namely bigger floorplates and a fair stock of recently renovated, heavily wired quality buildings after the dot-com bust. Their downsides are high upgrade costs if unrenovated. The locations may not be the most convenient for amenities and access or be safe at night.

For example, Manhattan's Far West Side, where much of its light manufacturing and warehousing occurred and where there are many converted prewar buildings, has no subway service. There are few restaurants or stores nearby.

When CMP was considering relocating *Call Center Magazine* to there from the Flatiron District around Broadway and 23rd Street, many employees balked. We would have had to walk at least two long blocks or more through not-great neighborhoods from the nearest subway. The office later moved just two block south, a baseball's throw away.

One compromise between both options is prewar commercial conversions, such as former department stores. These properties are rare compared with the other types, but they have the same appeal. Prewar commercial conversions often have better locations than industrial conversions. They sometimes—but not always—have larger floorplates compared with prewar conventional. If the property for lease is multifloor, you could gain revenues by subleasing the ground floor for retail or restaurants.

The major downsides of prewar commercial conversions are their rarity; ideally you want two stories at minimum. You don't want passersby to stare at your agents. But you don't want dark curtains or walls because they deaden the streetscape, which makes it less inviting. That in turn discourages business and other uses. Prewar commercial buildings often need expensive renovations.

✪ THE RECYCLED CALL CENTER

Vacated or recycled call centers are among the top picks of companies seeking to open new ones..

Recycled Upsides

The key reasons for these centers' popularity is that they enable a new one to move in quicker and cheaper compared to fitting out a building. Even conventional offices have to be adapted for call centers, with lighting, HVAC, noiseproofing, parking and break rooms.

These occupancies often require no new exterior or interior improvements, design, construction, furniture, or equipment. Many of them are quite new and well-designed. A number of them offer sound design features, such as large break rooms, large training rooms, indirect lighting, excellent HVAC system, soundproofing, and raised-access floors (more in Chapter 13). In some cases, the workstations and chairs are intact and are high-quality.

Modern applications and software such as CTI middleware, newer PC-based switches, VoIP and network routing, get around the older issues of technical incompatibility between the existing box and what you use.

Also, you're piggybacking on the previous tenants' or owners' site selection work assuming that they did it right, and their requirements were similar to yours.

"Many of these call centers are in good labor markets," says White. "They are just waiting to be discovered."

Recycling Downsides

Even if current, the design may be unsuitable for the new tenant, adding upgrade costs. For example, a tech-support center might not fit into a center designed for a former, lower-paying customer service operation. The furniture may be worn out or not up to modern standards. The cabling and switching equipment may also be obsolete. The HVAC and lighting may not be up to your requirements.

"Many centers are only one to two years old and up to date," says White. "But chances are that facilities older than that will be out of date."

You may also inherit another call center's labor supply, facilities, and technological problems, adds KSBA's Kingsland.

"The former tenant short-listed that call center for a reason," he points out. "So before taking it over, find out why they left it."

✪ REAL ESTATE STRATEGIES

Parallel Design and Planning

You can save months of planning and clumps of hair that would have been yanked out, not to mention possibly millions of dollars, in designing your call center right by figuring out what you want in the way of design before you begin property shopping. The next chapter covers design.

Roger Kingsland says call centers can shave weeks off the time required to open a new facility and increase the likelihood of its success by designing their center in parallel with site and real estate selection. The parallel approach can cut two to four weeks off initial planning and another two to four weeks of implementation.

"Parallel working also saves a small amount (one percent or less) by reducing contractor expenses or general conditions while on site," says Kingsland. "But the real financial benefits come by enabling call centers to make and take calls sooner."

By starting early and working in parallel with the location consultants, the design team can work out space requirements in advance. These include workstation design, amount of room per workstation, cubicle configurations, and circulation. If required, they look at shipping and receiving areas.

By working with IT experts, the teams determine the size of computer and mechanical rooms. They also work out voice, data, and IT needs.

There are many design features, like colors, that can be worked out before a decision to go into a building is made. KSBA works with clients to present and get their approval on design elements that improve productivity, such as sound masking, proper lighting, better air quality, and flat-panel monitors.

Improved performance has a great impact on the bottom line, Kingsland says. A three percent increase in productivity from design is equivalent to an additional $50 per square foot in construction cost.

"The beauty of what we call performance design is that the client gets to choose which options derive the greatest value under their circumstances," he points out.

The biggest variable, after the design and the property eventually selected, is the size of workspace. It can be larger, smaller, or split between floors.

"When there are differences, the site selection and design teams go over them with clients and analyze choices," says Kingsland. "But by knowing what clients want ahead of time, we can save a lot of time in the workstation design, floor planning, and computer room sizing."

Quality Design

It may pay to invest in better exterior design if it creates a better image and greater resale value, known as residuals. Examples include all-glass sides compared to finished concrete and glass. All-glass exteriors are 10–15 percent more costly than these other combinations. Tinting glass reduces the extra heating costs.

Build-to-suits and conversions can deliver better value for the money than conventional offices and vacated call centers if they use high-quality "infotech" design, points out Kingsland. The design typically includes raised access floors, high partitions, large floor areas, and natural light.

Kingsland says infotech design adapts to a wide variety of uses: offices (except executive suites), laboratory space, light assembly, and even warehouses. That adaptability makes infotech design desirable to occupants whose needs for space change over time.

Staged Occupancy

With staged occupancy you can add staff incrementally rather than opening an entire center at once. If you have a 500-seat center, you can add 100-seat sections over a period of time instead of having to commit to hiring a full staff by the time you open the center or resort to overtime or premiums.

A variation of staged occupancy for build-to-suit projects is phasing. Applicable in build-to-suits and renovations through lease options such as must-take and right of first refusal, phasing lets you test the labor market with a smaller center. You can then expand later.

The tradeoff with staged occupancy is that you get more time to build and staff your center. But, it is unattractive if your facility is not completely built out and you have clients and management going through it.

Hoteling

Hoteling is arranging for temporary facilities for certain activities like training for agents and taking and making calls, often with shared services such as printers, fax machines, and in some cases, conference rooms. There are firms such as HQ Global Workplaces and Regus that provide hoteling space, partnering with property owners. These third-party firms can also supply administrative support staff.

The key benefit of hoteling is that you test the labor force in a community before committing to lease space there. Hoteling space can be rented by day, week, month, or year, up to five years. Hoteling also provides disaster recovery.

One creative way of using hoteling is for training. If your firm tends to hire significant numbers of new employees in spurts, or if you have new products and you need to train staff—you may find it more economical to hotel it rather than provide for training rooms on your premises.

The downsides of hoteling are that you are limited to the locations where the hoteling spaces are available; (which may not be optimal to your labor market), you pay a premium for space that you do not control, and you use facilities, furniture, voice links, and data links chosen by the hotelier, not you.

Shrinkage and Hot-Desking

Not all call center agents will be there when you want them. Knowing that can help you save money by enabling you to lease less space and buy fewer workstations, if you couple it with workspace management in the form of "hot desks."

Call center agents take time off for breaks, lunches, training, meetings, absenteeism, and vacation. This factor is known as shrinkage, explained Gerry Barber, Secretary and Treasurer of the CIAC, a leading call center certification organization, and Bill Durr of workforce management software firm Blue Pumpkin, in a *Call Center Magazine* article. Shrinkage specifies the minimum number of staff you need to schedule over and above base staff in order to meet business and operational objectives, say Barber and Durr. Once you know your workload, you or your experts use Erlang C (see Chapter 3) or run a simulation to determine how many people you require. These calculations then go into call center workforce management systems.

There are two formulas to calculate shrinkage: linear and inverse. The details re not important for the purposes of this book but, says Barber and Durr, "If your center uses a workforce management system, be sure to find out the shrinkage formula utilized by the software," recommend Barber and Durr

Now here's where shrinkage gets interesting from a design standpoint. If you know that a certain number of seats are going to vacant during any one shift, why spend money for them?

Yes, everybody likes to have their own desk. But you're paying for that real estate, and you're not getting full utilization of it if it is empty.

John Vivadelli is founder and CEO of AgilQuest, which provides workspace management services. He reveals that there is high under-utilization of workstations in many buildings—from 15 percent to as high as 50 percent—because of vacations, sick days, training, and travel.

He recommends managing the workplace as a shared office environment. Agents should be able to select available workspaces, in advance or on the day of their arrival.

To accomplish this requires workspace management software, which tracks which workstations are available for a given day or shift, and then makes the assignment. You can tie this software into your workforce management system to show supervisors who is coming in, and who isn't, so you know how many workstations are available.

If your agents work in teams, you need to rely on communications tools like instant messaging and conferencing. That way they can stay in touch with each other and with supervisors no matter where they are.

If you have home agents or field staff who come into your call center periodically, you may also need workspace management. That way you can efficiently allocate free desks.

There is another benefit to this approach: business continuity. By knowing where agents are, you can shift resources in case of disasters. If any centers go down, you can route calls to agents at other sites, including their homes.

The only downside is agent morale, productivity, and turnover. Employees, especially women, like to post photos of their kids and other loved ones in their space.

If your call center has plenty of extra space and your leasing terms at that building or others or the local real estate market prevent you from disposing, subletting, or moving other functions into it, you may not need such practices because you're stuck with the property. But chances are, when your lease comes up, you will be looking for as small a space as you can get.

"If call centers can increase their utilization rates, and buy and lease only the furniture and space they actually need, the cost savings may tip the scales against sending that work offshore," says Vivadelli.

❂ CHECKING OUT THE PROPERTY

You know your requirements, the type of building you want and where. Now it's time to knock on doors and poke around. Here's what to look for.

Competition

If competition is an issue for you, check out to see who is around and determine by looking at their job boards and help wanteds if they compete with your center.

✳ Is the Building in Good Shape?

Just like buying a house or condominium, you need to find out if the building and its upkeep are sound. You should bring in your architectural and trades experts to inspect the structure, HVAC, electrical and plumbing to see if they are in sound shape, and if they comply with the building codes and with laws such as ADA.

Also see if the building can affordably meet your extra requirements, e.g., additional HVAC, more stalls for women and lighting, and raised access floors.

Don't overlook the small things, like the condition of the common areas, the entrances, foyers, elevators and stairwells. Sloppy painting and repairs, garbage and butts not being picked up and elevators frequently out of service are signs of trouble. Talk to the other tenants and see what they say.

✶ Air quality

Clean air is not a frill; it is a necessity. To ensure your prospective building is clean, check for the obvious, like the air changes, vents and smoking practices.

Track down the previous tenant and see if there were any problems (with this and with any other matter, like security). Raise the concern with the landlord/developer and ask them to give you records of past air quality complaints. *Business Week* recommends that you bring in your own indoor air specialist to measure air quality. If there are differences that cannot be resolved, go elsewhere. There's always another building.

Firesafety

Don't take a landlord's or developer's word that a building, space or property meets fire codes. They are after all minimum standards. Make sure of the little things that could cause fires and trap people inside to die, like improperly stored chemicals and locked exits.

There is a built-in incentive to cut corners, called profits. Also too many property owners, managers and supers are amazingly careless and stupid.

I once worked in a warehouse where I saw so many violations that I called the local fire marshal's office. They had a field day. Among the more egregious violations they discovered a lawn mower filled with gasoline stored next to the oil furnace.

As the building was located across the street from a residential neighborhood, it did not take too much of a stretch of an imagination to conjure the horrible scene that may have been prevented. Not exactly what people had in mind when they talked about "backyard barbecues," though I have been a reporter long enough—and my son is a paramedic—to know that's what those who respond and write about such ghastly scenes would call it.

Hire an independent building inspector check out the occupancies, just like you should do when buying a home. If your call center is in a building shared with other tenants, don't completely trust the supers to do their jobs. Visually check out the common premises, keep an eye out for the obvious and if something is not right, like frayed wires, call ASAP and demand that it be fixed. And if it isn't done, pronto, then call the fire inspector and your lawyer and put the property management firm on a spit. Better that they roast figuratively than your staff literally.

Safe surroundings

Check if the immediate neighborhood is reasonably safe. Look for the signs of trouble: graffiti, litter, smashed beer/wine/liquor bottles, condoms, needles, skid marks in parking lots, cracks and grass in pavement, damaged lights, bars on windows, snarling dogs. Who are the neighbors? Liquor stores, taverns, pawn shops, storefront churches, rundown motels and boarded windows are dead giveaways of trouble. So are kids hanging around.

Here's a test. Bring a female employee for a drive after hours. If she gets the creeps, scratch the building/property off the list ASAP.

First chat with the development agencies about the areas of eligible property. Then talk to other employers in the area. Contact the police. Check out the local papers. If what you find out contradicts what the development people told you, cross this community off your shortlist fast! Who knows what other b.s. they have spun you?

Surroundings also go for tenants in the building or complex. Are they upscale or downscale? Avoid those where there could be problems with your employees, like substance abuse clinics, escort agencies, marriage/relationship counselors, massage parlors and nightclubs.

Preferably pick a building with a flat surface parking lot. Garages offer too many hideouts for robbers, murderers and rapists. If all other aspects are equal and both competing options have garages, see if one of them has locked access after hours or patrolling security backed up by cameras.

* Secure/safe computer/phone placement

Have your design checked by your or an outside business continuity expert and select buildings that avoid critical flaws.

Examples: computer and phone rooms underneath hot water heater tanks; those tanks eventually leak or burst, says Dodge McCord, senior specialist-business recovery with consultants North Highland. He devised for one client a sloping tin roof over those rooms; the roof saved the equipment when the tank burst less than two years later.

Also, many outfits locate their voice/data and power switching in the basement, which is arguably the worst place to put them. Why? Guess which part of the building floods first? Electricity and water ***don't*** mix.

Make sure the phone room is properly connected to UPS and generator power. Though PBXes often operate on low-voltage DC like phone circuits, most high-end ACDs require conventional AC power, McCord points out.

Split your equipment, like having the phone room in one part of the facility and the computer room in another. If there is a flood or fire in one, the other is not affected.

"So many disasters are made worse by having one critical point of failure," McCord points out. "Having backups and preventative measures costs little compared with the cost of your business going down."

Check to see if the building you want is up to current building codes, such as earthquake and electrical. Authorities do not require buildings to be brought up to current code unless they are substantially renovated.

No matter the age of the building, check out its disaster response procedures. Does the building have an on-site super or one that floats between sites? Ask about the super's qualifications and experience. Who handles problems when the super is not there after hours? Are there security guards, what are their responsibilities and how well trained are they? How thoroughly are the backgrounds of the building's employees and contractors, like cleaning and security companies and their employees checked?

"You may be getting a good deal on a building or a lease and it may be in a great location," says Kurt Sohn, of call center business continuity planning firm 180cc. "But what is the price on your business if the building is not able to withstand a disaster and if the building staff has not taken steps to prevent and respond to them?"

Security

Check out the building security. Talk to other tenants. Be like an employee. Drop in when you expect your workers to be there. Use your street smarts. Look for dark alcoves for preps to hide, broken lights, blocked exits, smashed liquor bottles, condoms and graffiti. Sniff for British Columbia's best-known and "highly" lucrative agricultural export...and I'm not talking about fires with cedar, Douglas fir or spruce.

Companies hire guards usually to save money on insurance, but with little thought of what the guards would do in emergencies. I was assigned to a defense contractor soon after terrorists bombed a similar firm nearby. In the guard shack we had a book several inches thick on bomb disposal procedures. Were we trained and equipped on it? Do pigs fly? Only in an explosion...

In too many cases, the security companies only cared if the applicants were barely breathing. Too many of my comrades were drunks or were so old and out of it they worked up a sweat going to the toilet. Security work was for them an opportunity to sleep, or sleep it off, and get paid for it.

Even back in the 1970s when I worked security, we knew airport work was for losers; airports added security in the wake of skyjackings. We laughed at guards and guard companies that got those contracts. The hours, work and the risks for an extra 25 cents an hour? Forget it!

With the horror stories about the security guards assigned to airports in the wake of 9/11/01, and from what I hear from medics and Transportation Security Administration personnel not much has changed.

What can you do?

* Screen the security companies and the guards. Insist on **_no_** criminal backgrounds. Get that in writing from your landlord before you lease.

* Insist that the guards be trained, presentable, able to communicate crisply and effectively (verbally and in writing); and be reasonably fit, but no Rambos. Guards are there to observe, report and respond: in that order. "Rent-a-cops," as we were known, do not have the authority of the real thing.

* Spot-check the guards. Drop in on weekends, or at night, just to see if they're awake, sober, fresh (not fried) and on the ball. If your center is open after hours, don't assume your supervisors will do that for you.

* See if the landlord or the company varies the patrol and routine. They should *never* have guards make the same rounds at the same time. That sets you up for trouble. If there are two guards assigned per shift, have them randomly alternate their tasks.

✪ COMPROMISES AND OPTIONS

Unfortunately, you may be under severe time restraints to get a new call center set up. You have to hire and train staff, and then you have to buy and install equipment.

This may not leave you with much opportunity to prepare the ideal workplace. You might need that corporate-looking office, but your site analysis shows that the best places to locate your center, which you must have up and running in two months, lacks enough desirable conventional space. Or that you had found locales with plenty of great space that's ready for you to move into, but the labor supply, skillsets, and costs do not meet your needs.

In both cases, compromises have to be made on property and design. If the two or three candidate locations rank close together on labor, choose the one with the best building option. However, you should never choose better building availability over quality labor supply.

Think of this like being told you're being transferred to a new office or you change jobs out of commuting distance from your existing location, and you only have a small amount of time to pick, remodel, and set up in your new home. What would you prefer, a slightly older rundown house, say a 1950s-vintage ranch a 40-minute drive away, or 60-minute bus ride away from your office but in a safe neighborhood with good schools? Or a fantastic mini-mansion that is 10 to 15 minutes away but walled off with gates and pointed fences, with window grates on the buildings lining the access roads?

"You can always find property," says consultant Bob Mohr. "But is it where your desired labor force is and when you need it?"

If you are facing the time and space crunch and want to watch your wallet, here are some of the options you should examine:

Conversions

Architects and developers are not alchemists, but they do so much with old stores and factories that they can make you think you're inside a Class A, even though it was an Albertson's. (Albertson's worked for Dell; it can work for you.)

Multifloor Space

Such occupancies may not be ideal for managing a call center because you have to split up your operations, but they do give you greater choice in conventional offices. You may also find benefits in having your administrative and training functions and agents working for different and sometimes- competing clients on separate floors.

Property Near Mass Transit

You may find good sites near major transit terminals. Mass transit brings in people. The new and renovated older systems are safe, fast, and accessible. Because uses like call centers also generate transit ridership, cutting down on high-polluting car use, transit agencies and local governments may be quite willing to work with you. This alternative is limited, however, to those locations and sites that have good mass transit access by your workforce. There may be zoning and traffic issues to cope with.

Lease Options

There are several options you should consider and have stipulated in your lease.

Zero Net Leases

Zero net leases cover the space's apportioned maintenance and interest costs, with very small profit margins for landlords.

EXPLAINING OFFICE PROPERTY CLASSIFICATIONS

There are three classifications of office property: Class A, Class B, and Class C. CB Richard Ellis's Global Information Standard Reference Manual defines them as follows.

These are taken from the Building Operators and Managers Association (BOMA):

Class A—The most-prestigious buildings competing for premier office users with above-average rental rates for the area along with high-quality standard finishes, state-of-the-art systems, exceptional accessibility, and a definite market presence.

Class B—Buildings competing for a wide range of users with rents in the average range for the area. Building finishes are fair to good for the area, and the systems are adequate, but the building does not compete with Class A at the same price.

Class C—Buildings competing for tenants requiring functional space at rents below the average for the area.

Robert Marsh, senior vice president in CB Richard Ellis's Call Center Solutions Group, points out that buildings such as former supermarkets that had been classified as Class C can be renovated and upgraded to Class B or in some cases Class A, depending on the level of work.

"You can renovate and upgrade building classifications by making substantial improvements to the structure's interior and exterior," Marsh points out.

Consultants report that call centers rarely occupy Class A property because they lack sufficient parking and are too costly. Yet there may be some quality Class A space available at lower rates, depending on the location.

"Some Class A space can be good for call centers," King White, vice president at Trammell Crow points out. "Depending on the city, they can be in active, busy locations with access to labor. You may find them in enterprise zones, which means you can get them at lower rates. But we've never seen call centers move into downtown Class A buildings."

The cost and supply advantages of Class C over Class B space is often outweighed by having less-efficient HVAC systems, inadequate voice, data, and power systems and parking. That said, if you're interested in particular Class B and Class C spaces and you observe that they don't have sufficient room for parking, you may find the landlord willing to expand the lot if he or she has enough property nearby to do so.

"As companies expand, many of them are moving out of the Class Bs and Cs into newer buildings," explains White. "To fill this space, the landlords will try to accommodate new tenants. We've seen them expand their parking lots to ratios as high as 10:1."

Right of First Refusal

Consider right-of-first-refusal clauses on adjacent space. If your plans are to grow in that community, this stipulation can help you expand efficiently without the hassle of being on multiple floors or buildings.

Must-take

A must-take clause requires you to take on and pay rent on a certain amount of additional space within a year or so after you sign the initial lease.

"If you are not sure when negotiating and signing what your space needs will be, then you might want the right to first refusal," advises Susan Arledge. "If you know that you are going to need the space, then a 'must take' might be a better deal."

Exclusivity and Noncompete Clauses

If you are the first call center into a building and are concerned about another one moving in, see about negotiating exclusivity arrangements or noncompete clauses in your lease.

Lease Length

Where you get the best deals on property terms depends on the market. Trammell Crow's White says it is easier to get short-term leases (five years or less) on existing space in poor labor markets because the overall property market is so soft and many vacated centers are on the market. But where property supply is tight and a conversion or build-to-suit is required, the lease terms are typically seven to 10 years to finance all of the improvements.

Ensure before agreeing that there is space and land available on site, should you need to grow, if your labor market projections show you will have the people to make and take contacts at the additional workstations.

It is almost always easier and less costly to expand on site than going through the ordeal of finding another property and opening another call center. But finding a second site in the same area may provide disaster recovery protection, access to new workers, and leverage with landlords at lease renewal time.

Exit Strategy

Call centers, like their occupants, don't stay or live forever. They move on, change, or die. You must prepare for that eventuality. You should design and rightsize your call center to meet your needs and the market's.

Following are some methods to be prepared.

Make Your Call Center Flexible

You can prepare a new installation for flexibility. Rick Burkett recommends installing as many separate heating and air conditioning zones (which use thermostat-controlled air dampers to regulate temperature in each zone) as you can afford. Avoid placing private offices, large open areas, conference rooms, break rooms, and training rooms in the same zone.

SMALL CALL CENTER CHALLENGES

Many of today's call centers are smaller than they were in the past. Downsizing, IVR, and web self-service, outsourcing, and home working have made them leaner. Some firms are considering small satellite offices to minimize commuting.

The upside of this is that there are many more property choices than there are for larger centers, especially in small cities with cheaper labor. It is often easier to find 10,000 square feet to support a 100-workstation center than 100,000 square feet for a 1,000-workstation facility.

The downside is that small call centers may have to pay slightly higher rents per square foot than their large cousins pay. Property owners may be less willing to fund tenant improvements such as better air conditioning to support the increased density of workers in the space that call centers incur.

"Landlords often feel less secure with a smaller call center, unless it is being operated by a larger firm," explains Mark Collmar, president of design firm DM Communications.

Smaller call centers may have to make design compromises or put in more money to improve their space. Small spaces may lack adequate voice, data, and power for workstations; call centers may have to rewire the rooms. There may not be good-quality acoustical tiles in the ceilings. If the call centers do not take up the entire floor, employees may have to share washroom space with other tenants' employees.

Collmar recommends that smaller centers make environmental surveys of tiles, walls, wire, cabling, and power to find out where to invest more money to improve their space.

"When you are in a bigger, especially a newer facility, there has already been a lot of investment in it on things such as power," he points out. "Because you have more room, it is easier to retrofit than if you have small space."

"The call center floor will require more cooling because it occupies a bigger area than those offices," says Burkett. "If you don't have separate zones, the temperature set for the call center floor will be too cool for offices and vice-versa."

King White recommends selecting buildings with 200-foot to 300-foot bay depths to allow you to subdivide the space better for multiple tenants.

Consider spending a few more dollars to make it more attractive (see Quality Design above). Evaluate conventional offices from the exit perspective.

Build-to-suits and conversions can deliver better value for the money than conventional offices and vacated call centers if they use high-quality "infotech" design, says Roger Kingsland of KSBA. The design typically includes raised access floors, high partitions, large floor areas, and natural light.

Kingsland says infotech design adapts to a wide variety of uses: offices (except executive suites), laboratory space, light assembly, and even warehouses. That adaptability makes infotech design desirable to occupants whose needs for space change over time.

Because the infotech design is flexible and inexpensive to change, developers should be able to re-lease infotech space faster and for more money than typical, conventional office space," he says.

Be Flexible Yourself

Be willing to consider multiple floors, or better yet, a building with potential to be converted to apartments or other uses. That's another plus to downtown locations and prewar buildings. Many offices in cities like New York and in downtown Vancouver, British Columbia, are now condos. Converting postwar spaces in the burbs is more problematic but still doable.

Depending on the real estate market, you might find getting out of them much easier. The new owners may also like that you've wired them to the hilt, enabling superfast Internet connections.

Write in Exit Clauses in Leases

You need to get your outs in writing. Negotiate your lease to permit other sublease uses, such as office, retail, and industrial.

The terms should also allow the call center to "go dark" (that is, allow you to move operations off premises temporarily or until the end of the lease). You pay the minimum: base rent plus marginal operating costs incurred by that space.

"The dark clause obligates you to keep to the lease term—rent payments and upkeep—but without the penalty of leaving the space," says Equis's Ron Cariola.

These stipulations give call centers, especially outsourcers, considerable flexibility. If demand suddenly changes, or the centers gain or lose big contracts, organizations are not as easily stuck with space that's no longer suitable.

But if call centers had obtained government incentives to locate there, one of the conditions might be to stay there with the labor force committed to that site for a set number of years.

"You might find it cheaper to stay in that building for the lease period than to repay governments back their incentive money," says Cariola.

These methods usually have penalty clauses. But Susan Arledge says paying them is "certainly cheaper than paying rent for space you don't need."

Don't Expect Big Resale Value

When you close your call center, don't expect to get your investment back on furniture, computers, and phones, says space planner Laura Sikorski. The market is flooded with these items, and they quickly go out-of-date.

"You're better off donating them to schools or charity," she says. "Be sure to have your finance department review the possibility of a tax deduction with their accountant."

Chapter 12: Design, Ergonomics, and Safety

You've decided that you need a new or expanded call center. You know the functions it will have; and the property requirements. You're picked a building to meet them. Now you need to have the interiors outfitted so that your employees will work productively and safely. And you need have it ready yesterday and under budget. Factors include interior design, furniture, ergonomics, and equipment.

To maximize your investment, you also need to keep your call center clean, functioning properly, and protected. That includes ensuring the health and safety of your employees.

You can make nearly every space into a great looking call center. For example, you wouldn't know from this photo that this center, designed by Whitney, Inc and belonging to ABN-AMRO is actually a renovated downtown Chicago building erected in 1906.

✪ EFFECTIVE DESIGN IS KEY

The interior design of your call center needs to be focused on encouraging employees to be productive and on how to keep productive employees. Remember, 60 percent or more of your operating costs are labor. Working conditions affect your bottom line as well as theirs.

Workers who sit in chairs that support their bodies, use workstations that they adjust for maximum comfort and to avoid minimum glare, and whose surroundings are pleasant, will focus more on their tasks than workers who experience hassle and discomfort.

The agent who doesn't have to wait in line to use the washroom, who can grab a sandwich or salad, pick up the dry cleaning, and check up on their kid at the child care facility will likely stay with you longer than the one who feels they've been drafted and confined to barracks.

You certainly don't want to follow the example of a series of Dilbert comic strips that ran in late 1999. The strip's pointy-haired boss, in league with his firm's evil HR director, Catbert, cynically recommended keeping costs down by making the working conditions inhumane, with tiny cubicles, six minutes of bathroom breaks per shift, and incompatible goals, such as speed and customer service.

If the Dilbert strip touched a nerve, it was meant to. Too many call centers are operated that way, which could lead to poor customer service and relations, high turnover, and higher costs, and lower profits. In a boom economy your employees, especially your top performers, can tell you to "take this job and shove it!" and get another one easily. That could cause problems with customer acquisition, customer retention, and sales. If the only agents who will tolerate your poor working conditions are lazy, smart-alecky, or stupid that's not the customers' problem, it's yours.

WHY BOILER ROOMS DON'T WORK

When you set up a call center, give your employees adequate room—90 to 125–140 square feet. You'll pay more in the end if you don't.

Consultants Bob Engel and Gere Picasso cite the example of an unnamed service bureau that complained it couldn't get and keep agents at its New Mexico call centers even though they paid 50 cents above the local McDonald's wage. However, they crammed their agents into tiny booths.

"If you build a boiler room, don't expect to keep people," says Engel.

Engel and Picasso also point out that you will lose more money than you've saved because of furniture replacement costs, injured and out-of-work agents, workers' compensation claims and Occupational Safety and Health Administration (OSHA) complaints under the General Industry Standard. And as word spreads, you'll see staff leave with few others replacing them.

"About 90 percent of the operating and capital costs in call centers are associated with personnel, while only 10 percent are spent on facilities," says Picasso. "By disproportionately focusing on where you can save in that 10 percent, you'll pay much more in the long run."

Engel and Picasso cite a hypothetical 500-agent customer service or support desk call center where installing small, substandard non-ergonomic workstations can easily result in a 10 percent increase in turnover, which they say is an extremely low estimate. The center must now recruit and hire at least 50 new agents a year above its norm at $30,000 in recruiting, training, and starting wages and benefits, per agent, or $1.5 million annually.

"We don't think that any savings in furnishings or space rents could ever counterbalance this $1.5 million in additional personnel costs," says Picasso. "If the additional turnover is 20 to 30 percent created by facility 'savings,' then each year, $3 million to $4 million is thrown away, more than your total annual bill for facilities, furniture, technology and utilities."

State-of-the-art design is both interesting for the employees and highly functional for management. T-Mobile's call center, in McAllen, Texas, does both. It is located inside the Sustainable Technology Business Center, developed by CentraTek LP, a division of Hunt Power. Kingsland Scott Bauer Associates designed the center, which has advanced features, including raised access floors, large structural bays, high ceilings and so much natural lighting that lights are not needed on sunny days.

✪ JUSTIFYING EFFECTIVE DESIGN

Making the work environment employee-friendly is cost-effective. According to Roger Kingsland, managing partner of KSBA Architects, investing as much as $25 additional per square foot or $2,750 per seat is equivalent to only an additional 16 cents per hour or a 1.61 percent hourly wage increase.

Unfortunately, justifying many investments to senior management may be challenging and require more acceptance of concept rather than mathematical proof. Kingsland points out that although you can prove that some hard improvements such as sound masking can increase performance, conceptual improvements such as having nice surroundings are much more difficult to quantify and show to top executives.

"The difficulty is that the link between good design and employee savings is not direct," he explains. "With few exceptions, it is also hard to peg actual productivity improvements and lower turnover to improved facilities and design. There are many other variables that come into play, such as staffing, training, supervision, wages, and benefits."

Engel points out that call center building costs can greatly vary between $45 to $120 per square foot, depending on the type of building, location and finishes. He advises that the key to affordability is improved productivity, sales, and customer retention.

"For example, better lighting or a higher-resolution terminal costs $15 per month extra, yet it reduces fatigue to the agent so that, in the fifth or eighth hour of a shift, that person is 30 percent more productive," he points out. "The return on your investment is huge. Likewise, if the workstations are so tightly packed that most agents can't hear their customers due to the ambient noise, the overall facility cost savings will be more than spent in lost customer satisfaction."

There are many case studies to indicate the benefits are there. Herman Miller, a leading furniture maker and now a call center consultancy, found that practicing what it preaches pays off. *BusinessWeek* reported on June 5, 2000 that Miller witnessed a 1.5 percent productivity increase when it opened a factory that had 100 percent fresh air and daylight—enough to repay the structure's mortgage. Also, employees who had left the firm returned.

"If CEOs have half a brain they would start to pay attention to the fact that their employees are their main cost-and-benefit center," the magazine quoted architectural expert and consultant William McDonough.

✪ DESIGN CONSIDERATIONS

Corporate Traffic

Will your call center get senior corporate or client visitors? Is it part of a shared service center? If so, the consulting division at Hellmuth, Obata+Kassabaum (HOK), an architectural firm, recommends that you have it look more corporate. That includes higher-quality materials, chairs, and workstations. Additionally, you need audio, video, or design ads like large screens that show activity, say over an agent amphitheater, or install raised supervisory stations where an executive can look out over the call center.

Employee Appeal

If you are competing with other employers for quality labor, then consider having state-of-the-art design elements like lots of color to attract agents.

✪ EFFECTIVE DESIGN ELEMENTS

Following are the major physical characteristics and features for your call center that you will need to consider and ask about when working with your architect and designers, developers, and landlords.

Air Quality

As noted in Chapter 11, the standards are 20 cubic feet of outside air per person per minute. Also, the inside ambient temperature should be between 72 and 74 degrees Fahrenheit.

Achieving those parameters can be challenging. Climate and natural lighting impacts temperature; you may to pay more in capital and operating costs meet ambient levels. Air quality is therefore both a property and design issue because the solutions are expensive and require alterations to the building and equipment.

Following are some considerations and options.

Large Versus Zoned Units

There is debate over installing several large pre-manufactured, rooftop heating, ventilating and air conditioning (HVAC) systems for your entire call center versus many smaller or zone units. You save money with the former, but the latter array provides more temperature-control flexibility and individual comfort. You also gain more versatility in case you split up or downsize your call center.

When planning your HVAC system, balance the energy efficiencies with the air quality needs of your workers. For example, energy-efficient, variable-air-volume systems, part of HVAC systems, control temperature by varying the amount of chilled air delivered to zones throughout the building by local zone boxes, which have dampers that control the amount of air that is delivered to the zone.

"You want to make sure that when the ambient air temperatures outside the building are near interior temperatures that the air volumes are not so low that the air becomes stale," advises longtime design consultant Bob Engel. "If the inside and outside temperature are roughly the same, there is no cooling demand, and the air doesn't get changed."

Humidifiers

Consider having a humidifier if your call center is located in desert or cold areas, recommends David Meermans, Director of Product Management at EADS Telecom. Each unit, which costs about $4,000 to cover a space of 7,000 to 8,000 square feet, can address bad air, dry throat, and static problems. They work by injecting steam into the air ducts. They do require annual maintenance.

Air Amenities

There are portable air-quality amenities you can install on the call floor at agents' request. These include fans or small electric floor heaters, where permitted by code and law, to help accommodate individual temperature preferences, from the killer whale lookalikes to the fashionably tubercular.

The temperature is the number one source of aggravation between employees in office environments. Who hasn't seen people stripped down to what is decent or sanitary while others are bundled up for an Antarctic exploration for same ambient heat level? Some offices I've worked at have locks on the thermostats to prevent "temp wars" that give an HVAC system fits and ultimately breakdowns.

Having fans or heaters are also handy when the HVAC system goes down for repairs or squeezes its last cubic feet of coolant. I worked in one office where management shut the system off on weekends; it took half the week for the work areas to cool down or warm up (depending on the season) to adequate levels.

Draft Avoidance

Air drafts cause discomfort and at worse prompt muscles to tense up, increasing the odds of injury. If drafts are occurring, be ready to install air diffusers, advises Dr. John Triano, director of the Texas Back Institute's chiropractic division. To avoid window drafts you can fit workstations with skirts.

Lighting

Your call center needs to see the light. KSBA recommends 30 foot-candles of light at each workstation. But where you place lighting is as important as the quantity.

Engel and Picasso say that too many call centers are overlit; at the normal office level of between 70- to 80-foot candles. They recommend that you keep it at 60-foot candles if your agents are working with reference books and paper, 40- to 50-foot candles if they are primarily working on screens. You should also provide individual task lights to agents who request them, especially to those agents who need to read paper documents.

Call centers benefit greatly from glare-free indirect lighting, where you place or suspend fixtures to bounce light off ceilings or walls. It adds a soft glow to rooms and reduces glare. However, be sure the replacement bulbs that provide equivalent light output and energy consumption are not expensive.

Alan Hedge, professor of human factors and ergonomics at Cornell University, conducted a study on lighting at Xerox from 1989–1992 that tested direct and indirect lighting and found that indirect lighting caused less eyestrain and increased productivity by 2–3 percent. In the article "How to Design an Ergonomic Call Center" written by Randy Hayman in the April 1998 *Call Center Magazine,* Hedge was quoted saying that the lighting system costs about 1 percent of salary costs. The article reported that office space designers said the Hedge study ranked among the top in the field.

"If you can boost productivity by 1 percent you've covered the cost of the system in a year," said Hedge.

Careful Fixture Placement

EADS Telecom's Meermans advises that you arrange the lighting fixtures to keep the light off the monitor screens. Generally this is with the long axis perpendicular to the screen. You should also locate the fixtures in conjunction with the furniture, not just "every third grid."

Parabolic Diffusers

Consider parabolic diffusers. If you are using existing fixtures, throw away prismatic diffusers: the flat, white, or clear plastic ones with the bumpy surface. They only scatter light, filling the top of the room with white fog and creating glare on monitor screens. Instead, replace them with egg-crate diffusers. These look like a grid of small squares and direct the light down.

Less Is Good?

Your agents may prefer less light, such as those on a help desk where they spend a lot of time concentrating on the screen. A gloomier office may be less distracting than one that is brightly lit. On the other hand, older agents may require more light than their younger counterparts.

You should consider having adjustable lighting on the desktop, especially if your agents are reading manuals or doing paperwork, which require more light than staring at computer screens. While your employees can't control the indirect lighting, they can customize desktop lighting to compensate.

Where to Use Incandescent Lighting

Offices do not use incandescent lights because they create shadows, consume a lot of energy for the amount of light they give off, and have a short life. Yet there is a place for them, say Engel and Picasso, as secondary accent lighting, in visually bland areas to give some variation.

The Case For and Against Natural Lighting

Be careful how and where you place natural lighting. It makes a big difference in your call center's look and mood. Many studies have shown that people feel better when they receive natural rather than artificial light. How much natural light your center needs also varies by location.

Says Rick Burkett, president of design firm burkettdesign: "If your call center is in St. Paul, Minn. your day-shift agents won't see sunlight at all during winter. If your call center is in Jacksonville, Fla., agents will see it on their way in or going home."

But natural light presents two problems: glare and heat load, even in winter, making your air system work overtime. I once worked in a black-clad Manhattan office where my supervisors had the misfortune of having south-facing windows. It might have been 20 degrees outside but to them it felt like 200. Direct natural lighting is distracting. Burkett says some clients do not like providing it because having a view distracts agents.

Even so, employees like natural lighting. Burkett has heard of agents who said they'd prefer being by a window even with the glare because they felt better being able to see out.

The key is to figure out where and how to use natural lighting. Burkett suggests having work areas placed where there are north-facing windows that provide indirect light and your break areas where there are south-facing windows. If you can't have that arrangement look at providing shades if there is a glare problem.

You should also consider adding skylights if the building permits, because they provide light without overloading the air system. Exhaust fans can remove any heat pockets generated from the skylights.

"The easiest solution is to use lots of glass in the break and conference spaces and entryways where there are few if any screens which can be affected by glare," Burkett advises. "The natural light adds variety to the environment and enhances the employees' day-to-day experience."

Color

Color, like light, sets the mood in any business. Color reflects light in different amounts. You use color in walls, ceiling, carpet, and furniture. The closer to white, the more light it bounces off. Entire books have been written on how to use color to excite or calm people down. The stronger the color, the more eye-distracting, which is not good for workstations but fine for everywhere else.

Color breaks up the design uniformity in an open, "big box" call center. Color, like windows, compensates for dreary climates. You can put color on walls if you do not have a lot of windows or if the view from them is drab, like a view of another building or a parking lot.

Noise

Noise is a major issue in call centers. Sometimes the level can be so high that agents have a hard time understanding the person at the other end of the line and can't think. According to the KSBA's Kingsland, when the overall noise level rises above 50 decibels, people compensate by speaking louder, setting up a vicious cycle.

Dr. Triano of the Texas Back Institute likens the din to power tools. Research from by Professor Gary Evans and his student, Dana Ellis at Cornell University, showed that a typical office noise level of 55 decibels is high enough to cause problems.

Here are some methods and technologies suggested by consultants to enable you to reduce noise:

* Noise-absorbing furniture panels, ceiling tiles,workstation partitions and carpeting
* White noise generators, also known as sound masking. They can lower the noise of voices that employees hear by as much as two-thirds by masking the frequency of human speech. The sound is generated and distributed through ceiling speakers. KSBA's Kingsland say a study of an outbound call center that showed a 20 percent productivity increase over the same months in the previous year. Sound masking systems, also known as white noise, can achieve 4,000–8,000 percent return on investment based on performance improvements cited in several studies.
* Moderately high ceilings of 10 to 15 feet

Design out noise sources, provided you've put in sound-absorbing ceiling tiles and partitions. For example, locate the time clock and doors to the work floor away from agent workstations. Place copiers, shared printers, shredders, and busy fax machines in a separate room.

Another, much less expensive way to keep noise down is monitoring your agents for bigmouths. If you hear agents whose voices boom or sound like blackboards being scraped tell them to lower the volume: customers don't like to be yelled at either. Or as one middle-aged woman once sweetly told me when I was talking too loudly on a train: "Excuse me, but do you have a broadcast license?"

Flooring

How you design your floors affects where your power lines, voice lines, data lines, and HVAC is fed to your workstations. This will also affect the center's appearance and the ease with which you change the floor layout.

The conventional technique has been to have plain carpeted floors and run HVAC and cabling from ceiling ducts, dropping the wires to workstations in poles. A newer system, raised access flooring (RAF), is beginning to supplant it.

RAF is a skeletal floor that takes cabling and HVAC to the work areas underneath the walkway. Cable-only floors are about six inches high; those with HVAC ducts are 12 to 16 inches high.

KSBA's Kingsland, a longtime RAF advocate, cite the method's improved air distribution. Air rises past workstations to vents, instead of staying at the top. Raised access floors provide better esthetics by not having ugly poles.

Raised-floor panels can also be taken with you, like furniture, if you move your call center. Since raised floors are assets, like furniture you can depreciate them, which offsets its slightly higher costs compared to conventional methods.

But raised access floors also add to the leasing price, according to Dan Frasca, The Alter Group's vice president of development. "Companies that get the most out of raised access flooring are those that frequently change their floor layouts," says Frasca. "It gives them quick access to the underfloor cabling and duct work."

Raised access flooring, with the utilities underneath, such as at CompUSA's Plano, TX call center permits you to move workstations around more easily than conventional flooring that feeds voice/data and power from above-ceiling spaces. You must decide whether the flexibility and arguably improved appearance provided by this method is worth the extra cost.

Space planner Laura Sikorski of Sikorski-Tuerpe and Associates, says that raised access flooring is not necessary. She also doesn't like the feel when walking across it. How often do you move around workstations—which must be done by an electrician at up to $100 an hour—to change the cabling?

But conventional duct designs can work. Burkett points out that there are ways to drop utilities from ceiling spaces so they improve the look of the call center space. Partitions and closets that mask the wires yet allow access to these services break up the monotonous open appearance of many call centers. There are workstation designs like Herman Miller's "Resolve" that imaginatively feed cables from ducts to the equipment.

"You can also go for exposed cable trays, making the wiring a design feature, giving a fun industrial look that works in many call centers, like a help desk or where it serves young trendy industries like the Internet," says Burkett. "Such designs can help you recruit people by making them think 'Hey, this is a cool place to work.'"

Workstation Layout

Workstations are your call center's bottom line. They are where your agents produce. How you set them up the workstations will affect how efficiently the agents produce. And that depends on what the agents will be doing and how aggressively you need to supervise them. Are they working individually or in teams? Are they self-motivated and disciplined, or do you have to watch them like a hawk?

When you are setting up in areas lacking in labor, be very precise about how much supervision you provide and to whom. You may be forced to hire less-than-ideal people who need more supervision than you had planned. On the other hand, good agents may hate to be micromanaged and may go elsewhere.

Herman Miller's Resolve design not only accommodates conventional floor and cable tray wiring but it adds a contemporary, classy look for call centers.

Cubes, Pods and Teaming Spaces

Call center agents have traditionally worked alone with little need for interaction except with a supervisor. They have sat in stereotypical rectilinear cube rows, in straight or more contemporary zig-zag patterns with supervisor stations at each end. More agents are now sitting around cores, also known as hub-and-spokes or pods, where they face central cores separated by partitions. This high-density design helps reduce noise.

Unfortunately, these designs—especially the cores—are the workplace equivalent of the dreaded hospital gown, having the agents' backs exposed. They provide no privacy, permitting the supervisor to sneak up at any time. To check this, some call center agents have reportedly fitted rear-view mirrors at their stations.

"If you need more aggressive supervision, then rectilinear rows, hub and spokes work fine," says Kingsland. "If your supervision is less aggressive, then more conventional cubes, with three or three-and-a-half panels, offer more privacy."

If your call center works in teams, you should consider either a workbench design where more than one agent and desktop share the same primary work surface, or conventional cubes arranged around a supervisors' station. This way they can focus on individual and group tasks as need be.

Some workstation designs are highly flexible to permit a variety of configurations. Herman Miller's "Resolve" allows delta, single and multiple hexagonal, half-hexagonal, shell, and zigzag seating patterns and workgroups. It also has rolling screens, tables, and mobile bins.

Cost and Space Penalties with Pods and Zigzags?

Some design firms such as Workplace USA have pointed out that pods and zigzags cost up to 15 percent more. They also consume up to 20 percent more space because they require more circulation room between workstations. They claim that these designs have not proved more effective at retaining and attracting agents.

Instead, you can make the more efficient cube rows acceptable in regional or corporate headquarters. The options include more muted colors, like grays and tans, and using higher-quality fabrics.

✪ COMMUNICATIONS IMPACTS

Before deciding how you are going to set up your workers, look at how they communicate. Tools like email, instant messaging and conferencing are minimizing the need for co-locating agent teams by removing the requirement for in-person communication.

Emails and instant messaging are becoming the preferred tools of interacting with employees. Stud-

You can get very creative with high-density "cube rows" with the right kind of furniture. Interior Concepts' Swurv line uses gentle curves and contemporary styling while costing about the same as straight cubes. It also lets you house the same number of workstations in a given area.

ies show the managers often send them rather than meeting one-on-one. Management and staff have less time. When I worked in Manhattan, my colleagues and supervisors emailed or messaged me rather than walking over to my cube, even if they were one cube over.

Physically conversing with someone is more disruptive to work than communicating. You could be following a series of instant messages and emails while looking at a screen and talking to someone on the phone. But if an individual walked over in person you would have to give them your undivided attention.

If that is the case then you can use standard cube rows. Also, where your employees sit is no longer important. They could be at the other end of the floor, on another floor, or at home.

⊙ ERGONOMICS

Ergonomics is defined by the Oxford English Dictionary as the "study of the efficiency of persons in their working environment." It has come to mean making sure that the workplace is set up to enable employees to perform their tasks efficiently, with little risk of incurring injuries that impede individual and company efficiencies.

This chapter has covered air quality, lighting, and noise that impact ergonomics. This section looks at the immediate workers' environments: chairs, workstations, and appliances.

Workstations are where your call center's tasks take place. There are four key parts to these "machines:" the desks, computer, phones, and chairs, with the agents connecting them.

If any of these components are out of alignment, the agents will not produce efficiently and eventually will break down. That will cost you time and money in health care and workers' compensation, lost productivity and higher turnover and cause agent suffering, which will be heard far and wide.

Many people believe that injuries don't take place in sedentary office environments. Wrong.

Call center staff and others who keyboard a lot suffer from two major ailments, carpal tunnel syndrome and tendonitis, together known as musculoskeletal disorders (MSDs). Agents can suffer from head, neck, back, and leg injuries at their workstations. They also experience headaches, neck, arm, and mid-back pain.

Carpal tunnel syndrome is caused by compression of the median nerve as it passes through the carpal tunnel of the wrist that eventually results in damage to that nerve.

Call center injuries like tendonitus are treatable with rest, anti-inflammatory medications and injuries. Agents may have to wear wrist braces until the healing has finished. To prevent such hazards, you should have an ergonomist look over your workstations and at how agents work.

Twisted hand posture, repetitive hand movements outside the neutral range of motion (15 degrees), sustained pressure on the underside of the wrist, and forceful hand movement leads to carpal tunnel syndrome.

Tendonitis is the irritation and inflammation of the tendon. Twisted hand posture, high repetition, and force on the tendon cause tendonitis. Call center agents, administrative assistants, and programmers commonly get tendonitis in the wrists, forearms, and shoulders.

Dr. John Triano points out that such MSDs are insidious and often go unreported because one often can't tell whether someone has been injured or is in pain. Often, agents do not realize that their work activities are causing their pain.

"Unlike lower back pain, where someone who is injured on a heavy job can't do their work, upper arm, back, and neck pains are such for call center and computer workers that they can continue to work before the pain becomes severe enough to stop them," explains Triano. "This risks worsening their conditions and their health."

Fixing MSD hazards do not have to be expensive. A company where agents were complaining about wrist pain, particularly in the hand that operates the mouse, installed adjustable-height keyboard and mouse trays that sit atop desks and lower into position. The cost was $40 per employee.

Add such savings together and you could see a big payoff. Woody Dwyer, management consultant with ergonomics consulting firm Humantech, reports that Verizon's call centers saw a 38 percent drop in carpal tunnel disorders from 2000 to 2001, while its compensation claims dropped by $183,000 after it implemented a program designed by Humantech.

"The more enlightened firms are learning that good ergonomics makes good business sense through reducing workers' compensation cases, absenteeism, and health care costs, and improving productivity," he says. "Good ergonomics also improves employee morale, which is important in these tough times because it shows that their employers care about them."

Preventing such injuries reduces sick time, maintains productivity, and avoids workers' compensation claims. Though there are no federal OSHA ergonomics regulations employees can still file claims under OSHA's General Industry regulations.

These consequences and compliance benefits will increase as more agents chat with their fingers rather than with their mouths. Yet making call center workstations ergonomically sound is not expensive.

Interior Concepts designed its workstations with the advice of an in-house ergonomics expert. Its chairs, outsourced from other vendors, and its workstations are adjustable. That is key because agents come in different shapes and sizes. What is comfortable for the daytime agent who is 5 feet 6 inches tall is too cramped for the afternoon-shift agent who is 6 feet 5 inches tall.

By spending just $350 more per workstation, such as for adjustable chairs, keyboards, workstations, and wrist pads, call centers can prevent most carpal tunnel injuries.

"According to OSHA, the average claim from injuries like carpal tunnel is $35,000," says Interior Concepts' marketing director Christine Jacobs. "For the same amount, a call center can outfit a 100-seat facility with workstations that will help prevent that from happening."

Workstations

Look at buying workstations that permit agents to adjust them to fit their needs and to change positions to prevent getting stiff. Some models even allow agents to stand while they are working. Herman Miller's "Passage" line of freestanding furniture combines modular desks with storage and has adjustable-height work surfaces. It can also be reconfigured without disrupting other workstations. Individual stations are completely independent from each other.

The workstations you pick must be sized to meet the tasks, within your employees' reach radius, recommends David Smedley, principal with a facilities service and design firm, Wave. This arc space should be from 40 to 63 inches wide, with most tasks being done within the first 18 inches of the arc. Occasional tasks would be done within the outer edge of an overall 26-inch arc; this extended arc space would require a slight reach.

Debate About Desks

There is considerable debate about whether such desks or workstations should be fixed-height or variable height. Dr. Triano recommends variable-height workstations, which have manual or automatic lifts.

With variable-height workstations, agents can work while standing, making the tables especially useful, he says, for senior customer service or support desk agents who have to reach into bookshelves to access reference material.

The workstations shorten the distance from the computer surface to the books. Agents can still reach the mouse and keyboard without bending at the waist, craning their necks, and reaching too far.

"Your chairs and workstations should match the individual," he points out. "Some people are better off doing their work standing or semi-standing, while others are fine sitting at conventional-height (29-inch) workstations."

But variable-height workstations may also lead to more injuries than they could alleviate. Hedge says that you place more stress on your muscles when you stand up to work than when you are sitting down.

"Standing burns 20 percent more energy than sitting, and it increases strain on the lower back compared with supported, reclined sitting," he points out. "If you stand and work with poor posture and in a static position, then standing will have no benefit whatsoever. We recommend sitting to work in a neutral posture, then standing intermittently and moving around doing other kinds of work."

Laura Sikorski points out that variable-height workstations are disruptive, with agents yo-yoing in low-height cubes. She has seen call centers that put adjustable-height workstations in one area, and workstations with or without keyboard extenders in other areas, and have agents pick and choose.

"This is great for call centers where individuals work alone," she says. "It doesn't work where you have agents working in teams. You can't have them deciding one day they will work in an adjustable height area, then the next in a keyboard tray area. That makes it hard for agents to work together and with their supervisors."

David Smedley points out that most desks and tables are set at a standard 29-inch height. If that's too low or high to allow you to maintain the correct seated posture described above, an adjustable table or customizable wall-mounted work surface would be a better choice.

Worksurface Arrangement

The overall size of the primary work surface should allow most tasks to be done within your reach radius (an arc of reachable space to your front and side). Depending on your size, this arc space should be from 40" to 63" wide, with most tasks being done within the first 14"-18" of the arc space. Occasional tasks would be done within the outer edge of an overall 22"-26" arc; this extended arc space would require a slight reach.

When multiple work surfaces are used, it's generally preferable to have them arranged within an L-shape, or a curve, rather than along a wall. It takes less time and effort to rotate to a perpendicular work surface than to move to another part of the room. Using a separate work surface for the printer and fax machine will free up primary workspace for writing and other tasks. Storage accessories such as file holders and note displays will also help keep primary workspaces clear.

Adjustable high-quality chairs like this one made by Haworth are the single most important investment you can make in the health and productivity of your agents. Your returns are exponential compared with what you put in.

Chairs

Chairs are the literal backbone of call center design. If they don't fit or cannot be easily adjusted, your agents and your call center suffer. There's no such thing as a one-size-fits-all chair.

The chairs must provide adequate back and shoulder support. The armrests should support the arms while agents are typing. Yet because the chairs receive heavy use they must be strong, durable, and wear-resistant.

"Agents who work in call centers that 'hot-seat' (having the relieving agent take the same seat as the departing agent) don't necessarily have the same seat every time," Laura Sikorski points out. "A 5-foot-2-inch person could go into the same workstation previously occupied by a 6-foot-5-inch person. They have to readjust everything, but they don't know how to do it. They're not in a workstation ergonomically designed for them."

Wave's Smedley says that your agents will want to be able to adjust the chair height so that their hands can be placed on the keyboard with their elbows at a 90- to 120-degree angle with shoulders relaxed. Their knees should be bent at about 90 degrees or greater. Their hip angle should be 90 degrees or greater. Their thighs should be parallel to the floor. Their feet should be able to be placed flat on the floor, and their lower back should be fully supported.

Chair Feature Debate

There is some disagreement as to what you need on the chairs. Triano says chairs should have adjustable heights, separate foot rests, variable angled seat pans, adjustable back rests in depth and height of support for the lower back, and adjustable elbow supports.

Dr. Hedge says the most important adjustment is seat height, then backrest height. Other features like adjustable seat pans that slide forward or backward are often unnecessary. If the backrest can recline then a seat tilt is not necessary.

But Smedley thinks an adjustable seat pan is a worthwhile feature, especially when more than one person uses a home office space. This lets you move the seat pan forward or backward so that it stops short of the backs of your knees. This gives your upper legs maximum support while preventing direct and constant pressure behind your knees. A shorter person will require the pan to be closer to the chair back, while a taller person will need the seat pan to be slid away from the chair back.

Also, Hedge says if a workstation has a good-quality, negative-slope keyboard tray with a built-in palm rest, chairs do not need to have armrests. If a chair has armrests, they should allow users to lower them out of the way to type and control the mouse. Your agents risk injury if they place their forearms on the rests when typing and controlling the mouse.

Smedley says a curved or waterfall-style front edge is better for blood flow. Chair edges without the waterfall tend to place too much pressure on the backs of your legs, causing discomfort.

The chair's lumbar support should press against your lower back, allowing you to rest firmly against it. To avoid backaches, an adjustable lumbar support should be set so that it curves out to rest within and support the natural curve of your back at the waist. Your feet should be able to be placed flat on the floor or on a footrest.

"When choosing a chair, look for a feature known as 'synchro-tilt' that drops the rear of the seat pan when you lean back, allowing you to open up your hip angle and chest cavity," recommends Smedley. "Another chair option to consider would be a paddle or Flipper™ armrest. These soft arm pads rotate 360 degrees and are height-adjustable. They promote better working postures, particularly while using a keyboard or mouse."

Whichever chair design you select, make sure it is durable. Call center agents place a lot of wear on fabric supports, armrests, seat pans, and on lifting and lowering mechanisms. You're looking at spending about $400–$500 each in volume orders for top-quality chairs from top-of-the-line makers such as Haworth and Herman Miller.

Workstation Arrangement

Choose and arrange furniture so that monitors can be placed at least 25 inches from agents' eyes. It's better to place it even further than that because close viewing causes eyestrain.

The point where the eyes can rest between convergence and accommodation falls between 45 inches directly, and 35 inches with a 30-degree downward angle, Smedley points out. Having the monitor more than 25 inches away is usually better for most people and enlarging document print size is a better option than moving a monitor closer than 25 inches. Generally, the larger the monitor, the farther away it should be.

Cornell University's ergonomics department offers 12 tips, arranged around this drawing, for an ergonomically correct workstation:

1. Use a good chair and sit back
2. Top of monitor 2"-3" above eyes
3. No screen glare
4. Sit at arm's length
5. Feet on floor or footrest
6. Use a document holder
7. Wrists flat and straight
8. Arms and elbows close to body
9. Center monitor and keyboard in front of you.
10. Use a negative tilt keyboard tray
11. Use a stable work surface
12. Take frequent short breaks

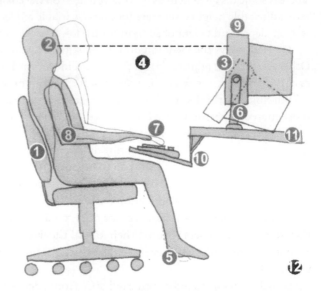

STRETCHING AND SITTING FOR PERFORMANCE

It isn't enough for you to provide excellent ergonomically designed chairs and work-stations to keep your agents and supervisors comfortable. They have to offer their share by stretching and sitting correctly.

The August 7, 2000 New York Daily News had a great article on how to stay limber while on the job, citing a new book "Stretching at Your Computer or Desk" written by Bob Anderson. The exercises include hand-wrist-arm stretch, shoulder-arm stretch, shoulder shrug, neck stretch, neck-upper-shoulder stretch, upper-body stretch, lower-back stretch and hamstring-lower back stretch.

Anderson also recommends that you and your employees should stand up and move around as much as possible, counting breaks and washroom trips. He suggests that you stretch every 45 minutes,

"To me the most difficult job in the world is to sit in front of a computer for hours and hours and hours," Anderson told The Daily News. "Your body atrophies, circulation decreases and muscles get tight."

The news story also had a box on how to sit right. According to Kell Roberts, owner of Real Fitness (Los Angeles, CA) and spokesperson for the American Council on Exercise, you should have a light but natural arch in the lower back. Shoulders should be directly above the seat and pulled down and back. The chest should be open and the head should be directly above the shoulders, not in front of them. If you're working at a keyboard, elbows should hang straight down from the shoulders.

"Sitting is one the hardest things to do correctly," said Roberts. "When you don't do it correctly you end up with back, shoulder and neck pain."

"When viewing close objects, the eyes must both accommodate and converge," explains Smedley. "Accommodation is when the eyes change focus to look at something close. Convergence is when the eyes turn inward toward the nose to prevent double vision. The farther away the object of view, the less strain there is on both accommodation and convergence. Reducing those stresses will reduce the likelihood of eyestrain."

Headsets are vital ergonomics tools in call centers. They are also becoming lighter, more comfortable and convenient. VXI's Associate S3 System is convertible between over-the-head and over the ear. It also has a compact amplifier.

○ ERGONOMICS APPLIANCES

There are many secondary appliances that may reduce call center injury risks. There are others that may be useless. These useful appliances include:

Headsets

The headset-bedecked agent is the trademark of call centers. Headsets have long been the leading ergonomics appliances and for good reason. They prevent discomfort and tissue damage to the neck and shoulders and are quite affordable.

Without them your employees will not be able to do their jobs, unless your center is online-only like eBay's. The good ones cost over $60 up to hundreds of dollars.

Headset manufacturer Plantronics cites a Santa Clara Valley Medical Center study that shows that headsets reduce neck, upper back, and shoulder tension by as much as 41 percent. An additional study by H.B. Maynard concluded that adding hands-free headsets to office telephones improved productivity by up to 43 percent.

Call Center Magazine editorial director Keith Dawson, author of *The Call Center Handbook,* provides a great checklist of features to look for, including one- or two-eared style, noise levels, sound quality, and universal amplifiers to make them compatible with different phone systems. With the growth of voice over IP, there are now headsets that plug into USB ports on computers.

Here's some other advice: If your call center is noisy, consider noise-cancelling microphones and binaural (two-eared) earpieces. People don't like hearing background noise when talking to agents.

Headsets must be adjustable; every head is different. Issue each employee their own headset and make them responsible for it. Give them cleaning and maintenance instructions. Have spare parts available.

Do not have your agents share headsets with others. The earpieces collect germs, earwax, and other detritus that could lead to serious infections, hearing loss, diminished productivity, and higher medical costs.

You don't force your workers to share their underwear. Don't make them share their headwear.

Another option is to buy bone-conduction headsets. The units don't transmit sound through the ear canal like conventional headsets but rather through the bone to the middle ear. One example is *Vonia,* from Dowumi.

Bone-conduction headsets are more expensive but may be worth it. They are more sanitary and potentially safer because they do not touch the ear canal. That avoids spreading infection and prevents hearing loss by not irritating ear follicles. The Vonia sets also have flip-up pads that cover the ears and deflect 30 percent of outside noise.

Ten to 12 percent of the population suffers from hearing loss, which limits them from wearing conventional headsets. As people age they lose their hearing. But the aging population represents a vast pool of smart, responsible, and mature call center workers. Bone-conduction headsets can enable you to accommodate that workforce.

Another benefit is that colleagues and supervisors can talk to agents wearing bone-conduction sets while on the phones. You can't do that easily with conventional sets.

Bone conduction headsets, like Dowumi's Vonia EZ-3000S are more expensive than conventional aural sets but enable people who are hard of hearing to work in call centers. They also permit agents to talk with supervisors and colleagues while on the phone and are more sanitary.

Desk Trays

Desk trays enable agents to place their keyboards at more comfortable angles. *Call Center Magazine*'s Chief Technical Editor Joseph Fleischer had found one desk tray design helpful: Allied Plastics' ComfortSlope. As the name implies it is a desk attachment that inclines up, with space cut out for the worker to slide their chair into and at the top for the monitor. There is enough room to rest the wrists at the front and place a keyboard at the top of the slope.

The appliance works by having the workers lean back in their chairs with their arms and wrists straight but somewhat elevated and relaxed, instead of the usual positions of leaning forward, tensed against the desk. The position prompted by ComfortSlope reduces stress-caused injuries to the wrists, back, and neck.

Then there are other appliances of debatable worth. They are:

Wristpads and Keyboard Trays

Wristpads are, as the name implies, pads placed on workstations that users rest their wrists on. Keyboard trays are trays that roll in and out from under the workstation.

Cornell's Dr. Hedge does not recommend wrist pads. There are sensitive nerves in the wrists that they could strain. If you position yourself properly, you have little need for wrist pads. Instead he recommends keyboard trays.

On the other hand, Laura Sikorski, a veteran space designer, likes wrist pads but dislikes keyboard trays because they place agents too far away from the screen. They also interfere with the legs. She thinks you should have the keyboard on the work surface and use wrist support.

Dr. Triano does not take a position on wrist pads or keyboards trays. Yet he says neither may be necessary with proper workstation design, equipment setup, and seating.

"All of these auxiliary ergonomic aids are useful to solve isolated problems that cannot be solved, for one reason or another, with the overall workstation set up," says Dr. Triano. "Under special circumstances they may be necessary. In many circumstances they are a crutch that can be eliminated by attention to proper work surface, seating, and equipment arrangements."

From my experience, I'm with Dr. Hedge on wrist pads and Laura Sikorski on keyboard trays. One of my former colleagues developed serious carpal tunnel injuries at their workstation because the attachments were located too far below their desks.

Wrist pads are nuisances; I bump my knees into keyboard trays. But my wife has one on her desk. It is individual preference.

Keyless Keyboards

Keyless keyboards are pads that combine mice and keyboards. The OrbiTouch from Keybowl lets workers type without using fingers or wrists by sliding two large discs in different directions to create letters. The firm claims OrbiTouch reduces strain injuries compared with standard flat keyboards. The device's built-in mouse capability allows users to keep their hands on the OrbiTouch at all times.

There are some downsides to keyless keyboards. According to CMP publication *InformationWeek,* which reviewed OrbiTouch a few years ago, the device has a much slower typing speed and a hefty cost of $399.

Keyboard Chairs

Several years ago at least one manufacturer came out with chairs that had keyboards built into the armrest. Those contraptions didn't exactly make a big splash. They were expensive, inflexible, and ignored the need of most employees to have work surfaces to write on.

Ergonomics experts recommend that call center agents take short breaks from using their keyboard and mouse. Magnitude's ERGOManager software reminds agents to do so; it also indicates how agents are working to avoid future problems.

AIROGYM FEET
THE RISKS OF SITTING

Sitting for long periods of time can hurt you. It can put agents at risk from deep vein thrombosis (DVT)—blood clots. DVT killed NBC correspondent David Bloom through pulmonary embolism.

DVT occurs when there is minimal or stagnant circulation in the ankles and legs, which can occur from sitting in workstations or in plane seats. Anyone sitting for long periods may notice swelling in the ankles or possibly pain in the lower legs.

Those conditions may lead to blood clots, which form far more commonly than once thought. They usually dissolve but sometimes they grow and break off and go to the lungs. Or they manifest themselves in swollen, discolored, or painful legs.

To help prevent DVT there is the Airogym. The Airogym, devised by a retired long-haul pilot, is a simple two-chambered foot cushion. The two chambers are connected by a restriction valve, which provides vital resistance to leg muscles when operated.

The appliance is proven. Clinical trials showed that the Airogym increased blood circulation in the deep veins typically by 500 percent; using it may be more effective than walking.

Airogym is easy to apply. The user inflates one chamber, places the device beneath their feet, and treads air from one side to the other. The action mimics walking, so pumping blood and fluids from the lower limbs and back to the heart. This drastically reduces the risk of DVT.

DON'T FORGET THE LITTLE THINGS

Don't forget to include provisions for the seemingly little items that are very big to agent convenience and comfort. One of them is where they can put their coats, umbrellas, and other personal items.

EADS Telecom's David Meermans suggests that if you don't have or want workstation storage, then provide small lockers, about a 12-inch cube, for headsets, handbags, and other small articles. Agents tend to want to keep their coats with them, so include a coat hook in each workstation.

"A coat closet should be included in the center, but do not be surprised if no one ever uses it for coats," he says.

There are methods to make sure your amenities like water and ice work more effectively. Meermans recommends having water filters or a reverse osmosis unit, serviced by the supplier to ensure that old dirty filters are replaced, which are less of a hassle to fiddle with than bottled water. For ice he advocates commercial icemakers because domestic models cannot keep up with demand. For keeping food cold a domestic frost-free unit works well and is far cheaper than sub-zero units.

TEACHING ERGONOMICS

You can buy all the great ergonomically-designed chairs, workstations, and appliances you want, but they won't do any good unless you teach agents how to use them, and train supervisors what to look for.

Simply training agents to use the adjustments in their chairs, desks and monitors, coupled with proper phone use, printer and in-box placement, and encouraging them to take breaks can reduce or eliminate pain, points out Woody Dwyer, management consultant with ergonomics consulting firm Humantech.

"You don't put agents on a computer or a phone and tell them to use it without training," Roger Kingsland, managing partner, KSBA points out. "So you shouldn't provide agents with adjustable chairs, workstations, and keyboard rests without showing them how to use these features."

Dr. John Triano of the Texas Back Institute recommends using the following guidelines to set up agents' workstations so they are individually correct:

1. Is the work surface elbow-high so if agents sit in their chairs with their arms bent 90 degrees, their hands rest on the work surfaces? If not, have them raise or lower their chairs.

2. Check pressure on buttocks and legs in the chair. Here is a simple test. Have your agents put their fists behind their knees. If they can pass their fists easily behind the calf of their legs then they have enough clearance to avoid putting pressure on the vital blood vessels and nerves behind their knees. "If they have too much pressure there, like a seat pan too deep for a short person, they'll end up with problems like swollen feet and aching and restless legs," says Dr. Triano.

3. Have them take two fingers and pass them under their thighs on the seats. If this task is too difficult and their fingers don't fit easily, then there is too much pressure on the nerves and buttocks. Dr. Triano recommends installing adjustable footrests to correct this problem.

4. When agents are sitting, they should have a back support that applies a little bit of pressure in the lower back to arch their backs. The backrest should be pushed forward 2 to 2 and a half inches more than if they had nothing in front of them.

5. Computer screen sizes differ, which means that the angles that people view them at also vary, potentially causing eye and neck strain if the angles are not adjusted properly. The easiest way to find a comfortable angle is to have agents sit in a properly positioned chair, close their eyes, and then, in a relaxed manner, open them slowly. Where they are looking—their direction of gaze—is their resting gaze position.

6. Have agents adjust their elbow rests so that they just barely lift their shoulders. If the shoulders are too high or too low, the elbows fatigue.

NETWORK OPTIONS

When the calls arrive to your switch, you have to get them to your agents. That impacts design, maintenance, and security.

The traditional choice has been circuit-switched voice, also known as public switched telephone network (PSTN) or plain old telephone services (POTS). But there are newer methods such as VoIP and wireless to consider.

PSTN Advantages
A PSTN provides the greatest level of voice reliability and quality. They are also secure unless someone puts a wiretap on the line. The technology is proven.

PSTN Downsides
A PSTN requires separate voice cabling in addition to data. That increases your installation and upkeep costs, adds to fire loads, and creates more work when you change workstations.

VoIP Advantages
Voice over IP in the call center uses the same network as data. The principal advantage within the center is that it reduces installation and changeout costs.

But the benefits are greater outside the immediate call centers. VoIP cuts costs by reducing the number of lines into the facilities. It also eliminates long-distance charges on calls routed from network switches or rerouted from premises switches to satellite offices or home agents. I am connected to my office's switch on Vonage VoIP phone.

VoIP Challenges
VoIP quality is still not as good as a PSTN, though it is getting much better. A PSTN uses dedicated lines to connect callers and called parties, while VoIP entails sending conversations in packets over many different routes that must be assembled in the right order and time at the receiving end.

✪ ENABLING EFFECTIVE DESIGN

Well, you might say: "That's nice but I don't have the time. I need to get butts in seats yesterday!" Or "I don't have the budget!"

You're not alone. Call center site selection consultants say they've seen the site selection window shrink. The day a company chooses to open a center to when the agents take or make their first calls has gone from 18 months to three months.

Yet, if you want your call center to be a success, you can't make employees unhappy. Especially when they can get a job at the local WalMart or at the insurance company call center down the road, the one with the neat gym.

And yes, like time, there's never enough money to do what you want.

But here are some ways to get more for less with your design and furniture.

Pre-Planning
What you need to do is pre-plan your designs, including workstation and furniture, and have the RFPs and quotes in hand along with the local dealers' addresses and numbers

Think of VoIP like shipping loads in railroad cars or in several trucks. If they don't all arrive or don't arrive at the same time you have problems. And unlike data, you can't as easily resend a conversation.

VoIP also requires extra steps to ensure security, like high-quality encryption such as through virtual private networks to prevent hacking, viruses, and spit (spam over Internet telephony). If your call center is mission-critical, you will also need to make sure your UPS or backup generator has the capacity to handle VoIP as because it relies on commercial or generated power. While a PSTN also requires such power to run switches, there are workarounds to enable some calls to go through using the low-voltage DC current in the phone lines.

Talk to your VoIP supplier about compatibility with existing call recording systems. A *Call Center Magazine* reader reported concerns about combining them. Call centers should ask vendors to guarantee to protect their investment against the security changes.

Wireless Advantages
Associated with VoIP is the option to have your voice and data routed to your agents via wireless. You eliminate the lengthy and involved installation of wires to every workstation. Instead they have antennas. Reconfiguring each station is also much easier. The clutter and fireload of cables is greatly eliminated.

Wireless Downsides
To date there has no demonstratable ROI for wireless in call centers. Wireless cards are much more expensive than Ethernet cards. If you need your call center to remain up and running during power outages, you have to ensure that the equipment is hooked into your backup system.

There are issues with potential interference, degrading voice quality, bandwidth, and more importantly, with security. Hackers wardriving outside of call centers and your employees can sniff and steal customer information that you are liable for.

If you go for the wireless option make sure the technology you pick is made to the standard that offers the greatest security. Better safe than...

so that when you need to expand you can literally roll them into place. You already have enough to contend with, such as staffing and training, building quirks, and inevitable hardware and software bugs.

When you do pre-plan and pre-design, involve your employees. Don't leave it all up to the architects, interior designers, upper management, and bean counters. They don't have to work in your call center.

Ask your agents and supervisors how you can make their existing call center more comfortable. Bring in different chairs and workstations and study their reaction. Try new lighting and noise-reduction techniques in portions of your call center. If the innovations work out, consider retrofitting your present call centers. As well as providing you with advice, you also empower your staff, making them feel better about working at your call center. The more agents feel valued as employees, and the more empowered they feel about their immediate environment, the more likely they'll be willing to stay and recommend your call center to others.

Kingsland recommends to his call center clients that they perform studies to measure the performance of different workstation types and layouts. This requires a minimum sample size of 25 workstations and rotating the staff through the alternative station types. You'll find that catching any flaws in the designs at this stage is a lot less expensive than catching them when you open a new facility.

He points out that much time can be saved by discussing and settling all design issues up front, instead of planning and making changes on the fly as the project gets under way.

"If the design firm doesn't give the information up front, including costs, then your executives may want to revisit initial decisions later," Kingsland says.

You can also fine-tune your design and specification to prototype blueprints to save time. Many call centers, especially outsourcers, have such blueprints. Prototyping shaves four to eight weeks off a 14- to 20-week schedule for a 300- to 500-seat call center.

"You'll begin to experience the benefits of blueprinting after the second or third center," he points out. "You're proving out the prototype in the first one or two centers."

Steel Indirect Lighting Fixtures

Rick Burkett, president of burkettdesign, has seen steel indirect lighting fixtures that are about one-third less expensive than the cost of conventional aluminum fixtures.

"The steel fixtures have very good illumination and quicker ship times," says Burkett "The process to produce aluminum fixtures is much more time-consuming: eight weeks compared two weeks for steel."

Fluorescent Lighting Tweaks

Call centers are often overlit, which costs you money. Space planner Laura Sikorski recommends removing two tubes and the ballast from four-tube overhead fluorescent lighting fixtures, thereby reducing glare and power bills.

Also buy task lighting designed for compact fluorescent bulbs. Those lights save power directly, by requiring less energy and indirectly by taking the cooling load off your HVAC system. But don't retrofit these bulbs into existing lamps. Compact fluorescents are usually bigger and longer than incandescent bulbs, and their ends may stick out from the lamp hoods, causing more glare.

One good example of such task light is Waldmann Lighting's Roma. They use 18W compact fluorescent bulbs that provide the same light output as 75W incandescent bulbs. The lamps' built-in parabolic louvers direct light evenly over work surfaces. The louvers also help eliminate troublesome computer glare and reflections, a key cause of employee eyestrain and fatigue.

Wallpaper, Not Paint

Another cost-effective technique is wallpapering rather than painting walls. Wallpaper can last 10 years or more, while each repainting lasts three to four years. Wallpaper also absorbs sound, reducing the high noise level in call centers, Sikorski points out.

While wallpapering is more expensive than painting, wallpaper's increased lifespan pays back the added up-front cost in three to four years. Not to mention from increased employee and customer satisfaction by limiting background noise.

Buying Pre-Manufactured Items

You can buy many items for your center literally off the rack. For example, Rick Burkett points to coat racks. You can often get great deals on ancillary items like bookshelves and garbage bins at office supply chains like Office Depot and Staples.

Spread out Workstation and Chair Purchases

Another way to control costs is not to build out your call center at once. Mark Courville, associate principal and facilities unit manager in the Houston office of Carter & Burgess, suggests that you buy enough workstations and chairs for the first year. If and when demand picks up, you buy more. You thereby spread out your costs to match volume. That first-year savings can be substantial. Workstations comprise up to 40 percent of the cost of a new call center.

"Because it takes six to 12 months to get new workstations built and delivered, you have to plan your call center ahead very carefully," advises Courville.

Avoid Being Penny Wise But Pound...

Some cost-saving measures can actually cost more money, even in the short term. For example, broadloom carpet is less expensive than carpet tile but often needs to be replaced after a few years in call centers where wear and tear is high, Kingsland points out. Cheap chairs and desks wear out, aggravating agents. Their prices are often directly proportional to durability. You get what you pay for.

Shopping Smart

When drafting your tender and reviewing your bids, remember that price is only one component. Other components that you need to examine are appropriateness of application, longevity, maintenance, finish, aesthetics, and accessories.

One key method for getting value for your money is doing your homework on what you're buying and shopping around, especially for items like chairs. According to consultants like Bob Engel and Gere Picasso, in many instances the dealers offer "free" design services and even maintenance that makes managers perceive they are receiving value in the transaction.

They recommend instead that you draw up your specifications, tender your furniture purchases, and get vendors to bid against each other. Engel and Picasso say there are over 250 office furniture manufacturers in the U.S.

"It is not uncommon with this process to yield 30 percent to 40 percent savings over one-source, no-negotiation buying," Engel pointed out. "In one recent project Sprint saved $700,000 on a $2 million furniture budget."

Pre-Owned Furniture

Check out quality used chairs and workstations either through dealers or from a vacated call center. Heavy-duty office furniture can and will last. You may get some excellent deals on top makes like Haworth and Herman Miller, better than buying a lesser grade new. You can get them as low as 10 percent off the original purchase price.

Be careful when buying them. Bob Engel warns that acquiring used furniture may cost less initially, but bringing second-hand furniture up to modern standards may be more expensive than buying new. Out-of-warranty repairs and component add-ons boost old furniture costs. While the frames made of steel may last, key features like adjustable chair lifts and arm and backrests do not. Nor does the upholstery.

As with buying a secondhand instead of a new car, make sure that all the components, such as seat pans, lifts, backs, and upholstery, are in good shape. Only buy ergonomically sound, adjustable chairs. Avoid purchasing older pre-wired workstations that have pass-through cable passages. They are more difficult to wire and fault-trace. Source from reputable dealers.

Lastly, the furniture may be outdated. If you have to add new or replacement seats you may not get a right fit of designs, color, and fabrics. The look is also important, especially to senior managers, top customers, and senior employees.

"If you don't plan to change around your call center and you don't care how it looks, then you can get away with older unrefurbished furniture," says Engel. "But if you do care about performance, appearance, flexibility, employee health, and total cost, then you should buy new."

✪ DON'T FORGET HOME AGENTS

If you have a portion of your employee workforce based at home, don't overlook them in your design and ergonomics considerations.

While they may be out of sight, they should not be out of mind. If your home working employees are suffering from ailments like tendonitis, you will find out soon enough, when they call in sick or say they have to go to the doctor.

Also, while authorities such as OSHA have as of this writing stayed away regulation of home offices per se, they are still covered under General Industry rules.

The CMP book *Home Workplace* delves into home design and ergonomics issues. Here are a few key points:

* Insist on dedicated, disturbance-free locations for their work. Consider separate rooms with lockable doors, especially if privacy is a major concern.
* Require that they purchase, or you buy for them, adjustable chairs and suitable-height workstations. Arrange them to minimize glare. Ensure that employees can perform their tasks safely, with minimal injury risk.
* Remind them to make sure that the electrical load in the workspace complies with electrical codes to reduce the odds of fires.
* Have them buy insurance to protect you (and them) in case of injury, such as to deliverypeople. Stipulate smoke alarms in the workspace.
* Set out your design and environmental stipulations in a home workplace policy that you and your employees sign.
* If the agents work from home on a regular basis, consider dedicated IT staff to support them on issues such as VPN lines and firewalls.
* If you require agents or others to come into your call center periodically, arrange for hot desks.
* Have them connect their equipment into a surge protector to protect them. If the

agents' or supervisors' work is critical, supply or have them buy small battery-powered UPS systems to keep them going through brief outages or enable them to safely power down through longer ones.

✪ MAINTAINING YOUR CALL CENTER

Do not forget to figure out how you will keep your clean and secure. Change air filters and lights often. Filthy microbe-riddled filters degrade agents' health and HVAC efficiency.

You will of course arrange for janitorial, maintenance, and security services. But there are other issues that arise. KSBA's Kingsland recommends obtaining agent buy-in to facility upkeep, such as by dedicating part of the vending machines' proceeds to maintenance, educating agents on why it is important for them to do their share, and involve them in housekeeping. This way the call center becomes theirs too, and they will help keep it clean.

Engel and Picasso conduct post-installation audits that use interviews, observations, and questionnaires to monitor the facility, work climate, technology, and job tasks. They track and document functionality, performance, and employee satisfaction. The audits survey 33 areas including furniture, safety, acoustics, lighting, job performance, training, and technology transfer.

Companies use these techniques to benchmark their call centers and establish quality standards and controls. Some use them to monitor the performance of their facilities personnel and evaluate how the changes that are made to the furniture, environment, technology, and training affect profitability and performance.

Here are some other key areas to look at:

Smoking

Unfortunately you may have to find some way to accommodate those employees that are nicotine-addicted and prefer to consume their drug by smoking it. I say "unfortunately," because permitting people to smoke tobacco where it could harm others makes as much sense as letting them release hydrogen sulfide gas and inhale it.

Smoking poisons people. It exposes employees and others to cancer and emphysema. It releases dangerous chemicals such as benzene and hydrogen cyanide and gases like carbon monoxide.

Smoking costs you money. It jacks up healthcare expenses from related diseases, hikes maintenance costs, and causes loss of productivity trough smoke breaks. Your firm may also pay the ultimate price from workers lighting up. It takes one idiot to toss a lit butt onto the grass on a hot dry day and you have a blaze that could spread out of control, costing lives, property damage, and tax dollars in firefighting.

Have your HR department actively discourage smoking. One method is to look for benefits policies that reward non-smokers, such as those that offer stop-smoking programs.

Keep It Clean, Folks!

You open this brand new, state-of-the-art call center, and six months later it is a dump. Yes, you paid for a cleaning service, but it looks and smells grungy, especially the common areas. Here's what you can do about it.

No Eating at Workstations

Allowing employees to eat at the workstations should be a no-no. Spilled food and sticky beverages, like coffee and soda, can ruin costly keyboards and chairs in an instant, while attracting real, live, creepy, crawling bugs and mice into your computer system. Individual garbage cans begin to reek of odors that are heaven to rats and naturally draw them in, just what you need to show clients and senior management when they tour your call center.

Hire Someone to Clean After Your Workers

One of the biggest lunchroom nuisances that can cause health problems is filthy dishes. Piles of them lying in a sink provide a buffet for bacteria, germs, and their carriers, such as flies, roaches, and rodents. Yum-yum. While employees should wash up after themselves, many do not.

Meermans advises that you hire and pay someone, a surrogate Mom or Dad, whose responsibility includes cleaning the lunchroom and the fridge. You should also consider buying a dishwasher—Mom or Dad can run it.

"This may mean the end of microbiology experiments in the coffee cups, but you are running a call center, not a pharmaceutical lab," he says.

Do Likewise for the Toilets

People are slobs, especially when no one sees them. That means the toilets.

If you have or will have a large call center, consider hiring someone to clean the bathrooms before and after break periods. Perhaps it can be the same individuals you hired to clean the break room.

Have Recycling Bins

One way to manage the mess is to require employees to recycle paper, cans, and bottles. You can buy these bins and have them in the break rooms. Many communities now have mandatory recycling because the dumps are filling up.

✪ SAFETY AND SECURITY

To function effectively your call center needs to be safe and secure. Here are some of the issues that arise and what you can do about them.

Personal Safety

As Chapter 11 outlined, you need to take steps to protect employees. That includes picking the right building in a safe area, with excellent lighting and few shadows. You may need security guards and electronic door access.

Yet the biggest danger may come within, violence between employees and the people they know. Call centers are at risk for such violence because it is a high-stress environment where low-paid workers handling sometimes-abusive customers or other employees are rigidly managed to maintain performance.

Another threat comes from the almost-always male spouses of female employees, who murder and assault their intendedvictims, but in most cases no one else.

SIGNS OF IMPENDING VIOLENCE

An excellent *USA Today* article that appeared July 16, 2004, discusses workplace violence. One of the sidebars lists some warning signs. The leading ones include:

- Past violent behavior
- Excessive argumentation
- Mental health problems
- Substance abuse
- Absenteeism

- Threats of violence
- Hypersensitivity to criticism
- Family or money problems
- Extreme changes in behavior

There is little call centers can do once the perpetrators have made their intentions known. Who is going to stand up to a gun? But relying on employees who report overheard threats is ineffective and unfair because it is nearly impossible to check the veracity or validity of such reports.

Bob Mellinger, president of business continuity planners Attainium, suggests training frontline people such as receptionists and agents to spot suspicious behavior and how to react.

"You can also create corporate policies banning such threats or conversations, like anti-sexual harassment policies," he says. "This may be a sufficient way for someone who breaks the policy to be disciplined."

Peace at Work, a non-profit based in North Carolina that is dedicated to workplace violence prevention recommends that you "target-harden" employees who have been domestic violence victims to lessen reoccurrence such by moving their workstations, job sites or their parking spaces. Also notify and consult with local law enforcement agencies.

More fundamentally the organization recommends a workplace violence policy that demonstrates support and protection for victim employees. All employee need to be trained to recognize domestic violence warning signs and dynamics of domestic violence, how to refer to available services and the importance of notify a supervisor if there is a potential danger.

"If an employee sees their workplace as a source of support they are more likely to disclose their predicament and the potential threat to management, " states founder John Lee. "This initial warning is the vital first step to any security planning."

Stealing

As with any office or workplace your equipment, property and employees' personal effects are liable to disappear. You would be surprised what can drop out of site when no one is paying attention or cares.

The last point is important. If employees don't feel their employers care about them, especially if there are rumors of downsizing in the wind, then they are likely to be temporarily blind. Call centers are at risk for property crimes because employees are low-paid and under high stress. There is rapid turnover in centers that inhibits the building of personal bonds between workers and with employers that enable each party to look after each other.

Your best bet is to follow the advice of police, and that is to mark and record every piece of equipment. Provide lockable filing cabinets or lockers for personal effects. Check to make sure all doors are securely locked and that the guards are doing their jobs.

Put in access controls like electronic card readers, requiring visitors to sign in and out and wear cards. Most important, educate and train agents to keep an eye out for each other and report suspicious behavior and persons who don't seem to belong. Better to be safe than sorry.

Data Theft

Data theft is another risk in call centers, because agents have access to this information. The temptation is there because employees are low-paid, which makes them vulnerable to debt and to those lenders with very high interest rates.

Experts like Jerry Brady, chief technology officer of Guardent (now Verisign), say call center agents can steal data on site because they have access to this information. They move it off site such as by copying onto outbound emails, re-entering it on cell phones and PDAs or printing data and sneaking the hardcopies out.

More computer-savvy employees install web spyware programs on their PCs that stream data out through encrypted tunnels on web browsers. They also install hard-to-trace keyword and network sniffers. They automatically look for names or keywords, capture that data, and retransmit it out the door without anyone knowing it.

Visiting family, friends, and guests who have access to the call center floor can eyeball, record, and steal information. They too could slip a sniffer on an empty workstation if no one is looking.

There are several ways to protect data. Require agents must to log in to applications that enable access to data, log out, and record what they did with that information. Agents must authenticate their ID. Data can be partitioned to give access only to what agents need, such as credit card numbers but no names.

Brady suggests setting up application-based rules, like agents can't do wild searches; they must enter account-specific information.

Another tool is storageless PCs to prevent agents from installing spyware. The total cost of ownership is much lower. The boxes are less expensive and so are software changes.

If companies allow agents to print or store information, they must back this up by checking employees when they leave the premises. Agents should lock up all printouts.

"You shouldn't be worried about people's memories, because there is a limit to how much specific information they can store, and there's no way you can purge them," says Anton Ragen, senior security consultant with Avaya's Enterprise Practice. "What you should be concerned about is personal information they can store and walk out with."

Fire Prevention and Response

Fires can happen anywhere, anytime. Equal if not worse than the danger of fire is poor response to it. Panic kills.

Arguably the biggest risk in office environments is from smokers, when discarding

DISASTER COSTS

How much will a disaster cost your call center? You can get an idea from this data from Contingency Planning Research , cited by 180cc principal consultant Kurt Sohn, based on 450 of the Fortune 1000 companies.

Cost per hour of downtime	$78,000
Hours per incident	4.2
Average incidents per year	9
Downtime cost per year	$2,970,000

Industry Average Hourly Impact

Airline reservations activation	$41,000
Credit card authorization	$2,600,000
Home shopping	$113,750
Online network fees	$25,250
Package shipping services	$28,250
Pay-per-view services	$150,250
ATM services fees	$14,500
Brokerage operations	$6,450,000
Catalog sales	$90,000

their matches and butts. Hence the stringent prohibitions smoking outlined earlier.

Employees need to be drilled on evacuations, to be taught that the last thing they do is go back for some trinket which puts their lives and others, such as those of emergency services staff, in danger.

All fire exits need to be kept clear and doors closed. Any employee caught violating that rule gets a dressing down and on second offense, suspension or dismissal. People have died from blocked or inoperable exits.

Business Continuity

Your call center needs a disaster response strategy, also known as business continuity. Laws like HIPAA require it. That can include:

* Offsite data backup
* Diversion of calls to home workers, other centers, disaster recovery rooms, or outsourcers* IVR, outbound voice mails and emails, and web pages informing people of the events and what to do

A designated 'panic room' that is resistant to storms, and has food and water for one to three days. Deploy this strategy if there is a risk that employees may be trapped on premises or when remaining stationary is preferred or safer, says Bob Mellinger, president of business continuity planner Attainium.

Formalized Disaster Plan

You need to create a formal business continuity plan. You look at whether call center functions are critical or not, rank their criticality, and devise responses.

Kurt Sohn, principal consultant with call center business continuity planning firm 180cc, recommends that at the very least, plans lay out evacuation procedures, training, and who to contact in emergencies. Call centers should also have some designated employees trained in first aid and CPR.

The plan should include an updated staff list, a vendor list with passcode information, procedures to declare the beginning and end of the emergency, a chain of command, a press information program, and purchase agreements with hardware stores and other vendors.

The procedures should set out expected response times from local police and fire departments and who to contact there.

Dodge McCord, senior-specialist business-recovery with consultants North Highland, recommends phone and computer system documentation and backup.

"Do not rely on what is in an IT or telecom guru's head," he advises. "In emergencies they may not recall, they may not be there at the premises when the disaster occurs, or be injured or dead."

Getting Buy-In

The key to such plans is obtaining top-level buy-in, especially when the plan requires building changes, investments, and new procedures. But managers may not have such plans, says Sohn, because senior management has not been convinced of how important the call centers are to their business. Many managers all the way up to the executive suite do not know what each call costs or produces in revenue.

"In business continuity, how much do we lose per transaction times downtime and how many customers go elsewhere never to return, is one of the first items to be considered," explains Sohn.

Your agents and supervisors also need to buy in. Yet often companies are not cognizant of employees' different needs. If they have families and houses, they will be more concerned about them than they are about your premises.

If you do get buy-in, that plan should have principal and backup incident managers authorized in writing by companies' senior executives to act on their behalf who will be listened to by all employees. They would be the go-to person in an emergency.

Incident managers should be bondable with no criminal records. They should have an interest and background in building infrastructure and technology.

There is business continuity training and certification available through the Disaster Recovery Institute International.

Bob Mellinger recommends having a knowledgeable third party, which could be a consultant or your local fire or police department, to review the plan and offer advice. A third party will see and catch items you may have missed.

"Think of how many times you walked by a door and not noticed it wasn't all the way shut," he points outs. "But if you have someone who is aware and trained to spot potential risks they will spot that and either correct it or draw it to your attention."

Chapter 13: Staffing Your Call Center

You now have your dream call center. The lights work, the switch is connected, the dust is off the workstations, the contractors' empty cola cans are out of the hallways, the paint is dried, the AC is cranked up, and the asphalt in the parking lot has a never-to-be-seen again deep black hue. You've got a rent-a-sign out front flashing "Good Jobs! Good Pay! Excellent Benefits! Child Care! Jacuzzi! Just Come On In!" Yes, open it and they will come. The "Field of Dreams" in the call center league.

Well, it doesn't quite work that way. You must have your staffing and training worked out and (ideally) the people lined up before you open a new center. Finding and retaining the right people is the most critical and challenging aspect of setting up a successful call center. All the labor-saving technologies in the world won't help you without qualified, friendly, intelligent, hard working, and reliable agents, managed by competent and dedicated supervisors.

If you want to keep your call center viable for as long as possible at the new location—which caused you to pull so many hairs out while researching, selecting, and outfitting it that the baldness-cures telemarketers are calling you—your attention should be focused on selecting the right agents. Which is what this chapter will explore, along with training strategies.

✪ STAFFING

The bottom line of a live agent call center is that you need to hire the right people to take those contacts. It's that simple, in theory.

Call centers already possess certain key advantages over other employers that compete for the same workers, like retailers. The pay is usually higher, there is less physical labor, security is greater, the environment is office-like and professional, and employees get to work on computers. Depending on the company, you may also offer flexible full-time and part-time hours.

"Even in a tight labor market, call centers will always draw people from other service jobs," observes King White, vice president of real estate firm Trammell Crow.

The downsides are that the work is high-pressure. Dealing directly with the public is often very stressful while the compensating pay and benefits are still low, especially when compared with employment in manufacturing, programming, transportation, and construction trades.

There is not the direct professional and personal satisfaction from working in a call center that exists in other careers. The hours are often irregular and inadequate to make a living, and it plays havoc with personal lives. Turnover, which costs anywhere from 50–100 percent of the agents' wages in hiring and training expenses, runs from about 25 percent to sometimes over 100 percent per year, depending on the local labor market.

Self-service, outsourcing, and offshoring have lessened the labor pressure somewhat by handling the minimal-skill calls like bank balances, dealer locators, and simple order taking. Call centers have become more professional, taking on more difficult cross-selling, upselling, service, and problem-solving tasks. Agents are interacting via email and chat as well as by voice. That has made the work more interesting and challenging, helping to retain agents.

LIMITS TO TRADITIONAL CALL CENTERS?

Call centers could reach their effectiveness limit. There is only so much you can ask someone to do at $12-$14/hour and find enough of them to fill your workstations.

Organizations are reluctant to grant call center agents widespread decisionmaking or problem-solving authority. To provide that requires these individuals to have business and technical expertise and experience beyond the ken of customer service, sales, and support requirements. You can't have every employee offering unrealistic deals or making promises that they can't deliver without understanding the impacts of their decisions and knowing what the organization can and cannot do.

Not when too many of those who you are hiring for the jobs couldn't be bothered to show up to the interview on time. Or lack the reading skills to understand this sentence.

At the same time can you afford to pay high wages to attract literate responsible people for seemingly and supposed-to-be basic skills? Put that case to your CFO and you'll hear "self-service," "offshore," or "outsource" faster than you can read this paragraph.

If you can't go up the food chain, then how about back "down" it?

There is an emerging school of thought that goes back to the basics of phone interaction. That alternative, telemessaging, implies that the agents cannot solve problems, that they can only triage them and connect callers to trained professionals. Calls are answered quickly with courtesy and empathy.

Telemessaging is proven. Agents have been answering calls for professionals like doctors since the 1920s.

With telemessaging agent skillsets are minimal, which enables call centers to reach out and retain a much larger labor force. Telemessaging firms have had surprisingly relatively low turnover rates. Firms report that some employees have been with them for over 20 years. Employees are often older and female.

Typically these call centers are small and friendly where everyone gets to know each other. Hiring is by word of mouth. It is not uncommon for mothers and daughters to be working there. Employees who go to college sometimes come back to work part-time.

Unfortunately, call centers may be facing a ceiling. The greater demands require agents with more skills like excellent grammar, comprehension, and multitasking; higher education levels or computer software or hardware certification.

Tracy Buelow, market planning manager for contact centers for staffing agency Manpower, told *Call Center Magazine* that call centers' evolution into a sophisticated customer service and sales channel has led to more-stringent agent qualifications. Managers who were once satisfied with basic call center experience and a high school diploma are now expecting specific job histories, such as five years as a benefits enrollment associate or a licensed insurance agent.

Take an applicant pool without skills advanced centers need and overlay that on a shrinking labor supply caused by baby boomer retirements and lower overall birth weights Then mesh that with underlying undesirability in the minds of many people of call center work. The net result is a shrinking group of applicants for the positions available.

"We're going to witness similar labor shortages, wage escalations and turnover issues as we did in the late 1990s, but worse," predicts Kathy Dean, partner at Banks and Dean. "The challenge for call centers will be to attract and retain sufficient quantities of these increasingly rare workers."

To make matters worse, applicant quality is reportedly plummeting. The American Diploma Project reported that a majority of college grads take either a remedial English or math class.

Call center staffing consultants say cognitive thinking, interpersonal skills and the ability to learn are also tanking. Jeff Furst, president of staffing firm FurstPerson, reported that many applicants lack even basic work aptitudes, like showing up and leaving on time, which are vital for call centers to maintain service levels. More students appear to be leaving high school with fewer skills and with no desire to perform challenging work.

This goes for bottom-level call center work, too. Furst said that his clients must weed through many applicants to fill entry-level agent jobs that require only voice with no deviation from scripts.

"To set up a hiring system for basic customer care positions and only be able to hire one in 25 people just doesn't make sense," he pointed out.

Who is to blame? A lousy educational system coupled with spoiled parenting that is creating a generation of brats is a start.

"[Young people] still feel that they have a sense of entitlement to better pay, working conditions, and promotions that aren't there," says consultant Elizabeth Ahearn, president of The Radclyffe Group.

Unfortunately as the employment market tightens, employers like call centers may have little choice except to hire slackers. No wonder why countries like India and China are so attractive, for those workers know the consequences of failure.

There is good news. Later on in this chapter I look at labor markets that may appreciate the opportunities to work in call centers, if you're willing to hire and accommodate them.

❂ DESIGNING YOUR IDEAL AGENT

Before you begin recruiting workers you should have a profile of a good agent for each position type. You then use the profiles to attract, screen, and select employees.

Ideally you should have the profiles created before you begin your site selection process. That way you can find out whether the labor force in the targeted locations matches them, and whether it has the quantity and quality needed for the life of your call center. The profiles should contain what Banks and Dean call non-negotiables agreed-on by call centers and HR. These include education level, experience, basic skills such as computer or keyboard skills, and availability.

There are some common characteristics that call center agents must have and should be in your profiles.

Job and Stress Tolerance

Working in a call center is very stressful. Few people call because they are happy and want to let everyone know it. Few people answer calls with smiles on their faces.

Agents have the toughness to withstand annoyance and verbal abuse, inbound and outbound. They must know how to take rejection and threats. Bullying is easier if you're not face-to-face with the person you're harassing.

Call centers are physically not easy environments to work in. Agents are confined to their cubicles except during breaks, punctuality is critical to help the centers meet service levels, the noise is constant, and there is heavy monitoring and supervision. For many centers there is shiftwork on weekends, holidays, evenings, and in emergencies.

Some people thrive in these environments, others will tolerate it, but others can't take it. It's best to see who can fit the first two by seeking attributes like self-discipline, ease at being at one place for long periods of time, stress tolerance, and acceptance of heavy supervision.

Personality Attributes

Required characteristics and skills can vary from center to center. If your agents sell products and services, inbound and outbound, then you should state that applicants must have the ability to offer and close sales. If they need to perform detailed calculations, your profile must stipulate math ability. If your center sells insurance, real estate and securities, your agents will need licensing and certification. If they will handle mainly customer service, the applicants should have listening, understanding, problem identification, and resolution skills.

Remember, it is always easier and more effective to screen people to see if they have these attributes than to train those who don't.

Service People Need Sales Skills

If your center is hiring for customer service or customer support, make sure you draft sales skills into your profile. That's because more centers are requiring agents to cross-sell and upsell products and services.

All customers have needs, Kathy Dean points out. It is critical for agents to discern those needs and understand if their product and service can meet them. You then need to look and screen for a different kind of person than those typically found on outbound and inbound sales and service. "You need agents who can ask questions, listen, assess, and come up with responses to meet customer needs, not just those who can sell or provide service," explains partner Berta Banks.

STEADY EDDIES INSTEAD OF STORMIN SALLIES?

It is human instinct to be wowed or intimated by strong, energetic personalities. We are seemingly hard-wired to equate force with success.

But what works at fighting off saber-toothed tigers doesn't exactly work for dealing with angry customers while chained to cubes.

The experience in India has been that the strong personalities do not work in call centers, reports the *Hindu Business Line*. They quickly leave and consequently turnover rates shoot up.

Instead, call centers should look for steady and hardworking agents who are dependable and can handle the conditions of the position. They are also good teachers and can easily be promoted.

"Sadly enough these individuals are not very impressive in an interview because they are mellow and conservative," says the paper.

Multimedia

Yes, most call center interactions are still calls, but slowly and increasingly less so. Some centers, like eBay's, are nearly all online.

That means you need to consider building into your profile the ability to communicate in other media such as email and chat. That means agents need to have superior written language and comprehension skills.

That's because in text communications you have to be painstakingly accurate in your grammar, syntax, spelling, and meaning. There is no opportunity to quickly correct mistakes with as there is with voice.

You are stripped of accent, emphasis, tone, and vowel length that gives context and meaning to language. Trainer Rosanne D"Ausilio, president of Human Technologies Global, says you can fall prey to homonyms such as "the farmer does produce" which can have two quite different meanings, but if you heard or spoke this line you will know which one. If the o is short, "produce" means agricultural goods; if the o is long it means that the farmer makes something.

Training consultant Stephen Coscia has seen agents with word usage problems, such as differentiating between "to," "two," and "too." Another common area of confusion is between the words "their" and "there." No spellcheck could pick up on those errors.

Poor language usage and punctuation coupled with informal abbreviations and symbols radically alters the meaning of a sentence. This leads to misunderstandings, lost business, and possible legal action if the misunderstanding caused damage. For example, a poorly worded email message from a help desk agent could lead a customer to make an error on a computer, causing a crash and either losing valuable data or disabling the machine.

The legal consequences are real. Emails and chat interactions are written evidence. Emails especially have been subpoenaed in court cases.

For these reasons text-based business style is and should be formal. It should have no slang or cutesy shorthand and symbols. If applicants who will handle email and chat interactions don't know to communicate properly by text, then they have no business working in your call center. You can't afford the consequences.

Unfortunately, today's frequently lousy American primary and secondary schools and colleges are spitting out too many graduates and students who lack essential grammar, spelling, and composition skills. Computer spelling and grammar checks make even the best students lazy.

This issue isn't limited to call centers. Publications like *Call Center Magazine*, which require a very high level of language competency, have seen some horribly written resumes from applicants.

Language Skills

If you serve domestic or international non-English speaking customers, your profiles should include other language skills, or you may need a separate profile for non-English speaking agents. In developing the profiles you need to look at whether your customers prefer native speakers and in which languages. If so, you need to decide whether the market is large enough to warrant serving them with native-speaking agents, whether you can get by with learned language-speaking agents, or whether you should outsource some or all of your language needs to an over-the-phone interpreting (OPI) service. Chapter 6 covers OPI.

Corporate Culture

Your profiles are also molded by your corporate culture. Dave Burdette, business development consultant with Development Dimensions International, uses the example of two call centers that provide customer service in competition with each other. One has a by-the-book process with a well-defined process and procedure, including escalation to succeeding levels, while the other has a less-rigid system by which agents are empowered to make their own decisions to solve problems accurately without escalation.

"The first call center would look for employees who can understand and follow instructions, while the second would look for employees who are creative and problem solvers," explains Burdette. "Yet they would have the same type of job serving the same type of customer."

Profile Caveats

Make sure that when you use profiling that your standards are high but realistic, so that you don't unnecessarily limit your labor pool by excluding good applicants. You don't want to leave seats vacant, which lowers your center's performance levels when you don't have to. Following are some common examples.

College Degrees

One sometimes-misused staffing profile element and screening device is a college degree. While it is nice to have the intelligence and prestige of college grads on your phones and keyboards, you may also find that they are not your best performers. They may view a job with your center as a temporary position and leave as soon as they accept an offer from an employer in their studied field.

BACKGROUND CHECKS?

Background checks can help call centers hire reliable, trustworthy employees, which minimizes the risk of theft and workplace violence.

Global staffing agency Manpower's extensive background checks may include verifying employment references and investigating multiple county criminal records. They may also confirm educational background, conduct credit checks, and administer drug tests.

But background checks, say consultants, won't stop acquaintances, strangers, or visitors from committing crimes on your premises.

Credit checks are especially problematic for two reasons. The information may be out of date or inaccurate, and difficult to correct. If someone had their identity stolen it could take months to repair that damage.

More seriously, why should someone who is otherwise qualified be denied a job if they were careless with their bills or declared bankruptcy? Any motive to steal to bolster finances is more than offset by the risks of going to jail.

Many excellent workers have had credit problems. Life happens. Events include emergencies like medical bills, supporting out-of-work relatives, student loans, and funeral expenses. Lower-paid employees like call center agents face a greater risk of that happening.

Denying someone a job based on credit is creating a de facto debtors' prison in one's home, or on the streets if they can no longer afford a place to live. After all, isn't the purpose of working to generate income to live, including paying bills?

"For example, one best-practice company we worked with found no correlation between success and significant prior customer service, but some correlation between success and a stable work history," reports Kathyrn Jackson, associate at Response Design Corporation. "They also found some correlation between success and college experience, but an inverse relationship between success and having a college degree. Therefore, it believes that the profile of an agent most likely to be successful is one with some college, between six months and one year of customer service experience, and a stable work history. "

Experience

Look hard at including experience in your profile. It is often not a good predictor for future. It is nice to have but not essential. Remember, agent work is not exceptionally high skilled. Also what works in one call center may not be the same in another, even though they may be serving the same types of customers on similar programs.

"Just because you have worked in a call center doesn't mean you're a good agent," Berta Banks of Banks and Dean points out. "Some of the best people have no experience, but they have excellent potential for the position. If you require experience, you will lose out on those people."

Jeff Furst says what is sound practice in one call center may not be acceptable in another. One example is the ability to get customers off the phone quickly to shorten average handle time. But that experience and attribute may not be acceptable in another center that prefers to resolve the matter, eliminating follow-up calls and resulting costs.

Brad Cleveland, president of Incoming Calls Management Institute, says the value of experience depends on the call center. Most people who can reason and communicate well with others can learn the requisite skills to be an agent.

"But you may have a call center where the agents' experience enhances their skills," he observes. "Examples include engineering or high-end technical support environments, financial services requiring licensed advisors, legal and medical services, collections, and complex sales. In those cases, you bet that hiring experienced agents can give you a big head start."

Multitasking

Be careful about screening for multitasking. You may be throwing away high-quality agent candidates unnecessarily.

Agents who excel at taking inbound calls may not do well making outbound calls. Or they're successful at service but not sales. Or they're great on the phone but not with writing emails or chat responses. And so on.

"I see many call centers struggling with whether to select individuals with these specific skills or require everyone to rotate responsibility for online communication, thus giving some job variety," says Anne Nickerson of The Call Center Coach. "In response, there are more call centers successful with a dedicated unit. They are offering this smaller group specific grammar and business writing seminars, along with establishing protocol, templates, and follow-up processes for more customized responses."

Finding Right Holes for Square Pegs

See that you're not excluding individuals who may not fit one set of agent positions but who are quite suitable in others. There are different tasks in call centers requiring their own skillsets and personality types.

"A detail-oriented person may not work best in general customer service," Burdette points out. "However they may be ideal in an upper-level desk where they're solving more complex problems."

Go to the Experts

When drafting your profile, be sure to consult with agents and managers about what makes a good employee. This method also makes agents feel that they are valued members of the team. When staff members leave, ask why during exit interviews.

"Front-line agents and supervisors know the jobs better than anyone else," Rosanne D'Ausilio points out. "Someone five steps removed doesn't really know it."

Remember...

Remember, profiles are not living human beings but your applicants are. Use profiles as a model of what your agents should be like and measure your applicants against them. However, when you interview these prospective agents, look at them as people, with bright and dull spots, to see if you can see them working for and with you.

"You can interview someone that has all the skills and qualifications, who fits the model, but if they show no life, no spark, no enthusiasm for the position, then what good are they to your center and to your customers?" asks D'Ausilio.

✪ AGENT CERTIFICATION

There have been and will continue to be attempts and suggestions for having agents tested on a given set of objective criteria, such as soft skills, sales skills, and multitasking, that show that they have the skills to work in call centers. Agents could then take these from employer to employer.

Certification is *not* to be confused with licensing. Agents undertaking specific tasks where are risks of financial and physical harm, such as insurance, medical/nursing, real estate and securities selling are usually licensed.

Certification Upsides

Certification advocates say having such testing and documents will help the call center industry image by making it more professional. That in turn will raise standards, better attract and keep employees, enable job portability by minimizing retraining at the next employer, and gain corporate respect.

Certification Downsides

So far such certification has not taken off. The main reason is that few managers, HR departments, or employees see any benefits. Critics point out that other high-stress/high turnover service sector occupations such as ticket clerks, bank tellers, restaurant staff, and retail salespeople are not certified. Neither are the managers. Yet they attract and keep staff. Few managers in any profession or trade are separately certified.

There are few serious financial or other consequences to customers if agents do not have general certification. There are doubts about how much of an agent's functions can be certified and are transferable. Teresa Hartsaw, president of outsourcer and training firm ePerformax, figures that no more than 35 percent of an agent's skills are transferable and 65 percent are specific to that call center. There are so few similarities among programs that she doesn't think that agent certification delivers value. Training must be customized to the specific products and services offered or supported to be effective.

Shifting agents from one program to another requires extensive retraining, she points out. Various programs require very different applications of agents' core competencies even if they share the same function and use the same hard skills, such as selling.

There is little agreement within the industry on standards. Kathryn Jackson reports that many call centers feel that soft skills cannot be quantified and that they are too prone to variation among call centers or from one caller to the next. Call centers need to agree on minimum standards for certifiable skills.

"To make certification work, we need agreement on competencies, such as what constitutes soft and hard skills and how are they defined to the behavioral level," says Jackson. "That's the only way certification will be transferable from one company to the next."

Roles for Agent Certification

There are two specialized roles for agent certification, where you should use them in your profiles. They are:

Technical Certification

Many hardware and software vendors offer certifications on their technology to support agents. That way these employees can fix problems on gear and applications that use these products, which avoids costly and time-consuming escalations to those suppliers.

Internal Certification

You should consider internal certification for your staff. Having such certifications enables you to set a more rigorous uniform set of standards geared to your processes. That way you can measure and track agent performance and career paths and offer them incentives to stay longer. You can also quickly screen people from other locales.

✪ QUALIFICATION

The next step is to put in place interactive questionnaires to qualify applicants based on the model developed from your profile. The methodologies and tools you employ can introduce the position, describe the work, and sort out applicants' skills before they ever sit down with you. That includes, if need be, their abilities in other languages.

The most common assessment means are IVR or web-based. They act as your 24x7 HR department, intaking applicants at their convenience.

An interested employee calls or goes through the web site to reach a call center and is asked a series of questions. The applicant is given a job description and is asked to see if they have the basic qualifications and interest. If the answers meet your criteria, the applicant is asked what dates and times they would like to come in.

Banks says all call centers should post openings online and use automated screenings, because call centers need to attract and capture agents who are looking for work around the clock.

At each step they let the applicants stop. This saves them and you from saying no to each other's face. You also save considerable time and money compared with having everyone fill out applications, play "pick the lucky resume," and meet them all face to face.

If you haven't gone to IVR and web-based qualification, do so. The benefits are well worth the expense. You avoid asking for and reading resumes, which Banks says is a waste of time.

"I've heard it takes 15 days to screen 150 resumes for five positions," she reports. "That is ridiculous. You can collect the same information and assess candidates much faster with online or computer-based screening in your office. The only human intervention required is in determining the screening criteria that will be used by the system."

Rosanne D'Ausilio is a big fan of qualifying tools. She says they can significantly cut down on time and costly human intervention. Because they are uniformly programmed, the responses are consistent, allowing for fair efficient evaluation.

"The great thing about these tools is that if I'm a call center manager, I can have an ad placed in the Sunday papers, and Monday morning I could have a list of pre-screened applicants and appointments," says D'Ausilio. "That's opposed to the old system where I get resumes arriving, or people stopping by at random to fill out applications, sorting them out, calling the ones I liked, then playing voice mail tag, and that's before these individuals come in for an interview. That's in between doing my job managing a call center."

Phone Assessment

The qualifying tool sets up the applicant for the next stage, which should be a phone assessment. This is conducted either by outbound or inbound at an appointed time. Allocate 15-20 minutes for each recruit. The assessments are conducted by professional recruiters to delve further into the agent's qualifications, availability, and understanding of the requirements of the position. The interviews also evaluate applicants' voice skills.

Phone assessment is the single most vital tool in call center recruiting. You can eliminate a lot of applicants—separating the wheat from the chaff—with this method. Unless your agents are doing nothing but email, which is still rare in call centers, your applicants will need to know how to respond by phone.

In your assessment, you should look for characteristics such as proper enunciation, grammar, timing of responses, and organization of responses, especially if those agents are going to handle online communication. D'Ausilio advises that you engage the applicant in a casual conversation and get them comfortable, then ask important questions such as "how do you handle stress" and hear how they respond.

"When someone calls a call center, in most cases it isn't to say 'thanks for doing a wonderful job,' " she points out. "You need to know if the applicants can take the complaints and the problems that customers present."

Screening

Once they've shown their phone skills, you then call them in for screening, although there are some highly detailed phone assessment systems that can combine that function. You conduct screening either through written tests or with sophisticated computer-based call center simulation tools. Recruits sit down in cubicles, are given sample role-playing situations, and are gauged on how well and quickly they responded. The process takes 30-45 minutes.

The screening usually tests potential agents in three areas: computer skills, math, and voice. The computer skills identify people who are able to navigate using the keyboard and mouse. The system will give a simple math test for addition, subtraction, multiplication, and division.

With the voice test, the applicant will receive a sample call from customers with a question and respond to the question. In addition to assessing problem-solving skills, this will help assess the agent's ability to use proper grammar, enunciation, and quality service skills such as empathy.

You can also screen people for email and web handling. You set up a terminal, transmit the types of email questions you get to the applicants and have them reply. The response doesn't leave your building. With this method you gauge not only how they responded but how quickly and accurately, like a traditional typing test.

Screening Tool Caveats

When seeking qualifying and screening tools, Jackson advises that you see whether they are customizable and if so, be prepared to spend additional money. The packaged scenarios, questions, answers, and scoring may not fit your business.

If possible, have your management team test them. This goes back to the corporate culture issue. The assumptions that went into the system's designs may not be shared by your company.

"If the vendor can't customize the scenario or range of answers, at least ask if they can customize the scoring to reflect the most relevant answers to your call center," suggests Jackson.

Accents and Dialects

The most important yet potentially the riskiest skillset a prospective agent must have is the ability to hear and speak clearly in the language they are hired to communicate in. If a potential employee can't talk and listen well, they can't take or make voice calls, which is what most agent positions require.

But be careful how you screen. Some individuals may overreact if you turn them down and could threaten legal action on grounds of discrimination. There are enough attention-hungry lawyers and political wannabes to jump at such cases.

Clarity and enunciation is essential in any language. Your agents don't have to speak the King's or Queen's English (or the more common variant in the UK known as BBC English after the droll monotone of the venerable broadcasting corporation's announcers) or Castilian Spanish or Parisian French. But they do have to be understood.

Most callers are tolerant of accents and dialects and have become more so as societies become globalized, but they must be able to understand the agents. Southern drawls are more acceptable to Yankee ears than they used to be and vice versa. The same holds true for foreign-accented English.

One of most amusing stories I heard about accent and dialect was about an inbound campaign in England being handled by Convergys.. In 1996 the firm, then known as Matrixx, arranged to overflow calls from its Newcastle, UK call center to its Ogden, Utah, Omaha, Neb. and Pueblo, Col. centers. In preparation, the bureau trained its U.S. agents on British dialects by having them listen to British soap operas. One of them, Coronation Street, is set in a mill town just north of Manchester, which is roughly the area where my family is from and where I lived on occasion.

I could just imagine one of my relatives in the UK, sitting in front of the television, likely with a fag (cigarette) in one hand, watching the infomercial, dialing the number, and talking to someone in America. "Bloody hell, did you hear that? I was on t'phone and some Yank answers it. They've sent th' bloody call to America!"

But the overflow program went spotlessly. Matrixx also tested UK agents handling calls from the U.S., where the British employees were taught how to recognize U.S. Hispanic accents. Americans and Britons understood each other quite well. No rendering asunder of the great transatlantic relationship, pesky little incidents like the War of Independence, the War of 1812, the Monroe Doctrine and that to do about a big chunk of rock, trees and snow known as Canada, notwithstanding.

Said Robert Ashcroft, general manager of Matrixx Europe: "We've found the UK and U.S. consumers were very accepting at hearing the different accents. [But] agents must not only have a good command of the language but know the cultural nuances of the country they're calling, such as form of address."

Yet, if the dialects are too strong and the words spoken too fast, most customers can't understand the agents then they cannot get the service or buy what they need from your company. Often customers will call back at another time, hoping to get someone else, which adds to your costs. Ultimately they may get so annoyed they will go elsewhere.

You must carefully screen out applicants that you feel your callers cannot understand. Make sure, however, of your grounds for not hiring them based on speech. While speaking and listening are essential job competencies, like knowing how to hit a nail if you're applying for work as a carpenter, the lack of which are grounds for not hiring, in this lawsuit-happy society you can never be too careful.

Be sure, though, that you don't fall for the nonsense that there is virtue in, as they say in northern England, "talking broad," that is, with accents. That is talking down to people, a big customer service no-no. Many people perceive strong accents in their own languages as lower class. And under *no* circumstances should your agents copy the twangs of callers or called parties. Such fakery will get them and you unprintable but quite unmistakable words, loud and clear.

Interviews

Those who pass the screening and assessment are usually given a conditional offer. It is then that you interview the applicants. It is at this stage that you must discover what kind of people they are.

The interview process has three stages. The first is online psychometric testing (quality service profile and sales potential report) for those attributes that will lead applicants to succeed and stay, such as customer orientation and potential to sell.

Second, applicants take an in-person behaviorally based competency interview with HR. Third, a supervisor conducts a fit interview to determine whether the applicant is suitable to work in the call center.

The last factor is crucial. Most call center turnover takes place within the first three months of an employee's hire.

RDC's Jackson recommends having interviewers ask questions that lead to an understanding of how well the potential agents line up with the profile. Interviewers ask questions around real situations and look for actions the applicant took and the results of such actions.

"An example of this sort of question may be 'Have you ever handled an irate customer? And, if so, describe the situation and how you handled it,' " she says.

Jackson also advises that you give leading applicants a chance to sit down with experienced agents to get the feel of being on the job. If your firm does not allow this because you're working with confidential customer material, see if you could permit your potential agents to watch interviews of experienced agents and client and resource specialists and then talk to them about what their job will entail.

She urges that you give them a picture of their long-term future with the company. The first interview should include a discussion about what the potential agent wants to do now and in the future. It should also include the commitment the center expects from the agent.

"All best-practice managers we studied demanded a commitment of the potential agent from Day One that he or she would stay on the job for (usually) 18 months before he or she would be eligible to move into other areas of the company," reports Jackson. "One manager was so philosophical about it that she took to introducing applicants to the managers of other areas of the company. She would say: 'This is Joe, he wants to be in your department in three years.' "

Don't forget to get to know the people you're going to be working and paying for. When interviewing, find out about their personalities, what they do in their off hours. That will tell you more than what they've done at work.

"What the manager should do when they see the applicant is find out about their attitude," advises D'Ausilio. "See what that applicant's hobbies are and what are their likes and dislikes. How alive they are."

Don't Bother with References

We all know and gulp when we get to that line on a job application that says: "Please list your references." Immediately we think of those people who will say nice things about us and hope they do. If we're really smart we'll contact them ahead of time and prep them.

And that's the point. For that reason Rosanne D'Ausilio argues references are meaningless. She also points outs that work-related ones can be difficult to find as it is rare for people, especially managers and their supervisors, to be in any one job in any field for a long period of time.

"Be realistic," she says. "Is your applicant going to give a bad reference?"

RDC's Jackson points out, for legal reasons, corporate HR departments tend to give nothing more than confirming that the applicant had worked there, and at the position, and the length of time that they stated on their application. However, if applicants give former co-workers and friends as references you can often gain additional insights on how that applicant worked if you ask the right questions.

"You have to pay close attention and ask behavioral questions to ensure that the reference is giving valid examples of the applicant's work behavior," says Jackson.

Analyze New Hires' Performance

Call centers rely on feedback to monitor and improve performance. That also goes for staffing and recruiting. Once you've selected and trained your agents and they've proven out in your call center, say 60 to 90 days after hiring, you may want to survey them on what they thought of the process.

"The value for you is seeing how the job fits the description you had given, the skills you tested to and the profile you've developed," explains Jackson.

✪ RECRUITING CHANNELS

There are four main channels for recruiting agents: word of mouth, advertisements, job fairs, and contract staffing. Each has their strengths and weaknesses.

Word of Mouth

Word of mouth is just that: employees telling family, friends, and acquaintances how great or terrible your place is to work.

Word of mouth makes the other recruiting methods happen; it prompts potential agents and supervisors to go to the job fairs, read the wants ads, scan the web sites, or go to a staffing agency.

It is also word of mouth that will make or break your call center when it comes to agent retention once it goes live, and govern its potential to expand. That's because word of mouth has long been proven as one of the most effective recruiting means. According to DDI's Burdette, about 39 percent of all employment occurs with this technique. Firms have recruiting and retention bonuses.

"One of the prime target call center employment markets are young people," he explains. "They are very social and they frequently network. Those that have good experiences with an employer will quickly tell their friends."

This ties back to developing and implementing agent profiles. If you hire based on them, these agents will recommend your firm to friends and family who will likely fit.

"Recruiters can prove recommendations (for example family, friends, next-door neighbors) from agents make the best potential agents," RDC's Jackson points out. "These referrals seem to understand a company's culture and expectations from the beginning."

Yes, the newspaper classifieds do work. When labor markets are right you use the best means at your disposal to find qualified staff.

Print Display and Classifieds

Newspaper ads are the other tried-and-true way of recruiting people. Many prospective employees still like to pick up and thumb through the rags, especially students, older adults, and people in rural communities.

If you are targeting these workers and looking to locate in a small city, make sure you run advertisements in the local papers. I live in a small, semi-rural city with two competing twice-weekly newspapers, supplemented by Saturday regionwide papers. Anyone with any job openings places their ads in all of them.

Also, if you are located or plan to open in a city of any size with a highly competitive labor market, consider newspaper advertising in those communities, too. You can't afford to miss potential applicants.

If your center is located inside a strip or indoor mall, make sure you take advantage of the owners' print marketing campaigns, such as flyers or inside sections. It is cheaper than doing this yourself.

Service bureaus are at a disadvantage here because their names are not on everyone's lips. The way around that is to get permission from clients to use their names in the advertising.

The downside is that print ads get costly. The reach is limited by the distribution. Major metro newspaper circulation has been going through periodic meltdowns because of strong competition for eyes and ad dollars from radio, TV, the web and direct marketing.

To get the most out of print ads is to put your phone number and hook it up to IVR screening. Also list your web site and do likewise.

Online

Of the recruitment methods, job boards and web site postings are gaining the most trac tion, especially with young people and web-savvy boomers. (I found my job on Monster.) You can hook these ads into screening and qualifying tools. The web recruiting office is open 24x7. Applicants pick the time when they want to seek jobs.

Web recruitment is becoming a very popular and effective way to staff call centers. West's Web page leads to an online application form. West's site also intakes self-employed workers for its West @ Home contracted agent program.

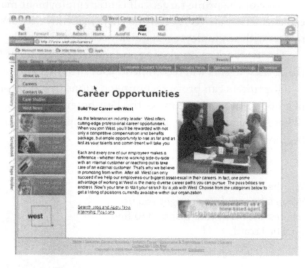

The big challenge with online is attracting eyes to your site and openings. That can get expensive with search engine and web marketing. The difficulties increase immeasurably if you are a service bureau and no one has heard of your company. Call center industry job boards are best aimed at supervisors and managers, not for people who have never or only sporadically worked in call centers.

Furst suggests active recruitment through web sites that have content related to your target audience. To do that, however, you need to understand where and when potential job candidates spend their time and money.

Job fairs are an excellent means to recruit not only traditional call center agents but contracted home agents as well, as shown by home agent contractor Willow CSN's recruiting staff in action.

Job Fairs

Job fairs are events either sponsored by you, a college, or a local economic development agency (EDA), to draw in potential workers to ask questions, fill out applications, and be screened.

If you are setting up a call center in a new location, your EDA is probably the best source of assistance. They want your call center to provide jobs to local residents, and the good ones will do what they can to help you. They know what channels people use in that community to find employment. They have links to all of the local colleges and staffing agencies and social services organizations. They will also arrange for call center training.

The downside is that job fairs require time and resources. You need to have own personnel there to meet, answer questions and intake applicants.

Broadcast and Nontraditional Advertising

Other ways to recruit for call centers include radio and TV ads, newspaper flyers and "Burma Shaving," that is, small billboards or signs spread out along a highway, named after the Burma Shave roadside ads in the 1950s.

Broadcast ads are very expensive but they are effective. They give a good audio and visual representation of the call center environment, including the smiling faces.

Community involvement is also an excellent way of getting known. Examples include sponsorship of events that are attended by your targeted labor market, like New Year's Eve.

All of these methods are especially valuable for outsourcers that do not have brand name recognition. One example of "Burma Shaving" I have seen was for West Teleservices outside of its call center near Victoria, British Columbia, which is located along a major road. RMH, which has a big call center in Nanaimo, helped sponsor a New Year's Eve celebration in that city.

Storefronts and Mallfronts

If you got the space, use it. List yourself on marquees and signs. Place ads in windows, or if you have a recruitment drive and you're allowed to, have your HR people up front with coffee and donuts. Also, if your landlord permits it, consider billboards or banners on building sides. That can help you immensely if you are located in the back or on an upper floor.

Also make sure you have information and applications immediately available inside your center, no matter where you are situated. During business hours (including mall hours), have someone inside who is trained to answer basic questions about available positions. Consider having a phone or computer connected to the IVR or online qualifying tool. That way people can get what they want quickly and leave, but if they are interested you can intake them right there and then.

Contract Staffing

Contract staffing is outsourcing your agent staffing to a staffing agency that handles all recruiting and training for your call centers. These individuals are their employees under your management.

Contract staffing is becoming a very popular alternative to staffing and training a call center. They will send in workers who perform their tasks directly for you. The vendor recruits, qualifies, screens, and pays for the agents under your management and supervision. If they don't perform or the tasks they've been assigned to accomplish are finished, they're out. You call up the agency and say: "We don't need so-and-so tomorrow."

Staffing agencies also offer "temp-to-hire" programs where if one or more of their employees is working out in your center, they can be placed on your staff, usually after four to six months. That gives the temps extra incentive to perform well and to get to know your company and customer needs. You can also insource with some staffing agencies, letting them run your call center day to day.

Staffing agencies rely on their large and frequently updated personnel databases. To fill them they effectively use all the recruiting means: word of mouth, ads, the web, job fairs and college and government job boards. Their core competency is people. The good ones know how to attract and keep workers. They also use an array of qualifying, screening, testing methods and incentives to select and retain the right candidates.

Contract Staffing Upsides

The biggest benefit of contract staffing is that the agencies bring in large numbers of qualified people to your center quickly. The ones that are very call-center-savvy have the latest qualifying, screening, and assessment tools. This is especially helpful if you are setting up a call center in a new community or your center requires high and fluctuating volumes of part-time and short-term employees.

These agencies are also very handy if you're setting up a center in another country. By knowing the cultural and legal lay of the land, these agencies quickly get your employees in place, and with less hassle than if you tried to recruit them yourself. By contracting with them to directly pay for your labor they may free you from being covered by onerous and often union-favoring labor laws that govern your vertical industry.

Contract Staffing Downsides

As with outsourcing to service bureaus, there are some downsides to contract staffing. You are leaving recruiting in the staffing agencies' hands. They may not screen as well for exactly the employees you're looking for as you could in-house. And they're still the agency's employees, not yours. Turnover is usually higher than in-house. People often go to staffing agencies when they're making career or life changes.

If you have a blend of staffing agency and your own employees in a call center, there may be friction between both classes. Agency people are doing the same work but are lower-paid. That might cause resentment by the agency workers. Employees may shun staffing agency personnel because they fear the employer is going to get rid of them in favor of the temps. They may also see that as a signal that the call center is about to close.

Making Staffing Agencies Work

For a staffing agency to work for you, your center must be in a community where there is what site selection consultant John Boyd calls a temp culture. In a temp culture there is a workforce that is accustomed to going to a staffing agency to find permanent or temporary employment, commonly in fields unrelated to their main area of expertise. Cities that have a large service industry have a temp culture.

When working with a staffing agency, ensure that they have an excellent understanding of your culture as well as your skills needs. Monitor their performance. Kathryn Jackson recounts that one company that hired a staffing agency for their call center had one of their HR personnel doing just that, including attendance at the same job fairs when they recruited for the call center.

"There are some very good staffing companies, devoted to quality and excellence," Jackson points out. "But the success of the relationship is dependent on the strength

Online tools like staffing agency Manpower's Netselect brings the employment officer to the applicants

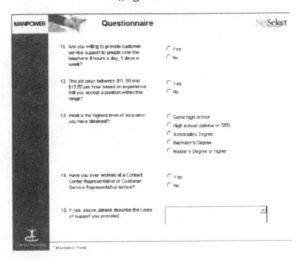

of the partnership between recruiter and the client, as in any outsourcing partnership."

You can also use these recruiting methods to test the local labor market, prior to making a final decision on a location. You can do this semi-covertly by placing a blind ad in a local paper. Drop a hint to local officials, who will spread it like wildfire to the media that you are looking to expand there but no decisions have been made. Or you can be open by holding a job fair. Even advertising it will flap the word-of-mouth network.

○ QUALITY IS KEY!

If word of mouth is going to work, then you must have a call center and a firm worth working for. If you followed the advice in the previous chapters and followed up with the consultants, you will have done much of that already by providing an excellent facility and work environment in a convenient location. Your company must manage people correctly by empowering them, treating them decently, and compensating them likewise. It should also have a positive image with its customers, delivering quality products and services.

Nobody likes working for a terrible firm, especially agents. They're the ones that deal firsthand with the angry customers who call up and send email to complain or who must try and sell to them. The verbal and written abuse is hydrochloric acid on the soul.

Service bureaus can piggyback onto companies with excellent images. Some clients allow bureau partners to use their names to recruit people. After all, the agents they hire will be working for that client, just as if they were hired directly. However, make it clear to applicants that they're working for and being paid by you.

"Money is not that high of a factor in recruiting people," D'Ausilio points out. "It is way down the list in what people are looking for in an employer. Flexibility, amenities, and motivation all rank higher. One day you won't be able to give any more money. What's going to happen?"

While in a tight economy you should try all recruiting means. Some work better than others for different types of call centers. If you have a basic low paying inbound order taking or outbound sales center, newspaper ads, job boards, and staffing agency listings may work very well. Listing such positions on the web or featuring the web site will not be very effective.

Call center work is not just for younger people. Older adults, whose numbers are growing as the Baby Boom ages, make for fine agents. They are by definition mature and responsible and are often willing to work part-time, which helps call centers meet employee supply with demand.

Also many services and products are now aimed at this demographic. Having someone on the phone that is in that market helps sales and service immensely. Credit: Brendan B. Read.

✪ TARGETED RECRUITMENT

When labor markets are tight you need to target your recruitment at those individuals who have traditionally been under-represented in call centers and in the labor force in general. The payoffs are obtaining highly-skilled and loyal employees. To tap them you may have to invest in additional training and needs accommodation.

Following are several examples.

Aging Boomers

People are living longer and retiring earlier, which have created a growing supply of potential workers over 50. They are ideal for call centers, especially for part-time work, because they are mature, calm, can handle most any objection or criticism, have strong ethics, and are very responsible.

The downsides are sometimes slower learning and time off for medical reasons. These are minor compared to the family emergencies and sudden time aways or tardiness by their younger counterparts.

Disabled

Applicants with mobility and sense impairments can make great agents. There is no shortage of such potentially excellent workers. Manpower reported that as little as 25 percent of an estimated 55 million such working-age adults are employed.

For the visually impaired there are Braille keyboards and special text-to-voice headset readers that permit them to be practically as effective in taking and making calls as other agents. And you can assign email and text chat to hearing-impaired agents, rectifying a critical skills shortage in call centers. There are often government grants to assist hiring and placing such workers. And it makes your firm look good.

The other benefit is that these workers can support callers who share their disability. One example is TTY service for the hearing-impaired.

You can accommodate these employees by designing and remodeling facilities to meet and surpass the Americans with Disabilities Act requirements, enabling those who require wheelchairs and walkers to get around. You can look for and correct simple things like removing plants that partially block entrances.

Outsourcer LiveBridge has worked with Oregon agencies and non-profit organizations to hire the long-term unemployed. A driving force behind this company's efforts has

OUTSOURCING YOUR HR

Outsourcing in staffing used to mean hiring temps. No more.

Outside vendors now provide employee agent and supervisor recruiting, screening, selecting, and training. You avoid investing in HR processes, technology, and staff in-house.

Outsourcing saves time, obtains quality staff quickly, and saves money. Moreover, outsourcers who are experienced with call centers can often provide better-qualified applicants than the typical corporate or organization HR departments that are unfamiliar with the call center requirements.

There is a vast array of outsourced offerings. They include HR consulting, staff sourcing, screening, selection, hiring and training. They also encompass hosted workforce acquisition and assessment firms.

There are firms that offer a blend of automated and in-person recruiting, screening, and selection. Others provide background checks and drug testing through partners. There are also recruiting web sites.

Staffing agencies provide a full range of services including HR consulting, insourcing, placement in individual positions, temp-to-hire and temporary-only help.

Manpower is seeing more call centers turning to firms like theirs. The company is increasingly asked to provide "onboarding" services, processing paperwork for new hires and conducting orientations for a call center's permanent hires.

Manpower serves as the employees' resource assisting them to become comfortable with their jobs.

"Companies want their HR departments to design their staffing strategy, look at ways to develop and retain people, and identify advancement prospects," explains Tracy Buelow, market planning manager for contact centers. "They want companies like Manpower to manage the day-to-day staffing, screening and skills training work."

been the experience of its co-founder, Patrick Hanlin. When he was younger, his father, a sportscaster, lost the use of his legs following an accident. In his day, announcers had to climb on ladders into boxes overlooking the stands to give the play-by-play.

"After the accident people would talk to him about their day-to-day experiences in the outside world, but this hurt him because he had very little of his own," recalls Hanlin. "Today our disabled employees have the same kinds of experiences as their co-workers. They can talk about the sales they've made and the customers they've dealt with. They have the life that disabled people a generation ago did not."

Minorities

There are many groups within countries that have higher-than-average jobless rates but who are willing to work hard if they have the opportunity, support, and training. Examples include urban African-Americans, Hispanics, non-Hispanic immigrants, and native peoples.

Minorities bring to the table special skillsets such as non-English-language profi-

MATCH THE JOB WITH THE DESCRIPTION

You should also ensure that the job description used in recruiting matches the job. When that doesn't happen, call centers don't receive the applicants they need. And the company has wasted money recruiting ill-qualified staff, says Launa Green, human resources strategist at Banks and Dean.

This may seem obvious. But call center recruiters are known to advertise and screen for qualified phone agents who, once hired, are told they must answer email and conduct text chat sessions. Many agents, however, lack the necessary written language skills. One result is turnover.

"If the candidate is given an inaccurate picture of the position and they do not fit the true opportunity, then there is a good likelihood of them leaving the job," says Green, adding that leads to retention and performance problems. "Either way, it costs the call center."

ciency in the case of immigrants. If you prove to be a good employer, you'll have a word-of-mouth supply of excellent, loyal workers.

In Western Canada, native peoples, known as First Nations, have historically been the most downtrodden of Canada's minorities. The history of abuse and neglect at hands of the conquering Europeans has only been slightly less shameful than in the U.S. Canadians may not have hunted them like in the U.S., but the effects are similar. The unemployment, alcoholism, abuse and crime rates on and off the reservation are appalling. You see that in the big cities, like Vancouver, and in small towns, like Tofino on the Pacific Coast.

Yet call centers have been successful at providing opportunities to First Nations people in partnership with local and native agencies. To help recruit staff for its 850-agent call center in Chilliwack, British Columbia, opened in May 2001, Stream International works with two local high schools and with the Sto:Lo First Nations Indians.

All of the high school recruits in this program are interviewed extensively and have completed pre-hire technical training. The Sto:Lo band informs potential applicants of the opportunities and arranges for any additional training to enable them to meet Stream's qualifications.

Another example is Istonish, which has a call center on the edge of downtown Vancouver. The outsourcer has been involved in a program in cooperation with The Aboriginal Community Employment Services Society (ACCESS) and the British Columbia Institute of Technology (BCIT) to recruit, screen and train off-reserve First Nations people.

ACCESS is a non-profit group that provides employment recruitment, support (such as child care and transportation), and training services to off-reserve First Nations people. ACCESS accepts recruits, gives them life-skills training, and provides job support. BCIT, through its Call Centre of Excellence program, trains agents on computers, phones, and common software. Istonish then approves recruits and trains them on the company's software, client programs, and procedures.

Working with the First Nations agents has had some challenges. Istonish program director Karim Farouki says ACCESS occasionally supplied suitable clothing to agents to meet the bureau's dress code because those employees lacked the money.

The program taught agents that if they have a problem they could talk to their supervisors; they will not be disciplined or fired if they do so. "Several people did not have much experience working, they had never had a job or haven't worked in one for so long they forgot how to work," says Quintana.

Some agents interviewed said their culture teaches respect for elders such as waiting for them to finish speaking. That made ending calls difficult.

In response, Istonish offered agents materials such as business rules and option statements to enable them to control the call professionally, quickly, and effectively.

Interestingly Annette Quintana, who is Istonish president, is the granddaughter of a San Juan Pueblo Indian. So when she heard about the ACCESS program she researched and began working with it.

"The ACCESS program offers a way for native peoples to leave that cycle and enables them to provide for their families and themselves," says Quintana. "Our hope is that we can grow the First Nations workforce with Istonish by opening more opportunities for them."

Chapter 14: Training and Retention

If you've done your staffing correctly, then you should have a fine crop of recruits to be trained on your front lines. They need to be taught the mechanics of the job: how to work the phones and computer, what are the contact-handling procedures and rules, and how performance is measured. You must also teach them about individual products and services, and about their specialties such as help desk, and insurance and securities sales.

You need to find ways to retain agents. And that can be difficult in call centers where the voluntary turnover rate runs from 30 percent per year to as high as 300 percent. To close the loop, effective skills training is one of the techniques recommended by experts to keep employees.

The length of new-hire training typically ranges from 40 hours to 12 weeks. You should also test periodically to ensure that the information is getting through; don't wait until the end. After the agents have been trained, they should be brought up to speed under close supervision. When they meet the same standards as the rest of your call center, then they can be given the same level of attention as the other agents.

✪ KEY SKILLSETS

There are several key call center skillsets most if not all agents must be taught.

Soft Skills

Soft skills are listening, reading, understanding, and responding appropriately to others.

Soft skills enable agents to connect with callers or called parties no matter what the transaction. With these skills, agents get the people on the other hand to buy into what they are saying by demonstrating empathy and interest. Agents can then begin to ascertain needs and limitations, such as when a prospect says, "I don't have the money but I get paid in two weeks".

"Training a new agent in the "art" of customer service should always be part of the new hire training," says Kathryn Jackson, associate at Response Design Corporation (RDC). "Customer service skills should be incorporated throughout the new hire training, regardless of the specialty. An agent who does not possess strong customer service skills is apt to alienate callers, and you face the potential of losing business."

Soft Skills ROI

The return on investment (ROI) exists for product and technology training, says RDC's Jackson, because agents are taught about products and how to sell them. But the ROI for soft skills is more difficult to show. She isn't aware of any studies that demonstrate either a positive or negative correlation between satisfied or angry customers and buying patterns.

"We know from customer satisfaction surveys that customers say if they had a pleasant experience then they will buy again and the opposite, but we don't know if they follow through," says Jackson.

Anne Nickerson of the Call Center Coach says companies can more easily quantify how new technology will reduce head count, or how training on sales skills will lead to increased conversion ratios. But it is more difficult to measure the impact of being empathetic with customers and asking appropriate questions to determine their needs.

Call centers might, for example, measure net changes in quality scores before and after training. Or they can determine cost savings resulting from fewer escalated calls to senior agents or supervisors. Or they might identify "once-and-done" completed calls compared with repeat calls on the same issue.

The long-term impact of soft skills training is on the net value of customer satisfaction and loyalty. To establish ROI, Nickerson asks companies to determine each customer's lifetime value. Let's assume that value is $600 and the cost to adequately train the agent is $200.

"By not providing that training you will lose at least $400 or more, because poor agent training will cost you more customers," Nickerson says.

Companies also find training in soft skills difficult to justify because they fail to adequately monitor and observe agents to learn whether the training is paying off, says Marcia Hicks, senior consultant at Kowal Associates. If you use customer satisfaction as a measure without monitoring and observation, you don't know where the problems are that lead to low customer satisfaction. Sometimes agents can retain customers by letting customers know they're sincerely sorry and are trying their best to ameliorate the situation.

"But you won't know that unless you monitor calls and survey customers," says Hicks.

The soft skills training ROI varies from firm to firm and by application to application, says Elizabeth Ahearn, president of The Radclyffe Group. Her firm has found that when companies boost their soft skills training, their customer satisfaction ratings and sales go up.

"[Because] every company and project has different requirements—some want to increase sales, others their customer satisfaction ratings—the ROI is very much on a case-by-case basis," she says.

Rosanne D'Ausilio's 1996 doctoral thesis, entitled "The Impact of Conflict Management Training on Customer Service Delivery," concluded that a utility whose agents she trained in soft skills saved more than $330,000 annually by slicing talk time by 22.3 seconds. She taught agents handling customer complaints conflict management strategies that enabled them to resolve issues on the first call.

"The company could handle more calls with the same number of employees," she explains. "The 22.3 seconds shaved off each call annually is equivalent to having seven extra full-time agents, according to the utility's calculations."

Cross-selling and Upselling

With the dying away of outbound cold-calling, enterprises are relying more on inbound cross-selling and upselling. If you have screened agents for this aptitude, then you can train them on it.

Sales is part of service; after all, service is about finding out customer needs and meeting them. That can include products or services your organization offers.

There are very successful cross-selling and upselling techniques. But the key to all of them is waiting until the customers are satisfied with what they were contacting the centers about before pitching.

For example, you can train agents on how to initiate and build relationships through engaging customers in a dialogue.

Kathy Dean, partner, Banks and Dean uses the example of a customer calling into a golf gear company to inquire about a brand of club. The agent has been taught or knows enough to ask the customer about his game and what equipment he uses. If the customer responds, the agent continues to build the dialogue. The agent then thinks like the golfer, listens, and identifies needs.

"If a customer reveals a need such as they want an edge in accuracy in their putt, then the agent supplies information like, 'We just got in this new version of putters with tungsten inserts that provides better roll. I could send or email you a fact sheet on them'," says Dean.

Don't make the mistake of focusing too much on product training at the expense of hard and soft skills training. It is the people who sell the product, not the other way around. You can have that practically impossible acme of an offering: air service that is on time, safe, with direct routes, ample and cushioned seats, and edible food at reasonable fares, or sheer, comfortable and affordable pantyhose that never runs, or software that never crashes and quickly boots up.

But if you don't have agents who can sell and service them and build relationships with your customers, you will get little initial or no repeat business. There are few proprietary products or services. If yours is popular, your competitors will soon mimic it. Those Asian prison factories and copy shacks can turn on a dime.

"The priorities, between product training and soft skills, which I call 'context training,' should be reversed," argues D'Ausilio. Today, the competition is just a click or a call away. The only way your front lines can keep your customers is if they have these skills."

The Radclyffe Group's Ahearn reports that call centers typically spend 80 percent of new hire training time on product training, 10 percent on call center systems training, and the remainder on hard and soft skills.

Instead, they should devote 50 percent to hard and soft skills training (Ahearn calls it "interaction training"), 35 percent to product, and 15 percent on systems. Companies would generate greater customer and employee satisfaction and agent retention if they changed their priorities, she asserts.

Every customer satisfaction survey she's done cites speed of issue resolution at the top of the list. Product knowledge and technology training rank far down the list.

"Customers discount their importance because they expect agents to know the product and how to use their tools," explains Ahearn. "It is the interaction training that

SOFT SKILLS FOR COLLECTIONS

Sounds like a contradiction? Not really. Call centers now seek collections agents with excellent customer service skills and who know how to touch late payers with the velvet glove.

"We're finding that with collections there is a greater importance on soft skills," says Teresa Setting, Kelly Services' vice president of marketing. "The agents must be trained to empathize with customers. But they must also make certain that they get that promise to pay."

matters the most to customers, how agents communicate in delivering world class customer service."

Accent and Dialect Neutralization

If otherwise excellent agents' have problematic accents or dialects, consider neutralization. Long used offshore in countries like India and to an extent in Canada, neutralization takes the broad edge off words, and reduces slang and idioms.

Some employees might object to this on cultural grounds and could make a media-grabbing stink. There's nothing like race and ethnic issues to get out the cameras and the lawyers.

Make sure you have evidence such as customer satisfaction surveys, call monitoring and positive performance scores of employees of the same gender and ethnicity but who have shown that they don't need the training. If all else fails you can remind your audience that the choice is between training people in America, providing jobs, or moving the work offshore. Take your pick.

ePerformax is a unique training firm in that it practices what it preaches; the firm operates a service bureau. ePerformax believes in gradually immersing its trainees. Its training floor is not in some basement or far off room but in a glassed off section of its call center. Agents pass through the firm's exacting standards apprentice with a 3:1 agent/supervisor ratio in its Acad-

emy Bay, seen here, on a nonpartitioned section of the call floor. Once passed, agents move to the main section. Credit: Brendan B. Read

✪ NEW HIRE TRAINING TECHNIQUES

Call centers must pay special attention to new-hire training. This process makes or breaks agents. Following are some good methods.

Breaking Up Training

Teach and train agents in one or two skills or call types at a single time, letting them get comfortable performing one skill before training them on another. You can route calls to an agent based on his or her existing ability to handle a particular call type.

In rare situations, the new agent may simply ask for a senior agent or supervisor to assist with the call. This technique also quickly weeds out uninterested agents as soon as they get a taste of the job.

"With this method, information retention rates are higher, agents feel greater achievement satisfaction, and you get agents on the phone quicker, which is important when turnover is high," says Rebecca Gibson, manager of education services with Incoming Calls Management Institute (ICMI)."Why pay to train agents for the full [training period] if some are going to leave right away?"

Use of Simulations

Gibson sees a place for computer-based simulations in new-hire training. She says it's better for agents to experience a simulation after a classroom lesson rather than the old model of supervisors standing behind them. Information retention is higher as agents learn more.

"A pattern of three classroom days, two simulation days, and three more classroom days enables agents to get the lessons, try out what they've learned and come back with the results of the trial and error, and train again based on what they've done," explains Gibson.

Apprentice-Styled Training

Adults also learn by doing, says Anne Nickerson of The Call Center Coach. Nickerson recommends apprentice-type training where agents get to listen in on calls but only begin taking a few calls, easing into parts of the call alongside the senior agent or coach.

Apprentice-type training is quicker than classroom-only or having agents in training bays with higher supervisor-to-agent ratios. "The apprenticeship reduces the anxiety and stress agents often feel on calls," she says. "That will improve performance and retention."

Integration of Skills and Product Training

Teach agents product and service information concurrently with techniques for handling calls and contacts.

"People don't learn how to apply knowledge just by listening to lectures," says Nickerson. "They learn by being taught what they're going to apply on the job."

✪ CROSS-TRAINING

Train your call center staff in other departments' functions if they interface with yours, recommends Peter Gurney, managing partner with CRM consultancy Kinesis.

If you do have stores near your call center, give your new hires tours and offer gift certificates. The more retail and the call center know each other the better the synergy.

Too many companies do not train agents in other channels, resulting in a disjointed image of the company, missed sales opportunities, and frustrated customers. Sometimes customers call asking about a web offer or a store sale, but the agents have no clue what they are talking about. And that will annoy customers and embarrass agents.

At the same time when customers go to a store, chances are someone there will know about your company's web site, catalog, and call center. Web sites will have store locations and telephone numbers.

You might be also wise to offer call center customer service training to non-call center staff such as salespeople, engineering folks, and webmasters. Why? Often, customers call your firm through contacts other than your call center number. Also, many people contact webmasters with customer service inquiries.

Some of these inquiries can be serious. On more than one occasion, notes a representative of a medical services company who requested anonymity, ill people emailed the company's Webmaster asking for help. Fortunately, the Webmaster had enough customer service skills to provide assistance.

"Every employee is potentially a customer service person," says Dianne Durkin, president and founder of training firm Loyalty Factor. "When customers contact you, you are representing the company, and it is your responsibility to help them or direct them to the right sources."

○ SOFT SKILLS TRAINING FOR TECH SUPPORT

Tech support is almost as different from call center customer service as both are from retail. Customer technical support requires agents to have two seemingly contradictory skills: customer service and technology problem solving.

The first requires people skills, such as listening and empathy; the second entails technical skills like puzzle-solving aptitude. Anyone who has been around engineers and techies long enough know they have the second skill down pat, but that all too many lack the first.

"Unfortunately what you have is a clash," says D'Ausilio. "You have customers who are very stressed out and defensive because they have problems they can't fix. And you have agents who are very knowledgeable and can solve the problems but come across as arrogant and condescending—just what the customers don't need."

While the agents may have the answers, the customers may not listen. The problems are left unfixed. The customers are ticked off, and if they're mad enough, they'll go elsewhere and bad-mouth your firm to others. Even if customers don't leave you, dealing with their issues may entail considerable escalation and productivity costs.

"Customers want to be treated with dignity and respect," says Durkin. "They want to be assured that the person clearly understands their problem and will be able to provide them with a solution. They want to feel understood."

But agents must have the technical skills to do their jobs. The solution lies in training agents and supervisors in customer service skills. Loyalty Factor's program, for example, aims to help companies improve customer satisfaction and service levels and lower turnover rates. In one instance, notes Durkin, Loyalty Factor reduced turnover from 65 percent to 35 percent annually.

The company helps agents learn about themselves and how they relate to others, during four weekly sessions of on-site training. Training introduces agents to four communications styles. Agents are also given profiling tools to understand their style and how to best interact with other styles.

Numerous exercises and role-playing gives agents the opportunity to put themselves in their customers' shoes, and to learn critical skills for managing customer calls.

One exercise demonstrates the need to have a clear, concise questioning strategy. Agents are called to the front of the class and asked to describe an image for classmates to draw. During the first pass, the audience is not allowed to ask any questions. They must rely entirely on the agent's description to complete the drawing. The resulting images are quite different from the original, says Durkin. Only when the audience is allowed to ask questions do their images begin to resemble the initial drawing.

"We teach agents about different ways that people process information and encourage them to listen for clues when the customer speaks," she explains "Then they can respond to the customer in a way that will make them feel comfortable and understood.

"For example, if a customer uses visual words like 'look' or 'clearly,' the agent knows to describe the problem in visual terms. We teach them to stay tuned in for word choices that are auditory or emotional in nature, and then to respond by using similar phrases."

The message that geeks have to be human is getting out. Rick Kilton, RWK Enterprises which provides support-desk training consulting, is seeing more companies look for customer service and people skills, not just product and technical knowledge from their applicants.

"Companies are now figuring out that when their reps deal with customers with polite respect, the customers become cooperative, which helps solve the problems," Kilton says. "Even if reps don't solve that problem, if they are respectful and friendly, customers are more likely to stay loyal than if reps are cold and hostile, even if they fix the customer's problem."

One motivator towards better soft skills training is the move by high-tech companies towards fee-based support, such as by buying service plans or by paying for service calls. Fee-based support allows firms to recoup their support costs. It can cost as much, if not more, to support a product per customer than the profit made on selling each unit.

Because customers are paying for support, they expect better quality support. And that includes well-trained agents.

"Increasingly high-tech products are purchased based on not just the product alone, but particularly the bundle of services and support associated with the product," Mia Melanson, principal at Performance Consulting, points out. "Quality support has become a key differentiator among competitors in the high-tech arena, especially in high-end software.

She says agents must be trained how to make judgment calls based on the value of the customer and the urgency of the request.

"Support reps now need diplomacy and negotiating skills," she says. "They must know the customer service level agreements and what can be done, even if this requires research or a referral to another employee, department, or vendor."

❂ REFRESHER TRAINING

This learning does not end when the agents begin to take live calls and emails. Periodically they need to be refresher-trained on existing skills and taught new ones. The type and frequency of refresher training depends on the company.

"Timing of refresher training should vary from center to center, based on each one's unique need," says Jackson. "What is triggering the need for training? Have the monitoring scores gone down across the board? Are we launching a new product? Has there been a process or policy change? Will the system be changing?"

Training Costs

Are you investing enough for training, especially for refresher training? Training consultants such as Elizabeth Ahearn say that training new hires costs $2,000 per agent per year. She pegs refresher training between $800 and $1,000.

Companies have been spending much less than that on refresher training, to their detriment. Ahearn reports that many companies don't want to pay more than $250 to $500 per agent per year. Some companies pay much less—nothing at all.

"If agents do not receive adequate ongoing [refresher] training, they will begin to feel unchallenged intellectually, and their skills will slip," she says. "Without that stimulation they will leave. Costs go up and customer satisfaction suffers."

New Skills Challenges

The toughest training challenges are not with new hires but in teaching existing employees new skills that are diametrically different from the ones they have.

One example is text-based communication. Many agents may not know how to write well, or they say they do but they use slang and symbols. You may then have to send them for remedial training to bring their writing skills up to standard.

Trainers such as Human Technologies now offer grammar and spelling modules. "If your agents don't know how to use words the right way, then they are misrepresenting your company to your customers," says D'Ausilio.

Another more serious instance is cross-selling and upselling. While agents experienced at outbound and inbound sales have little trouble transitioning to or from service—good salespeople know how to ascertain needs—the reverse is not often the case.

You are dealing with not only a skills issue but an attitude one as well. Many inbound call center agents, especially support reps, do not like selling because they see their mission as serving customers and fixing their problems. Selling violates the customer's trust.

Compounding agent resistance is the ham-handed manner in which many call centers script the sales pitch into customer service calls. Typically, the transition from service to sales is unnervingly abrupt. Agents are popped sales scripts even before the customer's original issue is resolved to his or her satisfaction.

There is not an inherent conflict between service and sales. When a customer buys a product or service, it is because the customer has a need that the product or service can fulfill, which is no different from when they call in for service. Therefore, when trained correctly customer service agents can sell or identify sales opportunities that they can pass on to sales agents.

Teresa Hartsaw, president of outsourcer and training firm ePerformax, teachers that all customer service must relate back to sales. Do customers buy or continue to buy as a result of the service they receive? She teaches her agents, and those who subscribe to her firm's training program, that every customer service transaction is a sales transaction, and vice versa.

"Customer service is selling solutions," Hartsaw points out. "Did the agents sell the solution to the customer's satisfaction?"

As noted earlier the key is waiting for the service issue to be resolved, the agents' prime motivation before marketing the other items.

Rosanne D'Ausilio uses the following analogy of a restaurant. "A fair restaurant will instruct the wait staff to say, 'Would you like dessert with that' as soon as the customer is finished eating. However, the wait staff in a top-quality restaurant will ask the customer how the meal was and note any comments, good and bad.

"If it is good, the waiter asks whether the customer has room for dessert. If the comments were bad, the waiter apologizes, speaks with the chef immediately, and offers something to make good, like a complimentary meal or dessert."

Training Tools

There are a variety of training methods and tools. You can train by people, such as your trainers or consultants, and by technology, through a growing array of computer, online, and video tools.

I break these up between instructor-led training (ILT) and technology-based training (TBT) also known as elearning.

ILT is where instructors teach students and obtain feedback by interacting with them live. Instructor-led training is supplied to large or small numbers of students, or one on one.

ILT can be in person, inside formal classrooms on site, offsite in colleges, or rented conference rooms, or conferenced at agents' desks through video or web hookups. Instructors can be supervisors, senior managers, product specialists, or outside training consultants.

TBT is where pupils obtain their lessons via media such as the web, CD, DVD, and video. The interaction, if any, is limited to answering preset questions. Agents typically undertake TBT at their workstations.

Conversations with instructors and colleagues are the key benefit of instructor/facilitator-led training. Photo courtesy Loyalty Factor.

ILT Advantages and Challenges

ILT's benefits are feedback and motivation. There is nothing like communicating directly with another person to learn from them.

Wherever you need immediate feedback for training from agents, use ILT. Your agent and supervisor can ask questions and take part in role-playing exercises and will be told the answers right away. Your employees cannot escape the eye or the ear of a good trainer; you can't switch or click them off. If your agents are not paying attention, the trainer, like a teacher, will notice.

Good live trainers also leave you, like a fine preacher or speaker, with an inspirational buzz to act, to do better. There is an emotional impact from being in the presence of a live person speaking that does not exist in TBT.

The downsides to ILT are cost and time. To minimize the time drain on call centers some trainers offer part-day as opposed to whole-day training. D'Ausilio's training takes 16 hours, but she breaks them in four-hour sets over one month.

It is also easier for someone to retain information if they come back to it two or three times, even if the material is new each time, because attending the sessions and seeing the trainers reinforces the lessons.

Animation is a cool training tool because puts trainees at ease. MaraStar uses animation to show how to defuse an angry customer: a key call center skill. The firm helps you teach this through customized 1-minute cartoons sent to agents.

TBT Advantages

The main advantages of TNT are convenience, retention, cost savings, and productivity. Each agent can tap into their sessions at their leisure, when they have a few minutes and, in the case of web-based training, at their desktops. They can easily review them to refresh their memory and go over key points, whereas instructors' words tend to fade with time.

Through simulations, TBT instructs and refines agents' soft skills, such as empathy with callers to build trust and help resolve issues. TBT also can help supervisors assess and improve employee performance.

Another TBT benefit is consistency of training and results. By contrast, the quality of in-classroom training will vary from one instructor to the next and even from one class to the next for the same instructor.

Still another reason to consider TBT is customization. You can pick the training modules that best apply to agents' deficiencies. Or in self-learning tools, agents select modules that will help them improve their skills.

TBT is great for product knowledge. Penny Reynolds, founding partner of The Call Center School, says that the just-in-time feature of TBT appears to be a good fit when your objective is to acquire knowledge or awareness of a topic. Examples include a product update, change in a procedure, or some facts to memorize. TBT also excels at learning new step-by-step processes, like handling credit card authorizations, points out ICMI's Gibson.

TBT saves money. It is approximately 50 percent less expensive to deliver than ILT. TBT is also more productive from a call-handling perspective. According to TBT vendor Knowlagent, in-classroom training engenders on average 30 minutes of downtime: 15 minutes each getting agents to and from the classroom. That one-half hour of downtime could be used for TBT training.

Knowlagent further estimates that call centers must overstaff by 10 percent to 15 percent to allow agents time for classroom training. In a 500-seat call center, that means 50 to 75 more agents who would have to be hired to maintain service levels. Call centers could eliminate most overstaffing by adopting TBT.

Typically 10 percent of all call center space goes for training. By going to TBT you can add more workstations into the same footprint or reduce the amount of space by the same amount.

TBT Limitations

TBT is not as successful when teaching frontline staff communications skills or call-handling techniques. "Technology-based training can easily be used to educate about transferring knowledge or, say, the proper rate of speech," says RDC's Jackson. "But an instructor is needed to help agents resolve anomalies between what is supposed to happen and what is happening during a customer interaction."

Reynolds points to simulations as an example. Most simulations are not truly interactive. It's hard to get a good measure of quality when students are simply picking from a list of multiple-choice options of how they would handle a call.

"Most centers find that communications skills are best taught and practiced the old-fashioned way, in a live classroom or one-on-one coaching environment," she says.

TBT is not good at handling out-of-the-box examples that will stump agents. It does not address all exceptions to company policies and rules.

Kinesis' Gurney sees companies shift back from total TBT to blended training. TBT is great at teaching mechanical skills and product knowledge, but not intangibles like teamwork, complaint resolution, and service skills.

"The great weakness of TBT is that it doesn't give a context to the training," he says. "It doesn't impart the company's or the brand's personality for agents to get interested in or enthused about. You can't do that on a screen, but you can with a facilitator in a room full of other employees."

Elizabeth Ahearn also fears TBT could lead to customer service deterioration. Customers constantly ask agents to make these exceptions. Agents must balance policy against customer needs.

A total TBT approach will tip the scales to policy, forcing agents to respond by the simple rote answers programmed into the TBT systems, or what Ahearn calls "rules-based service." Agents will give literal responses. The consequences are lack of flexibility and poor service.

"You could lose agents and customers with rules-based service by turning up the stress factor," Ahearn says. "Our studies show that agents who do not feel empowered to help customers are less satisfied. And there is a strong correlation between employee dissatisfaction and customer dissatisfaction."

Another TBT limitation is the software author's biases. For example, an author schooled in proper English may instruct an agent to offer help by saying: "May I help you?" The author might not permit the commonly accepted but grammatically incorrect phrase, "Can I help you?"

Also, web-based TBT apps, while offering cost savings compared with classroom training, can be expensive to buy and deploy. Some tools have been priced upwards of

$1,000 per agent though expect this to drop as technology improves. TBT implementations also require time and resources from your IT department.

Call Center Coach's Nickerson says call centers with 50 workstations or fewer may find the products cost-prohibitive, except if offered on a hosted or leased basis by an application service provider.

Even so, Ahearn is also skeptical about large TBT implementations, either on-premises or hosted, because they demand that you get usually overworked IT personnel involved.

"Even web-based TBT requires the IT department's attention, because they must find ways to work the applications through firewalls," she says.

Sourcing TBT

When considering a TBT product, Call Center Coach's Nickerson recommends that call centers ask themselves the following: Does the tool model the desired behaviors? Does the vendor offer implementation and post-implementation services? Are the skills taught behavioral, that is, measurable and observable? Is the product interaction-based or a page turner? Is content available? If so, is the content validated? And can the product support, and easily integrate with, the call center's quality goals?

The Facilitator-Led Compromise

A hybrid of ILT and TBT is facilitator-led training (FLT), where a staff member like a supervisor or team leader works with the DVD or online lesson and initiates discussions and feedback on each part of the lesson. With FLT, agents sit at their desktops or in meeting rooms, or in conferencing, and they can discuss the lessons with the facilitator or with each other.

RDC's Jackson explains how it works. An agent uses a computer and learns a lesson about how to serve an angry customer in 15-minute increments. Then the agent reviews the training content and tests with a coach or supervisor. The coach or supervisor may also role-play with the agent to help the agent hear what the content sounds like.

FLT is less expensive than ILT because you're not paying for trained consultants and experts, but the content is less customizable and fresh to your organization. Also, your workers may have questions that the facilitator can't answer and that only the package's creators can reply to.

The Call Center School's Reynolds says that instructor-led ILT web seminars provide instruction from industry experts without the expense of classroom seminar fees, travel costs, and time away from the office. Another benefit is that one seminar fee can be spread across multiple students, driving the price even lower.

"The organizations that are most successful with distance learning are the ones that implement it along with live classroom interaction with a facilitator," says Reynolds. "We find that test scores and successful application of knowledge on the job are higher when 90-minute web seminars are followed by a 30-minute facilitated discussion."

Kathy Sisk, president of Kathy Sisk Enterprises, urges call centers to track agents' performance after each session and provide hands-on training and coaching with those who are not improving.

"The typical type of instructor-led training on the market takes too much time out of the participant's day, and there is very little to no follow through," says Sisk. "The training needs to be conducted in a way where the participants are held accountable for implementation and results are measured to determine adherence and effectiveness."

Facilitator-led training, using TBT has its downsides. While it enables the classroom experience and interaction between agents, says D'Ausilio, the presentations are static. Often material is not customized to the call center and it's not often fresh. Some facilitators may not be very good motivational speakers. If a speaker isn't inspirational, students are less likely to pay attention and attend the session.

"With a good speaker in a classroom you will be more likely to get agents' attention and they'll retain the information better, which may not happen with a video and a facilitator," says D'Ausilio.

Conferenced Training

Alternately, ILT and FLT can be offered at call centers and at off-site locations like schools, seminars, and trade shows. Both instructor-led and TBT methods are frequently offered together. For example, the University of Phoenix offers advanced degrees taught through a blend of in-person instruction at satellite locations and online TBT.

Training consultants recommend that employers deploy a mix of ILT, FLT and TBT. ILT, via conferencing, enables interaction with instructors and colleagues, the throwing of curveball questions to see if trainees are on the ball, and provides motivation to trainees. The trends are towards ILT and FLT conferenced training and TBT training. The reasons are costs and productivity. Conferenced ILT and FLT as well as TBT avoids travel costs and minimizes the need for having training rooms that eat up real estate.

Conferenced training is the best and often the only feasible way to teach home agents. Some home-agent-using firms, like Alpine Access, never see their employees at all.

Equally important, conferenced ILT and FLT and at-desk TBT avoid productivity losses caused by having employees leave their work areas or the premises to be trained. Employees can take these lessons at their workstations or at home.

The only downside to ILT and FLT conferenced training is the lack of in-person interaction with peers and with the instructor. But as anyone who has watched and participated in chat and web conference sessions knows, those discussions can get quite lively too, and you can come away learning a lot.

Ensuring Training Value for the Money

No matter what you are training for, you should be prudent with your training resources and make sure you're training your staff correctly to get value from your investment. Training experts like Nancy Friedman of The Telephone Doctor point to a wide array of training tools and services: $10 books, CD-ROMs, videos, TBT, and instructor-led seminars costing up to a few thousand dollars.

"Price is not an excuse not to offer good training," Friedman says. "We ask 'What are your needs and your budget,' and we work with you to find the best mix."

To make the best use of training resources, RDC's Jackson recommends that companies quit "sheep dipping," that is, don't put every employee through every course. Those who listen well, for example, don't need instruction in listening.

Employers should also identify the skills they are going to hire for versus train for. Why train an agent to type 30 words per minute when the job only requires a 20-word-per-

Operational Factors

Center Size

Low turnover centers

High turnover centers

What makes a low-turnover call center? Low stress and burnout, supportive management and strong team performance reveals a study LIMRA International. LIMRA defines low turnover as 5% or less and high turnover as 45% or more.

On-the-job Training

Work Culture Attributes

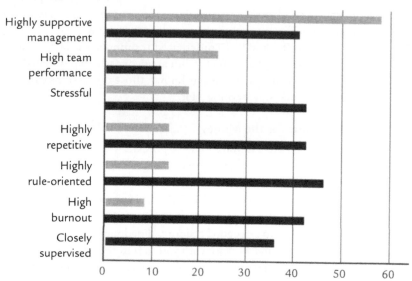

minute rate? Also, employers should determine which training methods—TBT, classroom, or both—best cultivate specific skills.

"The management team should be skilled in evaluating which employees need what courses and then deliver the training in the most cost efficient manner," advises Jackson. "For example, if a lot of employees need listening skills, is it more-cost effective to run a class, to coach individually or to deliver it electronically?"

Employers should also develop methodologies to assess and manage training performance. They can tie performance for example, to agent job description and skill evaluation.

Training, say experts, is only as good as what the company wants from their employees and how they evaluate their skills, and vice versa. Poor training affects job and recruiting performance. If the agents are poor-quality or take a long time to bring up to speed, call center performance declines as costs go up and efficiency, effectiveness, and customer satisfaction declines.

"If you don't evaluate the effectiveness of your training, you simply continue to train as always, which could be in a very inefficient and ineffective manner," warns Jackson. "You may be investing a lot of resources into something that does not give you a good return."

ePerformax's Hartsaw recommends that you train management and trainers to look at training as an investment that requires a financial return. Training isn't expensive, she says, if it works. Whether you're trying to reduce costs, increase sales, or improve customer retention, you must measure the training's impact.

"Training dollars invested during tough times usually have the greatest impact," says Hartsaw. "That's because everyone is focused on what it will take to build their individual job security. And people understand improved performance creates job security."

✪ AGENT RETENTION

You have selected and trained your agents. Now how do you retain them and for how long to make a return on your investment?

These are not easy questions to answer, because there is a wide range of retention techniques. Moreover, there are many circumstances where you may not want to retain staff for a long period of time. Let's look at them.

Retention Advantages

The key reasons why should retain your staff are to get the most value out of the money you put into your employees, not only in recruiting and training expenses but in lowered productivity until they get up to speed. Turnover can cost as much as 150 percent of each departed worker's annual salary. The payback period on agents ranges from three to four months to nearly two years depending on the skills required and depth of training involved.

More important, retention may help you keep customers. A study released by The Radclyffe Group found a strong correlation between employee satisfaction and customer satisfaction. The firm found, with no exceptions, that the companies with the highest employee satisfaction had the greatest customer satisfaction. The Radclyffe Group also found the obverse: companies that had unhappy employees also had unhappy customers.

How do employee and customer satisfaction influence each other? CEO and president Elizabeth Ahearn says that in companies where there is strong employee satisfaction when customers call to complain, agents are trained, for example, to respond with positive empathy. They're also empowered to give replacement items or discounts.

Agents tell customers that they understand how upset the callers are (without overtly or covertly criticizing their employers) and say they will try to resolve the issue.

But where there is strong employee dissatisfaction, the customers sense it. Such agents might tell customers that they would be dissatisfied, too, if they experienced what happened to callers. Supervisors tell agents that there is nothing they could do. And the agents, in turn, tell that to customers.

"Customers decide whether or not to make future purchasing decisions with a company, or to recommend its services to others, as a direct result of their experiences with the call center agents." Ahearn says. "Interactions with them can often be the first point of contact that a customer has with a company. A positive, enjoyable experience can lead to high customer loyalty and repeat purchases over time."

Retention Downsides

Yet there are downsides to keeping employees for long periods of time. The most practical if the most cynical rationale is to avoid paying them benefits, especially medical.

Having workers on the street before they are eligible saves money, but exacts a human toll on employees and their families in the U.S. where there is no government health insurance. It is also counterproductive. Employees who can't afford doctor's bills stay sick longer, which lowers productivity, come to work ill, which can spread disease causing more sickouts and they may leave on a moment's notice if they get a job that has decent benefits.

Slightly less nasty rationale to discourage retention is not paying wage increases with seniority. This is a fairer reason, especially with lower-skilled agent work, because there is often no real benefit in performance and productivity with longevity. Agents know all there is to know. Where this practice gets slimy is when companies hire people at lower wages at the start, promising to boost them after three or six months but who then find excuses to fire the workers before those dates.

Lastly, call center functions with naturally high turnover, like outbound business-to-consumer cold calling, tend to burn through their local labor forces quickly. There is little point of keeping those employees around.

Retention and Customer Reaction

A high turnover strategy may turn out to be penny-wise but pound-foolish. While new hires may have less expensive pay rates, reports Maureen Wolfe, principal at Legge and Company, they turn out to be more costly.

New agents' inexperience leads to longer call times, escalations, and return calls. They are less experienced at segue marketing, and training costs multiply. The churn creates termination, recruitment, selection, and orientation costs.

"Perpetually inexperienced staff leads to lower customer satisfaction, loyalty, and revenue," warns Wolfe. "That does not leave a positive customer impression of your enterprise."

In today's hectic economy, if you run your call center like this, you'll soon find your benches empty. There are many other call centers and other service sector employers that value good employees. Illustration by Travis Kramer.

Retention Challenges

Keeping quality people has its challenges, and as the labor market tightens, they will grow. Underlying the issue is that call center work per se is not a career for most people. No more than working the checkout line, ringing up purchases at the sales kiosk, flipping burgers, or guarding buildings. Organizations do not see any necessity to professionalize call center work and pay accordingly.

There are exceptions to this: higher-level work requiring experienced and/or certified individuals. These include inside sales, insurance, real estate, securities dealing, "dial-a-nurse," and senior-level technical support.

The traditional way forward in call centers, retail, and restaurants, is through supervision and management. This pathway is beginning to gather weeds for two reasons. Many agents do not possess supervisory skills, and the agent-to-supervisor ratios have been lengthening from 8:1 to 15:1 or higher.

But advancement is an important facet to offer in recruiting and retention. There is a direct correlation between lack of advancement and turnover. The Radclyffe Group reported that nearly 90 percent of agents who said they were bored with their jobs because they had no opportunities to grow their careers planned to leave within a year.

Retention Methods

There are several ways to retain your staff. Here are the key ones.

Providing the Best Center for Your Organization

Follow the advice in this book by providing agents with an easily accessible, pleasant, professional-looking, and safe place to work. That includes ergonomically sound workstations and chairs, glare-free lighting, excellent ventilation, even temperatures, low noise levels, and convenient break areas. Employees won't work in boiler rooms if they have a choice.

As consultants such as Bob Engel and Gere Picasso have pointed out earlier, providing substandard facilities leads to turnover and productivity costs that will quickly overwhelm any savings you initially achieved. You can also begin shopping for a subtenant for that call center real quick, because the word of mouth reputation as an employer will quickly dry up your labor pool.

If your site selection strategy includes co-location, you may have tapped into one effective means of attracting and keeping talented staff. By being in the same building as other departments you will be providing them with doors to advance their careers.

Unfortunately, the best call center locations may not be ideal for regional or national offices, and vice versa. Even so, make sure your call center employees are kept in the loop on new openings. A few employees may relocate to HQ if they are offered new positions there.

Treating Employees Well

Want to keep call center staff? The same three words that go for everyone: Treat Them Well. That means being fair and professional when managing their performance, keeping your bad day emotions in the black boxes where they belong when dealing with others and being accommodating to their needs as long as it doesn't impinge on your center's performance.

Treating staff well also means offering them affordable medical and 401(k) benefits and not canning them as soon as they become eligible for these necessities and for pay raises.

Empowering Agents

Call centers should make each agent's work personally rewarding. One way is to empower agents to make decisions, to become self-respected and customer-respected "go-to" people. Supplying such responsibilities gives agents pride and makes their work much more interesting and attractive.

The experience has been that when given the responsibilities, agents don't give away the store. Maggie Klenke, partner at The Call Center School, observed that agents who are empowered to decide whether to give refunds to customers and determine the amount of the refunds are stingier than the companies had expected.

Special Projects

Consider rewarding top performers with special training, and provide them opportunities to develop and succeed on special projects customized for them, recommends Klenke. These projects will help them win recognition, show talent, and grow their careers. The programs can also include community charitable work, which has tremendous publicity value. It can also attract new recruits.

Quality Supervision

Have agents supervised and coached by people who have the skills for those tasks. Chapter 15 explores this issue in depth. It costs nothing to say to an agent, "You did a great job!" Yet those few words are arguably the most important retention method call centers can deploy.

Additional Skills Training

Look at offering skills training. These include email and Internet response, monitoring, and technical support. If you have tasks that requiring licensing, such as insurance sales, offer it at a discount or pay for it completely for top-ranking employees.

Mentoring

Another means is mentoring, in which employees demonstrate and develop teaching and coaching abilities. That may open up advancement for top mentors.

Internal Career Pathing

Put together your own career path, such as creating more team leaders and sub-team leaders inside call centers.

"The recognition among colleagues that comes about from promoting agents to those roles counts more than any marginal pay increases," says Jeff Furst, president of staffing firm FurstPerson.

Paying per Performance

Coupled to career pathing is paying per performance instead of granting raises across the board. Outlay increases when agents meet certain milestones, like competing the end of mentored training, or when they reach certain attainment levels, provided you give them training. For example, in one company, a Level II agent may have email and chat training, whereas a Level III agent might be English and Spanish bilingual. In another company a Level II agent handles gold card customers and a Level III agent services platinum card customers.

"When agents attain these levels, you can offer them raises because they will be able to offer better performance than before," Radclyffe Group's Ahearn points out. "The agents will feel that they are getting ahead both in skills and financially, which will help keep them in your company."

Flexibility

Considering being flexible with your working hours in case employees have family responsibilities. Don't limit this to those that have families. Single and childless workers also have personal needs that are equally important to them.

Nothing like a truckload of prizes (or at least some candy) reports *Call Center Magazine* to keep agents motivated. Tinker Federal Credit Union filled a remote-controlled truck to deliver presents. The call center senior member service supervisor also handed out toy racing cars to encourage staff to "race for success" and watches to "watch out" for cross—selling opportunities.

Matching Jobs with Advertised Positions

Perhaps the best agent retention method is ensuring that the job advertisements and descriptions match what the agents will be doing, recommends Berta Banks, partner, Banks and Dean. Also, keep the recruiting and training teams updated on call center strategies and agent position requirements to ensure proper selection and training.

"This ensures that the qualities required for the job are properly identified in screening and selection and that training is aligned with the work environment and learning style of the class," Banks says. "When this process is in place, managers get the people they need to get the job done."

Incentive Programs

Everybody likes incentives, whether cash, gifts, trips, or "employee of the month." Or not have to come into the call center or office at all.

Brad Cleveland, president of ICMI, points out that incentives can dramatically impact call center, team, and agent performance. They also help maintain agents' interest in their jobs. Most employees will find higher pay for better performance appealing.

But, he adds, these programs must be set up right or they will fail. They should enhance quality and customer satisfaction, be tailored to the call center's mission (outbound cold calling agents are motivated differently than inbound tech support reps), and include non-monetary rewards (for example, agent of the month).

Also, no incentive program, no matter how well planned and implemented, can replace good leadership and management practices. "An incentive program is just one aspect of an effective manager's performance improvement arsenal," Cleveland says.

When deciding what incentives to offer, Manpower suggests that you consult with your team leaders. And when setting up your recognition and rewards programs don't forget the "Steady Eddies" who do most of the work. These are the employees whose performance may not be spectacular but who come in and leave on time, take few sick days, and meet expectations.

Also consider rewards for achievements such as attendance, on-time arrivals, and employment anniversaries.

Gifts programs

Rewarding agents and teams with gifts is a popular method to encourage excellent performance and agent retention. ICMI reports that gifts rank third in the incentives offered by call centers, behind recognition and awards but ahead of pay increases and job promotion.

How to supply gifts to deserving agents and teams? You can purchase the items, in consultation with the team leaders, and store and distribute the goods. Or you can outsource the gift program to incentives companies that partner with product or service vendors and handle fulfillment.

Here's how they usually work. You assign values (for example, dollars or points based on the performance you are measuring). Then you distribute awards directly to employees through certificates or tokens, or indirectly via the incentives firms that accumulate the awards in online accounts.

Employees can redeem awards directly through gift certificates or indirectly by, for example, purchasing items in program-specific catalogs with the points collected.

Gift certificates work especially well when they are for well-known firms with a broad range of products to appeal to a wide variety of employees. Because your agents redeem certificates on the spot, there is no waiting for redemption.

The Bill Sims Company champions gifts because, unlike cash and gift certificates, they are tax-free to employers and employees, with some restrictions. The firm offers a quality gift catalog to clients.

"Employees also tend to remember gifts more than cash or gift certificates," says President Bill Sims Jr. "Also my experience has been that 15 percent to 20 percent of all gift certificates are never redeemed."

Some incentives firms go beyond point-gathering and gift-distributing. Through experienced call center industry partners that assess hiring and performance, RYI Solutions analyzes your staffing, training, and retention issues. Once problems in these areas have been identified and corrected, RYI Solutions helps you devise and implement a gifts program tailored for your call center.

The downside of such gifts programs is that they have limited and short-duration results, but they can work with the right agents. Agents who handle customer service and support aren't likely to be motivated by them, says Jeff Furst. However, they can work for employees like sales and collections agents who are more goal-oriented.

If you do deploy such a program, you need to have employee participation in the setup so they buy into the incentives. "A golf outing is not going to appeal to the entire enterprise," says Manpower's Buelow.

Employee recognition

Recognition by supervisors and colleagues is the best motivator. Set up "employee of the month,," "top sales for the week," or similar programs. Reward them with their photo in the lobby. Take your top performers out to lunch. Reserve choice parking spots for this month's winners. Write them up and have their photos appear in internal newsletters.

Home Working

Many people love what they do for a living but hate the commute that is growing in time and stress and takes away from their personal lives. Home working gives that time back, along with money saved in commuting costs. Consultants estimate employees can save $4,000 to $5,000 per year by skipping the drive or the ride.

You'll likely find that you can keep agents longer and get more out of them as well as attract new employees if you offer this option. But make sure that you provide that to agents who have proven that they can work without direct supervision and you have the environments that permit home working.

Individualize the Incentives!

Whatever the retention method you pick, you must individualize them, recommends Maggie Klenke. "Different factors motivate employees differently," she points out. "Some are motivated by careers, others by experience, knowledge, recognition, and money. You have to find that motivator in each employee and tap it."

Detecting and Avoiding Burnout

Retaining your employees also requires that you be alert for problems with the ones who have been good workers. Call centers are high-stress environments, and the tensions get to everyone to varying degrees.

Kathyrn Jackson advises that you look for signs of burnout such as health problems, sickouts, short temper, bad co-worker interaction, and lack of responsiveness to corpo rate communications.

For agents, in particular, this is shown in lower monitoring scores and increases in escalated customer calls and poor attendance. In supervisors the symptoms are less monitoring being done, unavailability for questions, and decrease in coaching sessions.

To handle burnout RDC recommends you should consider:

* Measuring job satisfaction—conducting employee surveys, providing opportunities to serve on committees, working with management to improve the environment of the call center
* Rotating jobs—having agents and coaches work on projects from time to time
* Ensuring that appropriate training is delivered
* Offering access to gyms
* Providing defined career paths relieving the "dead-end job" syndrome
* Providing voluntary time off, allowing agents to go home (without pay) when call or contact volume does not require their services

"Managers noticing these signs should have one-on-one sessions with coaches, and coaches noticing these signs should have one-on-one sessions with agents," says Jackson. "Causes of burnout could include reaction to understaffing or personal problems."

DIALOGUE, NOT PITCHING

To gain revenue from customer service calls, many enterprises require call center agents to upsell on them. But the results have reportedly been less than stellar.

One key reason is that both agents and customers often feel uncomfortable at turning service calls into sales pitches.

"The old pitching model is broken," explains Kathy Dean, partner with consulting and professional services firm Banks and Dean. "Customers don't want to be aggressively sold. They would rather dialogue with agents."

Banks and Dean has a staffing, scripting, and customer follow-up program for business-to-consumer and B2B clients. The approach stresses assessing customer needs and building relationships with them rather than pitching on outbound, and upselling on inbound.

For example, on the outbound side, instead of calling with an offer, agents would call customers and ascertain whether that prospect has a need for the firm's products or service, like a bank calling people who just moved into areas near their branches.

On the inbound side, agents, after issue resolution, would create a dialogue with customers to hear if they have a need that the firm can meet and learn about the customer's interests, preferences, and decision factors.

"All customers have needs," says Dean. "It is critical to the enterprise for agents to discern those needs and understand if their product and service can meet them."

Revenue could increase significantly from customers calling back. Banks and Dean boosted sales for their customers by 45 percent to 90 percent using this method.

To enable this new method you need to look and screen for a different kind of person than those typically found on outbound and inbound sales and service.

"You need agents who can ask questions, listen, assess, and come up with responses to meet customer needs, not just those who can sell or provide service," explains partner Berta Banks.

Chapter 15:
Management Issues

You have invested considerable sums in your call center. How you get the most of it from beginning to end depends on how well you manage it.

This book is not about management per se. There are many fine tomes like Call Center Management on Fast Forward by Brad Cleveland and Julia Mayben. What this chapter will do is touch on the key issues that affect your center's performance and how to maximize your returns.

✪ MANAGEMENT STRUCTURE

Call centers are less like offices in how they are run and more like factories. Agents are almost always assigned to teams, with anywhere from six to 20 agents per team. You can have several teams handle the same types of contacts, like order entry. Or you can designate specialized teams such as for different products, like for "bismuth" or for "uranium" cards or services, like Latin language support.

Each team has a supervisor. The amount and level of call center supervision generally depends on the workload. In large groupings you may want to assign team leaders to assist supervisors, which are like corporals to sergeants. In the case of small call centers those supervisors can be the facility's managers. At the larger end team leaders may be designated assist supervisors says Maggie Klenke, a founding partner of The Call Center School. The team leader's role might be answering escalated calls, generating reports, real-time adherence management, or a variety of other roles that relieve the supervisor so that she can focus on the people-management roles.

Ultimately supervisors report to call center managers. If there is more than one center those managers answer to directors or executives. If there are designated or specialized teams and departments within the call centers, like sales, customer service, or IT support, their chain of command leads to the executives in charge of those functions such as vice president of sales, customer service director and CIO.

✪ SUPERVISORS: YOUR CALL CENTER'S NCOS

Effective supervision is key to call center manaement. Supervisors are your center's NCOs. They deliver your instructions and carry out your policies and relay back to you what is happening and what needs to be done to meet your goals. The second part is crucial. You need to know what is going on to make sound decisions.

Good supervision rewards your investment in staffing, training, and site selection, facilities, and design. Poor supervision breaks them.

Supervisors monitor and grade agents' performances. They see and suggest how agents could be doing their jobs better. That can include recommending additional training. If the work does not improve, supervisors can take steps to correct it, such as talking to the agents to understand what is happening and looking at shifts or tasks that better suit them. Supervisors can advise agents that they could be dismissed if the quality does not improve, and can have that threat carried out if necessary.

Supervisors praise agents when they do their work well and recognize them when they perform over and above what you require. Supervisors need to track the top performers and see how to get them in the call center through retention methods like career pathing, special projects, and training.

Supervisors also introduce changes to existing programs and new projects to the agent level, plus any change in corporate policies. That can include arranging for training. Supervisors take questions, listen to concerns, and bring them back to upper management.

Supervisors should also take part in hiring decisions, the final pass before new hires are accepted on the shop floor. They should meet with the people they are about to work with and see if they are a fit in the teams.

Most important, effective supervision can lengthen your call center's life span. A good set of supervisors can easily make your call center a favored place to work through word of mouth. You get less turnover, lower costs, and more productive employees.

At the same time poor supervision can shorten your center's life. Maggie Klenke points out that in most cases people don't quit companies, they leave supervisors. With supervisors being in charge of 10 or more employees those numbers add up—abetted by word of mouth accelerated by the observed 'fact' that bad news travels faster than good—and your center could be in trouble.

Supervisor Staffing

As necessary as they are, supervisors are overhead. They usually do not directly take or make contacts. Their salaries are higher than that of agents, $30,000 to $60,000 per year depending on call type (outbound often pays less than inbound) and local labor markets.

The key then is optimizing the number of supervisors needed for the functions to be carried out. Brad Cleveland, president of the Incoming Calls Management Institute (ICMI), pointed out in *Call Center Management Review*, the ICMI's journal, that most call centers have between 10 and 17 staff per supervisor, measured as a ratio such as 1:10 or 1:17. That rate varies by industry. For example, insurance companies, mutual funds, and catalog companies tend to be on the low end of that spectrum. Banks, utilities, and government-oriented call centers are generally on the high end.

"But there can be significant exceptions," Cleveland said. "Some reservations centers have 50 or more staff per supervisor. And software-support help desks can have as few as five staff per supervisor. Even within an industry, there can be wide variance. One well-known catalog company has 40 reps per supervisor, while another has 10."

Sometimes supervisors have assistants i.e. team leaders in larger (20 or more) groups.

Klenke says the team leader's role might be answering escalated calls, generating reports, real-time adherence management, or a variety of other roles that relieve the supervisor so that she can focus on the people management roles.

There are other factors affecting team size. Supervisors also have paperwork. The more they have to do, the smaller the team. For example payroll is four to six hours a week of work. Time away from the floor also eats into the ratio. Some centers draw the supervisors away for all kinds of meetings, such as with other departments, clients and prospects, and that affects team size.

Employees may need fewer supervisors if they are in self-sufficient teams or if they perform tasks that require little interaction or supervision.

Also helping to determine that supervisor-to-agent ratio is the experience and maturity of the workforce. The older you are or the more you know, the less you need someone to hold your hand. Outsourcer ARO cut almost by half the number of supervisors it needed, from 1:11 to 1:20, when its labor force changed from the 20-something baby boom echos to the 40-plus "big bangs."

The secret to ARO's boost in supervisor efficiency is that it dispensed with a traditional premises call center and went to home working, which means no cubes for supervisors to hover over. You can achieve similar results by limiting or doing away with face-to-face supervision.

Efficiency plummets when talking to someone in person. They can't talk on the phones or answering emails while listening to you. After you leave, they have to think about what you said. Their performance then must ramp back up.

Slowly supervisors and managers are relying more on email and instant messaging. They can get more done, which means fewer supervisors are needed.

Some call centers now have "supervisor centers" with help desks manned by supervisors, reports Roger Kingsland, managing director of architects Kingsland Scott Bauer and Associates. Because those agents get into queue too, there is now a subculture of knowledge, agents asking each other for advice.

"If the help desk concept proves viable, this could lead to virtual agents working at home and coming into the call center primarily for additional skill set training," says Kingsland. "[But] because the nature and complexity of U.S. call centers is constantly changing, that might be way off."

Recruiting Supervisors

The way to ensure that you have excellent supervisors is to recruit for them correctly. Unfortunately, many companies fail to do so. They will pick someone on the basis of seniority or because they are an excellent agent, or worse, because they're a nice person, without looking for specific skills. Not everybody is cut out to be a leader or wants to be one.

You can use most of the same recruiting tools as for agents: word of mouth, web sites, print classifieds, job fairs, and staffing agencies. There are some major differences.

Job boards, especially ones specialized to the call center industry are more viable. Supervisors, and especially managers experienced in the field, are much more likely to go there than are agents.

Also, classifieds in trade publications and web sites are now an option unlike with agent recruiting. The reasons are the same as above. Managers especially will have heard or seen of publications and will seek them out.

Talent Spotting

The best source for supervisors and managers is still within your organization. Focus on identifying, grooming, selecting, and training candidates who have people-management and organizational skills and enterprise knowledge.

Following are some techniques.

Pilot Programs and Special Projects

The Call Center School's Klenke suggests having pilot programs or special projects requiring teamwork, open to all agents. Examples include peer quality reviews and coaching sessions, or leading a team to prepare for the implementation of a new technology. That also gives potential leaders a chance to shine.

Extracurricular Leadership Examples

Look at agents who have shown extracurricular leadership, and encourage outside community involvement and leadership. Those candidates should be looked at because they have shown, on their own time and for no compensation, those skills and talents.

Develop a Management Preview Program

Maureen Wolfe, principal at Legge and Company, recommends setting up a management preview program where agents volunteer to learn about supervising and managing call centers, the challenges faced, and what is expected of them.

Not everyone is interested in advancement, and those who do become supervisors often do not have the skills to make an effective transition, says Wolfe. A preview program would give a realistic look at the supervisory role and begin to develop needed competencies before a person is promoted.

The preview program, which she has implemented successfully, enables agents who do not find supervision attractive to opt out. It also gives management a better look at the available talent.

Such a program would entice agents to stay, reducing turnover and creating a better-qualified pool of supervisory and management talent that is all too often missing from call centers.

"Joining the program would not ensure that they become supervisors, but it would better prepare them for the tasks ahead," Wolfe explains.

Screening and Profiling

Supervisor and management recruiting is less automated and systematic and more traditional compared with recruiting for agents. Resumes are typically submitted online or in person and screened. Applicants are called for pre-screening and in-person interviews. You can test online or in person for supervisory and management attributes.

Before you put out the opening on the web site or buy the display ad you need to put together job profiles for supervisors and for managers, as you should do for agents (see Chapter 14). These qualities include:

People Management Skills

What separates excellent supervisors from non-supervisors or incompetent ones is, obviously, people management. This is a "you have it or you don't" skill.

Supervisors must tell those that they are supervising what to do in a way that a person being instructed believes the order is in their best interest and that carrying it out helps them and their colleagues. They must watch for, listen to, and help resolve agent problems that might hinder the completion of your goals. They must hear without being personal and advise without bullying. Managers at each level must know how to identify potential leadership from applicants and from the ranks.

Demonstrated Leadership Abilities

Require your supervisory candidates to have demonstrated proof of leadership skills.

Look for past experience in similar fields such as hospitality and retail. These industries are very similar to call centers in that they entail dealing with the public, long hours, high stress, and high turnover. If an applicant has shown that they had successfully run an afternoon shift at a fast-food outlet, then they are quite likely ready to take on the call center at the same hours.

Successful volunteer experience can also show leadership experience. Examples include organizing a Girl Scout troop, coaching a Little League team, serving as president or vice president of a very active civic or fraternal organization, putting on fundraisers like a car wash for a nonprofit organization.

Some of the best supervisor or coach candidates are those who serve or have served their country in NCO, officer, or team leader roles in the military, Coast Guard, and in police and emergency services. They may be in reserves, the National Guard, with volunteer fire departments or on EMT/paramedic crews.. If these fine people are in your workforce, support them when they get called up. They make an important contribution to our freedom, security, and safety. They know what to expect, and what is expected of them, in *truly* mission-critical situations.

Negotiating/Networking Skills

Call center managers, much more than supervisors need to know how to negotiate with other departments and with senior staff for resources. Managers must also know how to corporate-speak: to communicate why they need to keep and improve service levels; and why they need the money to achieve this.

Managers need to know how to schmooze, network and negotiate so that they get what they need in the way of resources, cooperation from other departments and fight off budget cuts. They must understand, and be comfortable dealing in, corporate politics.

"Ideally these top call center managers should have both political and technical skills," says Donna Fluss, principal at DMG Consulting "But if you have to compromise when selecting managers, the political skills are more important. If top managers can't communicate effectively, they won't get the resources for their call centers."

Business Acumen

Your supervisory staff and certainly your managers need to have business acumen. That can include buying decisions and hiring staff. Berta Banks of staffing and training consultancy Banks and Dean, points out that they are spending a lot of your money—$25,000 or more—when they hire an agent.

Applicants who have trained or bought for organizations qualify. So would those who owned their own businesses, but be sure you find out what happened to their enterprise to see if they have good decisionmaking abilities.

College Education

You don't need a college degree to manage people, just like an NCO need not have attended West Point or Annapolis to lead troops. For supervisory or first level management positions do not have that requirement in your screening.

But having that education and piece of paper is a necessity to move up or enter the officer corps. You most certainly need to have that when hiring for most management positions.

Corporate higher-ups increasingly prefer supervisors and managers who have that college especially business education (and/or public administration in government) to succeed in their jobs and advance in the organization. There are no widely accepted degrees in call center management, but there are many training courses.

"A college education teaches managers critical thinking and reasoning that they need to apply in management," says ICMI president Brad Cleveland.

Donna Fluss, principal at DMG Consulting, believes that college instruction is not essential for day-to-day call center functions and for those who do not want to move up the ranks. But it helps build interpersonal skills and relationships. She recommends that supervisors who do not have a college education get one in their off-hours.

"If a soldier wants to become a general they have a much better chance of becoming one if they start at West Point than if they don't," says Fluss. "So too, a supervisor who wants to become a corporate executive will benefit from a college degree."

Certification

Call center management certification in open, employer-accepted performance standards has the potential to show to senior management that certified managers have mastery of the skills that may enable them to succeed at running centers. There are many more generic transferable skillsets in management than working as an agent.

There are generic call center management certifications. The Call Center Industry Advisory Council (CIAC) offers strategic and operations management certifications. CIAC tests for competencies in people, operations, customer relationship, leadership, and business management. The strategic leader certification goes into big-picture tasks such as establishing and achieving service levels and quality and site selection, design, and management. Applicants can take up to two years to be tested and certified.

But that's the theory. Call center management certification is still a way off from becoming accepted and the norm. So unless your senior manager has a call for it or your call center is highly specialized, think hard about requiring it in your profile.

"Certification could be useful to teach recognized skills but is not much of a screening tool," says. Connie Caroli, president, TeleManagement Search. "Employers want to see experience and proof of performance."

There are two specialized roles for certification where you should use them in your profiles. They are technical certification and internal certification.

Tech support management certification has taken off sooner than for call center supervisors and management. There is more of a certification culture in tech support. Many employees have vendor certifications such as from Microsoft and Sun to prove they are competent to fix products using their components. The key call center management certifications are the Help Desk Manager certification from the Help Desk Institute (HDI) and the Certified Support Professional–Supervisor and Certified Support Professional–Manager certifications from the Service and Support Professionals Association.

You should consider internal certification for your management and supervisory staff. Having such certifications internally enables you to set a more rigorous uniform set of standards geared to your processes. That way you can measure and track agent performance and career paths and offer them incentives to stay longer. You can also quickly screen employees who want to relocate from other divisions or who have been out of the organization and want to return.

Profile and Hiring Caveats

There are two main caveats when seeking supervisors and managers: don't place too much weight on understanding performance metrics and on product knowledge.

Jim Moylan, president of Call Center Jobs.com, recommends companies have behavioral testing to evaluate management candidates on their ability to work with people.

"Companies typically look for a manager who understands the performance metrics of a position, as opposed to what skills a person can bring in terms of building a team that will improve performance," he explains.

While inbound or outbound experience is beneficial, these skills can be learned quickly if you hire a person who can motivate people. Behavioral testing lets companies objectively evaluate people for their strengths and weaknesses.

Many companies err by recruiting for product expertise and not for skills and experience. TeleManagement Search's Caroli says this is unnecessary for most call center management positions. The exceptions are highly specialized fields like technical support and licensed businesses, such as insurance and securities firms.

Catalogers may need managers with similar experience, such as with product returns. But they could hire managers who have worked in high- and variable-volume call centers like those in financial services.

"Call centers prefer managers with experience and stellar performance," Caroli points out. "Call center skills are sufficiently transferable in most businesses to enable a call center manager in one industry to easily work in another."

Training

As qualified as your supervisors may be, they will likely need training. And that's an investment worth making.

With a quality management team focused on an integrated system of talent management, you can easily cut your annual turnover by at least 5 percent says Legge and Company's Wolfe.

"Average turnover costs are at a minimum $25,000 per agent per year," she says. "For a 100-agent call center that's a savings of $125,000. The money for quality supervisor and management training is a minor amount. The real savings are greater when you calculate higher customer retention and increased revenue."

Following are a few key areas to look at.

Leadership and Management

Applicants will have shown leadership potential or have some past experience. But that is often not enough.

Response Design Corporation recommends that successful supervisors utilize "full spectrum coaching" to train their agents. Supervisors' methods should include setting clear expectations and providing consistent, timely, and accurate performance feedback. They should also develop expert agent skill through educating, modeling, practicing, applying, and inventing successful behaviors.

Look for programs that teach supervisors how to manage their charges, including spotting signs of poor performance and motivating others. Seek out programs that teach supervisors how to build respect, create a fair and consistent work environment, establish credibility, and earn respect.

The best programs ensure that they are exposed to the same front-line conditions as agents are exposed to. Just as in the military, a "paper sergeant" who hasn't been exposed to incoming blasts is useless on the front lines and gains little respect from the troops.

"How will supervisors and managers be able to help their agents if they have not been able to be successful at it themselves?" asks trainer Kathy Sisk of Kathy Sisk Enterprises. "Managers and supervisors need real-world experience in order to effectively coach and train others."

It pays to train supervisors how to coach right. Such training helped reduced voluntary and involuntary turnover in ePerformax' U.S. call center to less than 10 percent per month.

"Many companies believe the front-line supervisors can have a much greater impact on turnover if they had further skill development," says Hartsaw. "Because the majority of front-line supervisors are promoted from within, and most companies don't offer call center supervisor training curriculums, the opportunity for improvement is great."

Many supervisors and managers don't believe they need training. After all, they're supervisors. How do you make that training stick? The tried and true way: the stick, as in carrot and stick. Todd Beck, service portfolio senior product manager of training firm AchieveGlobal recommends tying part of the supervisor's pay to customer satisfaction and agent retention.

"We've been preaching the importance of aligning everything, including pay, to the business outcomes the training is designed to achieve," says Beck. "And we've found that the companies that have picked up that concept have been the most successful."

Tools and Techniques

Your supervisors and managers will use sophisticated tools like monitoring and workforce management. They must know how to measure performance and glean through documents like ACD reports. Call centers are very numbers-driven. Therefore you need to train them on these tools and methods.

Relating Call Centers to the Enterprise

Managers will need to know how to relate call center performance to enterprise performance measured in terms of customer satisfaction, customer retention, sales, and costs. There are often tradeoffs involved.

One of the classic conflicts in call centers is between average handle time (AHT) and customer satisfaction. If you manage to strict AHT standards your center can process many calls but you might be annoying customers and causing lost sales that may outweigh the cost efficiencies gained by hurrying people off the phones.

Managers must also demonstrate that their function saves companies money by identifying problems early on through customers' complaints and suggestions on how to improve products and services.

BenchmarkPortal president Bruce Belfiore says call center managers must become "bilingual:" thinking and communicating call center results in call center and in senior executive "languages" - that is, they must think in terms of both return on investment (ROI), earnings per share (EPS) as well as first call resolution (FCR). Managers must learn how enterprises work, such as accounting, finance and in government agencies, public administration. They need to know for example how to do ROI analysis. They can take these classes at night or online.

SALARIES FOR BUSINESS-TO-CONSUMER CALL CENTERS

	Vice President-Inbound	Director-Inbound	Manager-Inbound	Supervisor-Inbound
Low	$94,100	$76,600	$53,200	$28,700
Average	$112,800	$94,200	$65,300	$30,300
High	$134,600	$99,700	$73,800	$33,900

Data Courtesy of Telemanagement Search (New York, NY)

There is a wide range of salary variations between supervisory and management positions even in the same field, in this case business-to-consumer. Data courtesy of TeleManagementSearch.

○ MEASURING PERFORMANCE

Managers need the right tools to do the job, and that means having the right performance-measuring metrics.

Call centers are plagued with too much data. Managers often can't see the forest for the dead trees that make up their printouts. As noted before, the measurements are so internally focused, using language that only other call center managers can understand, that those reading them can be blind to what is happening in the real world.

Here's what you and your managers can do to prevent this issue.

Avoid Bad Metrics

Bad metrics are those performance measurements that hurt call centers by focusing strictly on costs rather than customer satisfaction. Following are some of the leading culprits.

Raw Calls Handled

The number of raw calls handled says nothing about a call center's performance, says ICMI's Cleveland. A problem with a product will generate many more calls than if the product is working fine.

Cost per Call

Cost per call is almost as dubious. That is measured by dividing total costs for a given period by total calls. Conventional wisdom says the lower the cost per call the better. However, a climbing cost per call can be a good sign. For example, better coordination with other departments may reduce the number of times a customer contacts the center.

"If you're able to cut down on the volume of calls by good products and service, effective self-service, and first-call resolution, you actually often raise the cost per call," Cleveland explains.

Occupancy

Occupancy, the percentage of time agents handle calls versus waiting for calls to arrive, should be scrapped.

The reason, says ICMI's Cleveland is that occupancy in inbound call volumes is driven by random call arrival and is not within the control of agents. In addition, agents could be responding to emails, faxes, written letters, or conducting research.

Average Handle Time

Your center's AHT alone may be higher than for similar organizations. But that is not a bad thing. Many managers do not instinctively think that way because a lower AHT means you are processing more calls with the same number of agents.

"If you are using that time wisely, to prevent repeat calls or capture information that can improve products and marketing, your call load AHT times number of calls and associated costs may be less than the norm," Cleveland says.

Use Customer-Focused Metrics

Managers should not look at cost metrics like AHT in isolation but need to weigh them with the customer metrics, like customer retention and revenue and customer satisfaction scores, advises Jodie Monger, president of Customer Relationship Metrics.

"Instead you give ranges, like 125–200 seconds," suggests Monger. "If agents fall outside of the ranges, say 300 seconds, then managers need to look at other metrics and talk with agents to determine why."

Carefully Use Profit Metrics
First-Call Resolution

First-call resolution is a win-win situation for everyone. Customers get what they want the first time; the company gets fewer repeat calls and escalations; and agents have the tools and feel empowered to respond.

This data is in line with the precepts of CRM in that customer satisfaction is key to customer retention. You can't have a satisfied customer if your agents are trying to get them off the phone.

"Many line managers do not realize that CRM technology will increase talk time, hence those costs. But longer talk time will help lead to first-call resolution and improved customer satisfaction and retention, which outweighs the costs," says Jerry Barber, vice president with call center certification firm CIAC. "In revenue-generating operations it will lead to increased revenues as well."

Be careful how you measure FCR. There is no agreed-on definition of this metric (or others, as I will point out later) in this industry. Some organizations define FCR as when a customer hangs up. They don't count against agents' FCR statistics customer call-backs or drops in satisfaction.

Instead, Sarah Kennedy, partner at Service Quality Management, recommends they measure FCR by outbound phone or by simple IVR customer surveys taken no more than one to three days after the interactions.

"Call centers should measure FCR by what customers, not management, define as resolved," she points out. "Only the customers know that the issues have been solved to their satisfaction."

Customer Satisfaction

There is usually a strong correlation between customer satisfaction and retention.

Customer satisfaction can be ascertained indirectly, through monitoring agents' calls with customers, and directly, through surveys. Each method has its own challenges. Call monitoring infers but does not state customer attitudes and reactions. Jodie Monger notes that using the results from call center monitoring forms a quality indicator without surveying the most important component in the equation—the customer.

Managers or executives typically create monitoring forms that reflect how they believe the call should be conducted. The theory is that if agents follow the form's criteria, then customers will be satisfied and companies will retain them. But consciously or unconsciously, managing to the form deafens the call center and corporate managers to what customers really want, says Monger.

Customer survey information via traditional pen and paper arrives too late to be of much use to managers to correct problems. The challenge is asking the right questions, and that means with immediacy and impartiality.

At minimum, satisfaction surveys should help you find out why the reasons customer purchased the product, in rank order, their experience with it, and whether they would buy it again, and why. The forms should ask about access to a live agent, the agent's ability to solve a problem, and first-call resolution.

The surveys take should take no more than three minutes to fill out. You may need to incent customers to take part—no one works for free—such as offering such as a paper or online gift certificate.

You should conduct these surveys shortly after their interactions to obtain actionable information. Peoples' memories fade and risk becoming inaccurate over time.

If firms do them with in-house live agents, customer satisfaction surveys can be skewed if the same agents make the survey calls as those who were on the original calls with the customers. The agents might be aware of a pending evaluation, which would prompt agents and supervisors to be on their best behavior.

"The consequence is that managers will make decisions based on invalid results," warns Jim Rembach, vice president at Customer Relationship Metrics. "A manager may focus resources to address an issue that does not need attention. Or the issue can mask problems that need to be addressed."

A solution to this problem can be found in services like Customer Relationship Metric's Completely Automated Telephone Surveys (CATS). The product invites customers to participate immediately after the interactions but while still on the phone. CATS can also make outbound calls to customers and ask them to participate soon after they hang up their inbound calls.

Another more expensive but potentially more effective solution is to outsource this to a bureau that will use live agents to make the calls. The benefit is that customers may be more willing to talk to people than to machines.

The best way is to combine both call monitoring and customer satisfaction survey calls, recommends Monger. That way you avoid any gap between internal scores and customer feedback.

A caveat in customer satisfaction is if your organization is blessed with a natural monopoly, your customers may think your service stinks but they may still buy from you. But don't take them for granted. If another enterprise sees an opportunity—customer service is a marketplace differentiator—and can offer an equal at the right price then you have problems.

The same goes for government departments and agencies. If the public-facing service is lousy, the press will feed on that story, and the elected officials responsible will look bad—they hate that with a passion—and you have problems, too. Heads will roll. There's always an outsourcer ready to take business. There is always some other office wanting to horn in on your budget, and you can bet that their chiefs will make that known to their elected official champions.

Employee Satisfaction

Companies are also looking at measuring employee satisfaction to cut turnover, reduce costs, and boost customer retention.

"Studies have demonstrated that customer satisfaction increases as agent job satisfaction increases," argues Cleveland. "Retention, productivity, and quality often have a definable, positive correlation to agent satisfaction."

Monetizing Metrics

To make business cases, managers need to find ways to monetize their metrics argues Teresa Hartsaw, president of outsourcing and training firm ePerformax.

For example, if the cost per transaction is $6, what does that transaction mean in customer retention and revenue? If the cost per transaction can be reduced, say by less talk time, what will happen to customer satisfaction and revenues?

Wolfe suggests deploying a new measurement: total service cost as a percentage of revenue generated, rather than low cost per call. If revenue goes up that may free up more resources for call centers.

Encouraging revenue will prompt managers and agents to lengthen calls if they obtain revenue, which increases other results such as customer loyalty and satisfaction.

"Who cares if service cost as a percentage of revenue adds extra costs as long as return on total service costs continues to grow?" Wolfe points out. "A low cost per call strategy only works on the denominator of ROI, whereas return on total service cost works on both the numerator (revenues) and denominator (costs)."

Call centers should also look at and measure turnover costs. Call centers can assign a standard cost per turnover transaction, based on current costs.

"Experienced, high-performing agents are more economical than new agents, even if they make much higher wages because they work more efficiently and can effectively cross-sell and upsell," Wolfe points out.

Performance Measurement Software

DMG's Donna Fluss suggests adopting performance management software that enables all departments—call center managers, marketing, sales, and senior management—to see call center performance from their perspective through scorecards and dashboards.

Performance management software can identify issues that impact customer attitudes with speech analytics and text categorization searches that identify core causes in emails and phone calls. It could point to new revenue and competitive situations. It gathers operational data to measure agent productivity including average talk time and new revenue.

Managers should measure and evaluate agents using this system on a mix of productivity, quality, and revenue measurements. The data enables senior executives to determine whether the call centers are meeting corporate goals.

"Performance measurement tools will help senior executives appreciate the value that call centers bring," she points out.

○ DEMONSTRATING VALUE TO THE ENTERPRISE

Call centers managers can show worth by demonstrating business skills. For example:

Do ROI Analyses

Managers can show their business savvy by doing ROI analysis on any major change, such as buying new hardware or software.

They can find ways to help their employers cut costs, which shows the value of their centers.

Sarah Kennedy suggests launching a short survey periodically, as a follow up to the customer's call, to find out if the callers used the IVR menu or the web site first and, if so, what problems did they encounter? The data, including customers' suggestions, goes to the IVR and web teams.

"By asking customers you can find out what the problems are that necessitated the calls, resolve the issues, and head calls off in the first place," she says.

Engage with Other Departments

Too often other departments do not communicate with the call center, leading to instances such as a new product or campaign is launched with the call center team being the last to know, and left scrambling when the phones ring.

When that happens call center managers should meet with their counterparts ASAP to show how those actions create long hold times.

"Don't gripe, but act," Klenke urges. "Work out methods to avoid such problems in the future, like how much lead time to give the call centers to ensure the added workload can be handled effectively."

Be Proactive

Managers need to take steps to minimize or deflect the spray when the fertilizer hits the fan, as it inevitably will.

One method is to propose arranging additional in-house or outsourced agents in advance of a new product launch. Managers can demonstrate the ROI including additional net sales, sales per hour, customer satisfaction, and customer retention.

Another technique is to have agents listen and probe when customers are dissatisfied as an early warning system for enterprises. They identify marketing and service problems before these lead to much bigger customer retention problems. Much like scraping the paint and hearing the creaks. Managers can relay the results to the other departments. "In many cases, customers will tell you why they are leaving before they do," says Donna Fluss.

✪ BENCHMARKING

To help ensure performance, organizations benchmark their operations against those of other firms. Executives, shareholders, and analysts like to see it. A gap between your company and others is a red flag that there may be problems.

✳ Lack of agreed standards

Benchmarking is challenging for call centers because of the arcane performance metrics the industry uses.

You can't measure anything if you can't define what it is you are measuring and how to measure it. You can't have a standard in an industry unless there is agreement on what constitutes those standards.

That is the bizarre state of affairs in the call center "industry." I put it in quotation marks for a reason. As demonstrated in Chapter 1, there is no agreement what makes a call center or the industry for that matter.

There are no as of this writing industry-wide standards on how to define and apply call center metrics. There are no governing bodies, trade associations, or dominant vendors that set them. There is no enforcement, either by operational incompatibility, moral authority, or legal requirement.

Customer satisfaction is a key measurement, because it relates to customer retention and revenue. But it is difficult to benchmark. Jon Anton, director of research with BenchmarkPortal, points out that customer-satisfaction survey questions and scoring change from survey to survey within companies' call centers and in the same industry.

First-call resolution is a seemingly straightforward metric because it relates back to customers. But it can be distorted.

Richard Fredrickson, managing partner of Unisys' CRM practice, pointed out that FCR could be defined in many different ways. It can mean solving the call, passing it on to another agent, to resolve, or marking the call to investigate.

Jodie Monger, asks, "How do companies define a call as resolved? When the agents say it is resolved or when the customers say it is resolved?" The customer, she adds, must be the one to answer this question.

Cost per call is an easily understood but bad metric because it does not equate back to customer -facing performance. It can also be easily distorted.

For example, if the definition of cost per call includes agent labor costs, then managers and consultants must understand all the variables affecting labor costs. These might encompass geographic area, agent seniority, union agreements, and benefit packages.

Call centers in the same labor market and industry could have full-time employees receiving benefits. Others could have part-time agents with no benefits. Still others may use staffing agencies, which can also affect cost per call.

"If each secondary factor is measured elsewhere in the survey, then I can cut the data to represent a segment of the sample most representative of my call center," says RDC's Kathryn Jackson. "But each time I cut the data, I end up with a smaller sample size. At some point, I may go below the number necessary to have a significant sample."

Customer Relationship Metrics's Rembach points out that companies have different accounting practices.

One company he's familiar with assigned total telecom costs to the organization, with no allocation given back to the call center. Some firms may also miss expenses, such as security guards, janitorial services, or office supplies.

"While these items may seem small, they add up very quickly and can result in a considerable variance when comparing this metric," he notes.

Statistical Reliability

Even if measured right, benchmarked data may not give an accurate picture of the call center's performance.

Shashi Jasthi, managing partner of Eumotif says that because call center benchmarking statistics are snapshots of a center's performance during one day, the data may not be representative of the center's overall performance.

"Is that day average, below average, or above average?" he asks. "Was the day in the center's peak season or off-peak season? Was it after you or the other centers just announced raises, or the day after a long weekend? There's no way of knowing."

Such snapshot statistics are often unstable, he explains. Without stable statistics the metrics may be trending upwards or downwards. Taking or comparing such measurements would not give meaningful information about the state of the call centers.

The call centers in benchmarking firms' databases may not be representative of call centers in aggregate or by peer. Response Design's Jackson observes that call centers often voluntarily opt in to benchmarking studies rather than being randomly selected.

Benchmarking Recommendations
Benchmark Against Peers

Select firms that are of similar size, function, industry, and location. Anton says often trade organizations (such as the American Bankers Association) that offer benchmarking as a service to their members.

"Companies need to establish peer group comparisons," Anton points out. "But when benchmarking within that vertical they should select further to get more accurate comparisons, such as of similar size companies, call centers, and geographies."

Get the Whole Picture

ICMI's Cleveland says call center managers should go beyond benchmarking operations metrics and instead compare customer loyalty and retention programs. They should look to see if call centers are doing the marketing and R&D and what have been the impacts.

"Senior executives want comparisons with other companies to gauge performance," he says. "But that comparison should be the total picture."

Get Stable, Valid Statistics

You need stable statistics. To get them, take frequent enough samples of their performance to account for variations like demand spikes, seasonal slumps, and compensation changes. If more centers and benchmarking firms gathered and contributed stable statistics, the greater the odds that call center performance data would be more accurate and reliable.

"Stable statistics gathering is like your doctor asking you to fast before giving blood to a lab, to account for the sugars in the meal you've had," says Jasthi. "If you eat, you will throw off the results."

Also, gather data from a sufficiently large sample of the population being measured. The sample size should be as close to, and as representative of, the population size as possible, for example, tech support agents if you are benchmarking tech support agents. You should also randomly select the sample.

With the caveat that much of the data obtained from benchmarking may be inexact, if there are gaps between the benchmarking data and your performance, then you should conduct what consultants call gap analysis or process benchmarking. You determine why these gaps exist and what you can do about them.

If the benchmarking looks at best practices the analysis should examine how the best-practice firms operate. How do these companies train, monitor, schedule, and supervise agents? How do they conduct customer-satisfaction surveys? What tools are available to help call centers close the gap?

Gap analysis can pay off by avoiding wasted time and money on unnecessary programs and investments. Let's say your agent training time is three weeks, but the benchmarked training time is five weeks. Should you add two weeks' training time for every agent? It may be that you could have higher-quality agents, screening, and training than the centers benchmarked against.

Response Design's Jackson recommends that you link your benchmarking data with what your customers say in surveys and how they act with sales reps.

"A benchmarking study could tell me that my call center's hold time is higher than average," she explains. "But when I survey my customers I find that hold time is not correlated to their satisfaction or loyalty. If we will not lose any customers by not lowering hold times, why should we invest in additional agents, lines, and IVR to cut hold times?"

To link benchmarking data with customer experiences, Unisys' Fredrickson advises that call centers look at why callers call rather than the raw volume. For example, a high average speed of answer and first-call resolution is more vital to a caller complaining about an unwarranted $1,100 charge on his or her credit card than to a caller who has been told to pay a $17 shipping charge.

"There is no reason why you should trade off additional costs to reach that best-practices level when customers do not perceive any value to it," Fredrickson points out.

✪ RUNNING THE "24" IN 24X7

Managing call centers is challenging enough. But there are unique issues in providing the third or midnight shift to supply that "24" in 24x7.

A chief concern is keeping agents awake. Maureen Wolfe cites circadian rhythms that naturally cause people to feel tired around 3 a.m. Often there is no food service available on site or nearby to supply energy or caffeine boosts to employees.

Other issues are a lack of IT support on hand in case there are problems and a lack of access to HR staff and training which are typically only available during the day.

"The midnight shift often feel like third-class citizens," says Wolfe. "They feel they are not given the time and attention from management that other shifts get because their hours don't overlap traditional management hours. They have less visibility with management and are more likely to feel more like a number."

Not surprisingly, enticing agents to work those hours can be difficult. While some people like it, most others do not.

Some people have more fears and security concerns regarding working the night shift. Sleeping during the day can be difficult because of outside light and noise.

To ensure those workstations are staffed, some companies rotate shifts, which are unpopular with employees. While rotation treats everyone the same, the changing hours cause havoc to sleep patterns and personal lives.

To work around these issues, Wolfe suggests making sure there is food and drink (like coffee) available for overnight shift workers. Those agents should be allowed longer or more frequent break times to wake up by walking around and stretching to recharge their brains.

There should also be an HR person there at least once a week for night shift issues. Also consider training agents to fix some of the likely-to-happen technical problems.

Make sure you have good security and ensure that all doors are locked. Consider having a "buddy system" to escort employees to and from their cars.

She also recommends building an esprit de corps among overnight workers. For example, give them more authority and responsibility to solve problems without relying on supervisors.

"If it's done right you can make the midnight shift the elite shift, independent-thinking and able to solve problems," says Wolfe. "That will help attract top-quality agents and better prepare them for management careers."

✪ UNIONS

There is one word that is guaranteed to frighten many businesses, especially those with

BEST PRACTICES?

I wrote an article on benchmarking for *Call Center Magazine*. Some months after it appeared I received an email from a reader who pointed out that the story referenced "best practices." But the individual wanted to know which are the best companies with the best practices, and why wasn't this information in the article?

That had me scratching my head. This individual clearly didn't get it. By asking it revealed in my opinion a dangerous lack of business acumen.

The answers were obvious. The best companies are those that are eating your lunch or threaten to steal it from you. The best practices are how they are doing it. If managers don't understand that, then it won't be long until their operations and jobs are off-shored. Because without understanding one's competitive and financial environments that senior executives work in, then call centers are going to be treated like cost centers—best minimized if not eliminated.

call centers: "unions." To many firms they raise the frightening specter of higher wages and benefits costs, grievances and strikes, and, underlying them, the biggest evil of them all, another powerful force telling them how to run their business.

Many site selection consultants are often told by clients to avoid locations where there are union cultures where the communities support strong unions. Even if a company or operation isn't unionized—most call centers are not—there is social and economic pressure to raise wages and benefits to near-union levels. In union towns there is often also a poisonous us-versus-them attitude between workers and management that could seep into agent attitudes toward your managers.

On the other hand there are good reasons *why* there are strong unions and union cultures. Many industries that have been unionized, such as manufacturing, resource extraction, and transportation, exploited their employees by making them work long hours, under tough backbreaking conditions, for little pay and no benefits, laying them off at a whim, and giving them no respect.

Management often regarded their workers as less-than-human —"wage slavery"—to rationalize their ill treatment of them. Unions emerged, after many bloody battles, as protectors of workers' rights. Their wage gains laid the foundations for today's middle class. A union wage, such as the increase in the average hourly base pay to $29.25 per hour from $26 per hour for Verizon's customer service agents in 2000, is much more livable for a family facing the Northeast's horrendous house prices and rents than the $10–$15 per hour typically paid by non-union companies.

Yet, just because call center agents and supervisors work in nice clean offices, many think that conditions can't be tough, perhaps not enough for them to consider joining a union. It is difficult to sit electronically tied to a cubicle taking a never-ending barrage of often abusive calls and emails and making calls to similarly impolite people, knowing that every nanosecond of your behavior is being measured, recorded, and judged remotely by management in supervisory towers or behind walls. Call center agents do come down with painful injuries caused by repetitive motion, including carpal tunnel syndrome and tendonitis.

Then there is the low pay and often the lack of benefits. Some firms will close centers before employees gain too much seniority. It is blue collar work in a white collar environment.

On top of that are the hours and shifts, which vary radically and change at a spur of the moment and in doing so can wreck an agent's personal and family life. Sometimes the wage and the hours worked are not enough to pay the mortgage or rent. There is also the usual Dilbertian' capriciousness of top management, ignoring workers' needs and opinions and deploying new hardware and software and rule changes with little training or notice.

According to the August 21, 2000 issue of *The New York Times*, Verizon's workers won several important issues in their strike: the ability to organize non-union wireless division workers more easily, limits on the right of the company to transfer out jobs like call center agents, and the lowering of mandatory overtime. They also receive up to five 30-minute periods each week in which they can do non-call work. But does it have to take a strike to make these changes?

Your company may or may not have Verizon's clout or can or cannot afford those types of wages. And yes, you can relocate your call center anywhere there is enough cheap, loyal labor, like India, if you can find qualified local managers or have some adventurous types on your team who'd like to move.

But you might be better off if you made your call center a decent place to work, designed and well located, with excellent amenities, paying decent but affordable wages and benefits, and employing fair, productive work practices including listening to employees and acting on their suggestions. That's how many other companies have avoided unions, if that's what you want to do.

Employees have the right to be represented by unions. The propensity of workers to join one is, therefore, up to employers like you.

○ THE END (LITERALLY)

Yes, everything comes to an end. That includes this book, this writer, and call centers. How to consolidate, downsize, and close an operation and lay off staff, and which centers and employees to let go when it isn't an outright closure are arguably the toughest decisions executives and managers have to make.

It sure isn't easy on the employees. I have been laid off twice: once when the publication closed and second through a downsizing. Not long after I agreed to rewrite this book *Call Center Magazine* eliminated my position. Fortunately a month later I was hired by one of the companies I had written about, The AnswerNet Network of Princeton, N.J.

In the years since I wrote the first edition I saw the magazine's staff peak to the dotcom bust. Then slowly my colleagues began to disappear, with empty cubicles and later, after I had relocated back to Canada, vacant floors.

In many respects finding out that an operation may be closed is like being told you have a terminal illness. You go through all the steps but finally you have to get your affairs in order and accept it, closing the door behind you.

Can This Center Be Saved?

Oftentimes the rumor or threat to close or downsize is like a mild heart attack, a wake-up call to get your act together. Here are some steps to examine to avoid a shuttering or lower the number of staff to be let go.

Cut Costs and Boost Productivity

Many of the solutions are outlined in this book. They include effective self-service, home working, outsourcing, relocating to less expensive real estate, and improving employee output by hiring, training, and managing them smarter.

Make the Business Case to Senior Management

You should also make the business case for your call center to senior management. Show that it is a revenue-making and retaining center. Take a look at where your center could add revenue and cut costs.

"If I was a call center manager I would demonstrate to my manager the value of the call center in how much revenue we generate and how much is at risk if they closed or severely curtailed it," advises Teresa Setting, Kelly Services vice president of marketing.

Talk to Government Officials

The financial advantages of closing may be slight. Your reason for closing is that you cannot afford to open a new building in that community. Therefore you could talk to economic development agencies and governments about incentives such as tax breaks or free buildings to keep you there, especially if you provide a large payroll in a small city.

Involve Your Staff

Make sure you include your staff in call center cutback discussions. To preserve their jobs they may be willing to work more hours or take on new projects. "By getting their buy-in you may avoid problems like badmouthing your company to people and to customers, and theft if people have to be laid off," advises Rosanne D'Ausilio, president of Human Technologies Global. "At least they would leave feeling that they tried."

Cut by Attrition

When facing staff cutbacks, do so by attrition when possible. This method has less impact on morale, hence productivity, service, and customer retention. It also sends the message that the remaining team members need to work harder to save their jobs.

Consolidations

Consolidations, either internally or through mergers or acquisitions, save money. The downside is that they are often horrendous to implement, and the benefits can take a long time to achieve.

There are many questions such as what to do with the real estate, furniture, and people? Where do you put the smaller teams and the new agents? How do you mesh formal and informal dress codes and cultures? Or do you open a new building and throw everyone together?

HOW NOT TO MANAGE A CALL CENTER

Any good manager must know how to listen, to what is being said and equally important, what is not being said directly to them or behind their backs by their employees.

A case in point is this email *Call Center Magazine* received in response for awards candidates. The names and other incriminating evidence have been omitted to protect the accused.

"I am not sure if you are the correct person to send this to or not.... It has been brought to my attention that manager X from my company has been nominated and is in the first round of competition for the call center manager contest that your magazine is having.

"I am asking my identity be kept secret as it may jeopardize my job. I need to tell you as one of the employees in this department, X is the worst manager I and most of the team has ever seen. He does not follow up with employees on issues, rarely answers our emails, and leaves policy up to team members, thus alienating half the team at once.

"There are many more examples of the ways he is an extremely poor manager. While this call center is very efficient and may keep some customers very well taken care of, this should by no means be inferred to mean that X is in any way shape or form a good or even decent manager. To give him an award for this would be a slap in the face to our entire team."

Then there are the nightmares of re-leasing, packing, cleaning up, moving, unpacking, and setting up, including hiring and familiarization. Together they threaten to throw your productivity into the toilet.

Here are some suggestions to help you cope.

Do Your Site Selection Carefully

Make sure you are consolidating in the right places for the right reasons.such as reducing labor costs rather than shrinking facilities expenses. Don't underestimate the expenses or the hassle.

Consider a New Location

If you have new employees or are meshing in existing staff, it is easier and more conducive to productivity if they are on neutral ground rather than expecting them to move into a center inhabited by the other teams.

Don't Count on Agents to Move

People don't relocate for $10-an-hour jobs unless unemployment is very high in their area. But they will pack their bags for managerial positions that pay much higher.

Look at Home Working

Consolidation is an excellent opportunity to switch agent jobs to home offices, saving you a bundle in real estate costs and achieving significant productivity gains (see Chapter 5). Using home agents avoids the facilities and staff integration issues altogether.

✪ GETTING THE AFFAIRS IN ORDER

Sometimes cutbacks and closures are inevitable. You then need to get your affairs in order. Here is what you need to look at.

See If You Need to Close

Once you decide to pull the plug on a call center, ensure that your diagnosis is correct. If you draw the sheet over it and it revives, your reputation as an employer has just gone down the drain. It does not look good to give your agents pink slips and then buy huge want ads looking for workers three months later. You may not get them back because they won't trust your company's management.

While this may work in high-paying factory or mining jobs, where people will swallow their pride because of the excellent money and where furloughs have historically been part of the work pattern, this is not necessarily the case in low-paying call center employment. Manpower says call centers that lay off are going to have a much harder time rebuilding their image. People will fear that if they are hired they will be laid off again.

To give you some maneuvering room in the event of a future downsizing it may be worthwhile to outsource a certain percentage of your workforce, either off-site to a service bureau, or on-site through a staffing agency, from the beginning. This is an especially valid strategy if your business is successful but seasonal.

To avoid potential us-and-them hassles you should be certain that agency-staff and in-house agents have a clear understanding of who their employer is. The agency must take responsibility for all personnel issues related to their staff.

Consider Closures Instead of Layoffs

Sometimes a call center is worth more to another company if it is closed completely. A new call center may be more willing to pick up the lease, cutting your losses if these assets are still fresh and available to use than if the staff had gone to work for other employers or if the equipment had been sold and ripped out. There are specialized site selection firms such as Arledge/Power Real Estate that have databases of just-closed call centers.

This approach may be fairer to employees, allowing them to obtain other employment instead of hanging on vainly in hope. Call centers do not pay enough to merit such wishful optimism.

Property

Hopefully you will have negotiated exit or sublease clauses in your lease (see Chapter 11). Otherwise you may have to pay to keep space that you don't use.

HR Policies

You will need to have a fair redundancy policy in place to withstand legal challenges. The most common method is seniority—last hired, first fired. The problem with that method is that it doesn't measure quality. The remaining workers may be poor performers compared to the ones terminated.

The other fairer but trickier method is by performance. Fair in that you let the worst performers go. Tricky in that you need to have a reasonably objective and consistently applied method of evaluating agents and managers to avoid discrimination suits.

Look at where in the organization you can place them. Typical jobs that utilize similar skills as call centers are reception, employee benefits, collections, sales support, and research. Staffing agencies can retrain or outplace laid off workers.

Vandalism and Theft

Be prepared when you do give out layoff notices for vandalism, including computer viruses. Some people will be angry and not go quietly into the night.

Make sure you follow all necessary steps to protect your network and facilities, like password changes. Also to avoid these dangers, communicate why you are downsizing and offer as much assistance as you can with other employment.

Bad Publicity

Layoffs and closures are the corporate equivalents of motor vehicle accidents and homicides. They make for great media stories. Reporters are ghouls (I've been one) and love to get their hands on even a rumor of a downsizing.

Therefore be ready for the questions from journalists and from elected officials and for the negative coverage, especially if the reason why you are cutting back is because the top executives decided to offshore, particularly if your center received government grants to locate there. Have the armor-lined shorts on—maybe literally—if your call center is in a small city with high unemployment. Being a major employer in a small burg makes you part of the community family. Closing down is like a divorce, leaving many hard feelings and resentment.

There are no easy ways to respond to this. But there are steps you can take such as:

* Consult behind closed doors first with civic and community leaders. They may come up with some tax breaks to help you stay.
* Demonstrate that you have done everything you can to avoid downsizing or closing.
* Show that you are taking steps to accommodate employees such as home working, offering jobs that they will take elsewhere in the organization, and placement with other employers. Also, work to find other tenants who will hire your people.

Peoples' needs change, as does the means to meet them. Customers want products and services that are better, more convenient and less expensive. The development of call centers is one such effect because they meet these needs. If call centers should diminish or go extinct that's because they fell victim to the same inexorable forces that created them.

Chapter 16:
Resources Guide

There are plenty of sources of information about call center planning, setup, operations, and management that I recommend you to check out.

✪ OTHER BOOKS

This book focuses on putting together and expanding the means to deliver call centers such as through in-house or outsourced facilities, home agents, and self-service. It also touches on staffing, training, and management. There are several books that go in depth on operations, technology, and management issues.

CMP Books www.cmpbooks.com publishes these titles. Among them:

Call Center Dictionary, by Madeline Bodin & Keith Dawson

Call Center Handbook, by Keith Dawson

Call Center Savvy, by Keith Dawson

The Complete E-Commerce Book, by Janice Reynolds

The Complete Guide to Customer Support, by Joseph Fleischer and Brendan Read

Customer Service over the Phone, by Stephen Coscia

The Experience, by Lior Arussy

Home Workplace, by Brendan Read

Logistics and Fulfillment for e-Business, by Janice Reynolds

Maximizing Call Center Performance, by Madeline Bodin

Newton's Telecom Dictionary, by Harry Newton

A Practical Guide to CRM, by Janice Reynolds

A Practical Guide to Call Center Technology, by Andrew J. Waite

Telecom Handbook, by Jane Laino

Tele-Stress, by Stephen Coscia

Other great books not published by CMP include:

Call Center Management on Fast Forward, by Brad Cleveland and Julia Mayben

Call Center Staffing—The Complete Practical Guide to Workforce Management,
    by Penny Reynolds

Business School Essentials for Call Center Leaders, by Maggie Klenke

Wake up Your Call Center, By Rosanne D'Ausilio

✪ TRADE MEDIA

There are several trade media that provide a wealth of how-to advice and news on call center applications, setup, site selection, staffing, training, management, operations, technology, and issues. The sponsoring firms often put on or endorse conferences and trade shows.

The leading trade media outlets include:

Call Center Magazine
www.callcentermagazine.com

Call Center Times
www.callcentertimes.com

Contact Center World
www.contactcenterworld.com

Contact Professional
www.contactprofessional.com/

Customer Inter@ction Solutions
www.tmcnet.com/cis

Customer Interface
www.c-interface.com

CRM/Destination CRM
www.destinationcrm.com

There are media that cover call center applications, issues, and methods in leading user industries or at specific fields (such as site selection) that apply to call centers. Some of these titles include:

Area Development
www.area-development.com/

Bank Systems and Technology
www.banktech.com

Catalog Age
catalogagemag.com

Catalog Success
www.catalogsuccess.com

CommWeb
www.commweb.com

DM News
www.dmnews.com

Direct
Directmag.com

Disaster Recovery Journal
www.drj.com

Electric Light and Power
uaelp.pennnet.com/

Internet Retailer
www.internetretailer.com

Managing Offshore
Managingoffshore.com

Operations and Fulfillment
www.opsandfulfillment.com/

Plants, Sites and Parks
www.bizsites.com/

Site Selection
www.siteselection.com

Target Marketing
www.targetonline.com

If you have operations in Canada:

Contact Management
www.contactmanagement.ca

Direct Marketing News
www.dmn.ca

There is also a unique publication for telemessaging as well as traditional call centers that is worth reading. Connections Magazine www.connectionsmagazine.com is published by Peter DeHaan, a well-respected industry consultant.

Note: *Trade media come and go, almost like call centers. The titles I've written here may not be there by the time you pick up this book. Check the sites frequently or do web search to stay on top.*

✪ ASSOCIATIONS

Another great set of resources is business, professional, and trade associations. They often have their own trade books and media, plus conferences, seminars and trade shows.
Some of the leading associations are:

American Teleservices Association (ATA)
www.ataconnect.org

Association of TeleServices International (ATSI)
www.atsi.org

Call Center Networking Group (CCNG)
www.ccng.com

Direct Marketing Association (DMA)
www.the-dma.org

Disaster Recovery Institute International
www.drii.org

Electronic Retailers Association (ERA)
www.retailing.org

Help Desk Institute (HDI)
www.thinkhdi.com

Incoming Calls Management Institute (ICMI)
www.incoming.com

International Customer Service Association (ICSA)
www.icsa.com

Service and Support Professionals Association (SSPA)
www.thesspa.com

Society of Consumer Affairs Professionals (SOCAP)
www.socap.org

Society of Workforce Planning Professionals (SWPP)
www.swpp.org

The Telework Coalition
www.telcoa.org

And in Canada:

Canadian Call Management Association (CAM-X)
www.camx.ca

Canadian Marketing Association (CMA)
www.the-cma.org

Canadian Teleworking Association
www.ivc.ca

These organizations may have links to their counterparts in other countries, which is an extremely important asset if you are planning to offshore or serve other foreign markets. One example is the Federation of European Direct Marketing (FEDMA) www.fedma.org, which is an umbrella group for direct marketing organizations in individual European countries. The Direct Marketing Association works with FEDMA.

Note too that there are frequently local chapters of national associations. In addition there are call center equipment owners groups. One example is the International Nortel Networks Users Association (INNUA) www.innua.org. Also, many regions, states, and provinces may have their own call center organizations. .

✪ SPECIAL ASSISTANCE TO THIS BOOK

There are consultants and firms who have generously donated their time in interviews for past articles and for this book. I highly recommend contacting them; I have included their contact information. Note that these individuals and companies provide services in more than one category, so check out their web sites for a complete range of their offerings.

These fine people and firms are, by field:

Outsourcing and Offshoring

The AnswerNet Network
800-411-5777
www.answernet.com

R.H. Oetting
954-949-0616
www.oetting.com

TeleDevelopment Services
(330) 659-4441
www.teledevelopment.com

Home Working

Jack Heacock and Associates
303-841-8799

Tanner Group
801-538-2320
www.tannergroup.com

Telemessaging

Peter DeHaan
866-668-6695/269-668-6695
www.peterdehaan.com

Site Selection and Real Estate

Arledge/Power Real Estate
214-696-4800
www.arledgepower.com

The Boyd Company
609-890-0726
www.theboydcompany.com

CB Richard Ellis
800-368-8976 ex.5599/602-735-5599
www.siteselectionservices.com

Equis
215-568-4330
www.equiscorp.com

Trammell Crow
214-979-6193
www.trammellcrow.com

Design/Facilities

The Alter Group
847-676-4300
www.altergroup.com

Burkettdesign
303-595-4500
ww.burkettdesign.com

Kingsland Scott Bauer Associates
(KSBA)
412-252-1500
www.ksba.com

Laura Sikorski
631- 261-3066
www.laurasikorski.com

Wave, Inc.
972-387-7555
www.wave-fcm.com

Business Continuity

Attainium
571-234-4752
www.attainium.net

North Highland
404-233-1015
www.north-highland.com

180cc
800-550-4180
www.180cc.com

Management, staffing, training

AchieveGlobal
800-456-9390
www.achieveglobal.com

Banks and Dean
888-241-8198/262-240-9400/
416-385-1611
www.banksanddean.com

BenchmarkPortal
805-614-0123
www.benchmarkportal.com

CIAC (Call Center Industry Advisory
  Council)
615-373-2376
www.ciac-cert.org

The Call Center School
615-812-8400
www.thecallcenterschool.com

Customer Relationship Metrics
336-288-8226
www.metrics.net

Davis, Trotter and Giblin (Customer3
  [cubed])
314-569-0990
www.customer3.com

DMG Consulting
973-325-2954
www.dmgconsult.com

ePerformax
800-925-1974/901-751-4800
www.eperformax.com

FurstPerson
888-626-3412/773-353-8600
www.furstperson.com

Help Desk Institute
800-248-5667/719-268-0174
www.thinkhdi.com

Human Technologies Global
845-228-6165
www.human-technologies.com

Incoming Calls Management
  Institute
410-267-0700
www.incoming.com

Kathy Sisk Enterprises
800-477-1278/559-323-1472
www.kathysiskenterprises.com

Kowal Associates
617-892-9000
www.kowalassociates.com

Legge and Company
585-242-9700
www.leggecompany.com

Radclyffe Group
973-276-0522
www.radclyffegroup.com

Response Design Corporation
800-366-4RDC/609-398-3230
www.responsedesign.com

Service Quality Management
800-446-2095
www.sqmgroup.com

Telephone Doctor
314-291-1012
www.telephonedoctor.com

Appendix I:
Call Center Gallery

Here is a selection of interesting and imaginatively designed and located call centers. You may learn a few tips from these companies and designers for your facility.

✪ CORE COMMUNICATIONS (FREMONT, NEB.)

June 2001

When most people think of steel skinned-and-framed buildings what comes to mind are plain, ugly sheds and warehouses. Yet as one call center and its designer has shown you can create an elegant, highly functional facility that costs less and is quicker to erect than conventional concrete-and-masonry structures.

A new steel building may turn out to cost no more and take no longer to finish than a retail conversion. This is an important benefit if you need to open a call center quickly in small cities that do not have many suitably-located convertible properties with adequate parking.

You can make metal buildings attractive. This is Core's innovative call center with high-quality interior furnishings.

Core Communications is a new full-service inbound multimedia outsourcer, founded by Brent and Sheryl Diers. Brent Diers, a former executive with West, a large service bureau, is quite familiar with the challenges of setting up new call centers.

The Diers selected Fremont because many employees for Omaha-based call centers live there. They felt they could attract many top-quality employees who preferred not to commute.

When they looked for a home for their new business they chose not to lease and renovate an existing structure, which most bureaus do, but decided to build their own call center when they found that there were no suitable buildings.

Core contracted with Prochaska and Associates to design a new building. They selected a steel structure laid out to give a handsome, naturally lit entrance and high-quality office interiors difficult to re-create in retail conversions. Construction work began in August 2000 and the center opened in February 2001, after some weather delays.

The 13,500-square foot call center and headquarters has capacity for over 160 work-stations, including 25 training stations that could also be used for full call and contact handling. There is also a break room and human resources area. There is room for up to 300 parking spaces. The call center is on a main east-west highway north of the down-town, with shops and restaurants within walking distance.

To make the building more appealing to employees and clients, Prochaska added an aluminum entrance with skylights that pour natural light into the reception area and con-ference rooms. The feature gave the conventional structure height and airiness that it would not normally have.

The new metal call center cost just $40 per square foot, compared with $80 to $100 per square foot and saved 90 days for conventional masonry construction and about the same or less than for renovating an existing building.

Though the steel skin is thin, the interior equipment, training, and break rooms can withstand tornadoes. These rooms are constructed with concrete blocks and covered with a poured concrete cap. Employees and equipment can be protected here during a severe storm.

The site has multiple power and fiber feeds plus a UPS system.

"It was quite by chance that we ended up on the property we did and building a steel facility," explained Diers. "Through contacts we met the landowner who erected steel buildings. He worked with the architect to design the pre-engineered steel building. When our business increases to where we need to open a new call center we will proba-bly go with a steel structure."

⊙ INFONXX (SAN ANTONIO, TEXAS)

Call centers have been located in some very interesting buildings and under extremely tight timeframes.

Directory assistance outsourcer Infonxx has done both by opening a new 500-work-station call center in a former movie house and in less than four months over the Christ-mas holidays.

Infonxx opened a new call center in March 2003 in a former 1960s-vintage movie house that had been converted into a two story office by the landlord. The movie house has landmark status. It is located in an older suburban part of San Antonio near a pair of major expressways.

Infonixx's agents head to the center's break room to relax and recharge. Photo courtesy of laukgroup.

Infonxx picked San Antonio because it found the metro area had an excellent labor market. The outsourcer has a 250-workstation call center there.

The movie house has several first-run features: It had been recently renovated as an office, it is well known and visible so potential employees could easily find it, there is plenty of parking, and there are restaurants and shopping nearby.

Equally important, Infonxx's first call center is close by. That call center is in a conventional single-story office building.

"We needed 500 workstations to be ready quickly," explains John Chell, director of real estate and facilities services at Infonxx. "We looked at many different buildings, and we found this one met our needs. Because it was near our existing call center we were able to save time and money by not duplicating administration functions in the new call center."

Infonxx needed every edge it could get. The firm had a tight deadline to meet client needs. It decided on December 1, 2002, that it needed a new center, and the facility was open on March 14, 2003.

To expedite the project the company split it into two phases, with two separate permitting processes: the computer and training room, and the primary work floor. It hired lauckgroup, an interior design firm, to undertake the design work.

Infonxx paid a professional expediting service to get the permitting achieved faster; expediters are professionals who know how to prepare, file, and track building permits.

Infonxx also ordered long-lead-time components like HVAC systems, switch and electrical gear, including a 90kW backup generator at the get-go. It hired Schmidt & Stacy Engineers because of their in-depth expertise with highly technical call center projects.

The project splitting paid off. The training room received its certificate of occupancy in February. The center had agents trained when the rest of the center got its certificate in March.

Infonxx was able to make an attractive looking call center quickly and with a tight budget, explains Brigitte Preston, design principal at lauckgroup. The call center outfitted the work areas with Herman Miller Equa chairs. And the outsourcer's design team used colors to brighten up the atmosphere and to give it a sense of youth and playfulness. It also kept the break room ceilings exposed to keep a cutting-edge, warehouse feel to the space.

"We wanted to give the agents a sense that they were going on a break so they could relax and recharge," says Preston. "So we created a trendy, hip-looking room that had a completely different feel to the employees."

○ NEXTEL

Yes you can build attractive shared-use and multiple-use call centers. You can also take the same design and adapt it to very different climates.

This is Nextel's highly versatile call center. This design enables shared use such as for shipping, receiving, and retail.

The Alter Group built in 2001 for Nextel, a nationwide cellular firm, two very different new call centers in Temple, Texas, and Bremerton, Wash., using its customizable CallCore design with that functionality in mind. These centers illustrate how you can accommodate different needs and geographies while taking advantage of economies of scale in a single build-to-suit design.

The CallCore series is designed for high-density users operating multiple shifts around the clock. The models generally accommodate eight employees per 1,000 square feet and a more sophisticated wiring system than is typically found in office buildings to handle heavy voice and data traffic.

The structures generally have a single floor plate, 10-foot-high ceilings, windows on all four walls and a parking supply of eight spaces per 1,000 square feet, which is double the supply in conventional offices. They have on-site cafeterias, outdoor smoking patios, diesel generators, and UPS backups.

Where the two Nextel call centers differ is in size, adaptability, and in design reflected in their locations. The Temple center has 108,000 square feet and will eventually have 600 workstations.

The Temple building has been designed to withstand an F2-strength tornado. The F2 rating has been achieved through larger concrete footing and foundation designs, additional reinforcing placed in the exterior tilt wall system, along with 1-inch laminated glass to resist projectiles often created by these devastating storms.

The Temple call center can be converted into a light industrial or flex building, taking in and shipping out products. The rear of the building has a thick slab foundation and provision for truck bays. There is a 24-foot clear path for trucks to turn around and back up. The call center has been built in an industrially zoned area.

The Bremerton center is 60,000 square feet and will hold up to 400 seats. The foundations, exterior tilt wall, and structural steel systems are compliant with earthquake standards.

The Alter Group has included more storefront glazing in the Bremerton building for future retail use. Parking and sidewalk access around the building is also included in the site design, anticipating this future use.

✪ AMERICREDIT (PETERBOROUGH, ONTARIO)

One challenge of locating call centers in small cities is finding adequate property when the centers need it. This is especially true in Canada, where the country's more conservative business culture does not encourage speculative building to the scale seen in the U.S.

This is AmeriCredit's custom-built call center. The community where it is located, Peterborough, Ontario, enabled the new construction by agreeing to lease the property from the developer and sublease it to AmeriCredit.

Answering that challenge, the Greater Peterborough (Ontario) Area Economic Development Corporation (GPAEDC) asked local developers in 2000 to suggest ways to provide new buildings quickly.

Signum Corporation responded by investigating building methods and processes that could be pre-engineered and manufactured off-site, such as factory-made acrylic exterior panels. Manufacturing off-site takes less time. It also permits building installation during cold winter months. The developer contracted with an architect, Petroff Partnership to design the exterior, and with Stinson Design to design the interior.

Signum submitted its response to GPAEDC. When AmeriCredit, represented by Arledge/Power Real Estate Group came to Peterborough to locate its call center, GPAEDC brought Signum's proposal to AmeriCredit's attention.

AmeriCredit and Arledge/Power had earlier sought, without success, a facility with adequate space and parking that could be renovated into a call center. The developer convinced Arledge/Power and AmeriCredit that it could construct a new call center quickly.

The firm developed a design and plan for a new building in one week. The proposed building had 89,400 square feet with 500 workstations on two floors with plenty of parking: eight spaces per 1,000 square feet.

Arledge/Power and AmeriCredit settled on Peterborough and the Signum building after examining 60 other cities. AmeriCredit signed the lease on December 1, 2000. Agents began taking calls 28 weeks later, in July 2001.

"It is not the design that makes the difference when you are planning a new building on a tight deadline, but the building process," says Signum's president John MacDonald.

❂ EBAY (BURNABY, BC, CANADA)

Imagine walking into a call center where there are no calls and no voices except brief conversations among agents and supervisors. There is no near-deafening din of talk. Instead, you hear the occasional clacks of keyboards. Instead of having headsets clamped to their skulls, agents sometimes don earphones to listen to personal music.

This isn't imagination, though it may be the future of call centers. This is eBay's center in Burnaby, British Columbia, Canada, where nearly all of the contacts come in by email, making it a true "contact center."

eBay opened the Burnaby center in April 2003. It joined centers in Salt Lake City, Utah; Omaha, Neb.; and Dreilinden, Germany. The Canadian center handles Canadian, American, Australian, and UK customers.

eBay picked Burnaby, which lies next to Vancouver, for the area's high PC usage, large number of customers, and a culturally diverse labor force.

eBay's silent call center, except for the clicking of the keys. As more people communicate by email and text messaging, expect more call centers, or portions of them, to be like eBay's.

"We wanted to be in a location to ensure that we could hire people who were already familiar with and had a positive impression of eBay," says Lynn Hardin, senior director of customer support.

The center is in an inner-suburban office park with low-rise multi-floor buildings nestled between a major expressway and a tree-lined stream that shades a busy rail line. Close by is a large shopping mall.

Unusual by American standards, about 50 percent of eBay's workers come by mass transit. There is frequent bus service and a rapid transit line nearby. There is limited on-site parking.

The center occupies 75,000 square feet on two floors. It has 550 workstations and plans to grow that to 1,000 workstations. There are training areas and huddle rooms.

There are seven coffee and drink stations on the work floors plus a break room on each floor. Off-site there is a McDonald's plus many other nearby restaurants and shops.

What sets the center apart is the nearly noise-free environment. It exudes a literally unheard-of blend of activity and serenity, a calming, yet very aware "humm."

The center, designed by Valerio DeWalt Train, is outfitted with carpeting, partial fabric and glass-topped partitions, and specially designed wood wall panels in strategic areas to reduce noise. The partitions let in light and create a feeling of openness.

eBay eschewed cube rows for zigzag workspaces of two agents per shared table.

Agents sit back-to-back from each other in these mini-rows. Steelcase supplied the chairs and workstations.

Agents work in customer teams based on nationality, identified by flags at the work-stations. They know American, Australian, and UK slang.

The center has a 14:1 supervisor to agent ratio. Supervisor workstations have higher partitions.

eBay has a strong, socially aware corporate culture. In keeping with it, much of the material in the center is recycled, such as the carpeting and the wood wall partitions, which are reprocessed wood particles.

The center was designed for voice with Ayava phones on every desk. But the phones are used only for interaction between team members regarding questions, and to communicate with eBay's other sites.

Instead, eBay's agents use instant messaging (IM) to communicate with colleagues and supervisors. IM allows agents to ask questions while having the customer's email on screen. Agents can also forward a copy of the email to the supervisor to check draft responses.

Many of the features in eBay's contact center came from experiences with its Salt Lake City center, also email only, which opened in 1999.

"We learned, for example, that we needed larger rooms for team meetings, smaller rooms for coaching, and consolidated training center for trainees," says Hardin.

"We also learned that we needed separate areas off the work floor to meet with vendors and eBay members, which we also provide in Burnaby, off our reception lobby."

Appendix II:
Locate in the Big Apple? Don't Say 'fugedaboudit!' Just Yet!

Open a call center in New York City. Are you crazy, you might ask, in New York-ese.

Yes, New York City is one of the most expensive locations to do business in, with high labor costs and rents. Not surprisingly many call centers have moved out, to New Jersey, Connecticut, Florida, and Canada.

But don't toss the Big Apple out of your site considerations just yet. There are many reasons to look at locating there.

New York City has the world's best sales culture with people who know how to hustle from cutting deals (and everything else) in the boardrooms to selling batteries while walking through the cars of the uptown subway B train. The education level of the workforce is very high, bolstered by throngs of college students from all over the world.

New York City's multilingual and multicultural workforce is unmatched. Lady Liberty in New York Harbor still beckons immigrants worldwide, speaking a United Nations of languages and dialects, though they now land at JFK and nearby Newark Liberty Airport instead of docking at Ellis Island.

AT&T had a famously successful long-distance marketing campaign by featuring ads placed above the interior windows on subway cars assigned to the No. 7 line, which is the "United Nations" of subway lines. The ads had dialog bubbles of people speaking in many different languages. No. 7 rumbles above and below Queens neighborhoods with nearly every ethnic and linguistic group on the planet before tunneling under the East River, by the UN building, stopping at Grand Central Station before terminating at Times Square in Manhattan.

Call centers that sell securities, provide customer service to unions, local government agencies and HMOs, or which require access to that multilingual labor base can justify locating there.

New York City and the surrounding area is a market unto itself. Agents inside New York have invaluable local knowledge (and attitude) that customers come to expect. Agents from outside of New York can be intimidated when talking to New Yorkers.

New Yorkers hate false sincerity and time wasters. They don't suffer fools gladly, and they are not hesitant to let others know what they think in the clearest possible terms. *The New York Times* reported that agents have been paid "danger pay" when they called New York City numbers.

New York Downsides

When examining New York City as a location note that the traffic is often horrendous, though most commuters get around on a mass transit system that is surprisingly safe. Yes there is crime. But the city is also one of the safest of the large metros even before 9/11/01.

The city's bureaucracy can be slow on the permitting process. Call centers are not exactly big on their lists of businesses to attract and move quickly on.

Facilities Issues

New York City has a property menu as vast as its dietary variety. There are occupancies ranging from midtown and Wall Street offices, and in Brooklyn's MetroTech, to renovatable and renovated older buildings in the Flatiron and NoHo districts and in lower-cost borough neighborhoods in The Bronx, Queens, Brooklyn, and Staten Island. Staten Island also has room for large-floorplate buildings just minutes from Newark Liberty Airport.

Mark Collmar, president of DM Communications, reports that many of the buildings within the budgets of call centers are older, which may require extensive retrofits. Mold is sometimes an issue in older buildings.

One common problem is that many of these buildings lack sufficient riser capacity to bring cabling to upper floors; the floors are solid concrete. In response, firms must do core drilling between floors to run conduits. Fire stairs are also used.

Some structures are designated landmark buildings. There is a lengthy and process by which you need to gain approvals to extensively renovate them, such as changing walls, floors, and ceilings.

"Landmark building landlords don't even like you touching the walls," says Collmar.

Almost inevitably, your call center will depend on elevators, which the building management is responsible for. In some buildings the elevators will be in good shape. In others they may be slow and prone to breakdowns. Avoid leasing in a building with only one elevator.

"If you're a small call center with, say, 25 agents, and the tenant knows growth is possible, sign a short-term lease agreement. One to three years," Collmar advises. "If the elevator service doesn't meet your requirements, you have the opportunity to find better space to accommodate your growth."

Whether in new or older buildings, chances are good that you will have to use union contractors to do the remodeling. Unions, and their higher costs, are an accepted fact in large cities. The landlords or the building management firm will let you know if you have to use union labor.

Rodents can be a problem in some older buildings. While the rats stick to the basement, mice often head to upper floors. Collmar recommends that you ask existing tenants if there are any pest problems within the building, check enclosed areas like bathrooms and hidden closets, and ask management how they exterminate all pests.

Once you set up your call center, establish a policy forbidding eating and drinking at desks to reduce attracting potential pests. Always call an exterminator to help combat the problems.

"Mice, and other pests, are a fact of life in a city like New York," he says.

Give New York this much, there is life everywhere you look. Even if it is unavoidable.